AF559704

Practical Manual of ENTOMOLOGY

Published by
Sumit Pal Jain *for*

New India Publishing Agency

101, Vikas Surya Plaza, CU Block, L.S.C. Mkt.,
Pitam Pura, New Delhi-110 088, (India)
Ph.: 011-27341616, Fax: 011-27341717, Mob.: 09717133558
E-mail: info@nipabooks.com
Web: www.nipabooks.com

ISBN : 978-93-80235-90-5

Composed and Designed by NIPA

PREFACE

The term 'Pest' in English, derived from the Latin word 'Pestis' means 'causing destruction'. As the name indicates, pests cause extensive damage to cultivated crops and other vegetations, besides destroying household articles and other possessions and attacking livestock and even humans. Some pests are responsible for causing or spreading a few serious diseases in man and livestock. Pests include insects and several other non-insects belonging to invertebrates and vertebrates. Pests are responsible for causing enormous economic losses. According to a survey conducted in India in 1976, it has been established that a total loss upto 18 per cent is caused as a result of pest attack every year. Of this about 33 percent loss is due to weeds, 26 percent due to diseases, 20 percent due to insect and non-insect pests, 6 per cent due to rats and 7 per cent in warehouses. The monetary loss had been estimated to exceed Rupees 5,000 crores at that time. Among the pests, insect pests are more destructive and are responsible for causing huge losses.

Insects constitute slightly over 50 per cent of all living organisms viz, other Arthropods, plants and animals put together. They are ubiquitous and are found in every place on this world where life can be sustained. There are about 1.0 million species of insects, of which comparatively a small percentage of about 1,000 species are pests in the true sense, while the others are wild and harmless. Many species are beneficial because of their role in our ecosystem. Some are a part of the food chains of wild creatures, while some others are either predators or parasites. A vast majority of insect pests are predominently polyphagous. Many of them feed on the foliage or other plant parts. Some of them bore into plant parts, such as stems, branches, roots, flowers, fruits or grains and feed on the inner contents. Some pierce the plant parts and suck and feed on the plant sap, while some others are vectors of virus and mycoplasma diseases. In nature there is hardly any plant species, which is free from attack by insect pests of one type or other. Even plants, which possess insecticidal properties, such as *Azadirachta indica, Calotropis procera, Allium sativum* etc. are also subjected to attack by some pests.

It is an established fact that insects came into this world more than 250 million years ago, while man came into existence only about 1.0 million years ago. Over the several million years, insects have survived because of their innate adaptations to live under any adverse environmental conditions. Actually it is man who has intruded into the domain of the insects. In order to meet his requirements of food, clothing and shelter man started cultivating crops and in doing so, the natural habitats of the insects were destroyed. The insects on the other hand have gradually changed their food habits and have started attacking the crops raised by man to meet their own requirements of food and shelter. Thus many insects have attained the status of crop pests leading to large scale destruction of crops. From the time man stepped in, a stifle has started between the insects and man and the struggle to gain supremacy has begun. Man considers insects as his enemies and to safeguard his interests tries to destroy them by various means. However, inspite of his vast potential, supremacy, intelligence and ingenuity, it has not been possible for man to eliminate these much inferior creatures completely and the struggle is going on as a never ending ordeal.

Unlike plant pathogenic microorganisms, such as fungi, bacteria, viruses, mycoplasma etc., which cannot be seen with naked eye, almost all the pests can be seen with naked eye or with the aid of a simple lens and the damages caused by them are also quite clearly visible and so appropriate control measures can be taken up at the proper time to control them. Knowledge about the nature of damage caused, life cycle and dispersal of pests goes a long way in adopting suitable control measures at the proper time. In this context this book, which has been compiled as per the syllabus of B.Sc.(Ag.) Degree course of the Agricultural Universities will be of great help and guidance to the students of Agriculture. It consists of three chapters. The first chapter deals with 'Pests of field crops'. The second chapter deals with 'Pests of horticultural crops' and the third chapter deals with 'Pests of household and livestock pests'. Detailed account of each of the major pests and possible measures to control them have been given. Besides the major pests, list of minor pests which may assume serious proportions when environmental conditions become favorable for them has also been furnished. The book, I hope will also be of immense help to the officials working in the Department of Agriculture, people interested in Scientific farming as well as the General Public.

Numerous illustrations have been given to amplify the nature of damage and the life cycle of the pests so that the reader can understand and follow the text easily.

In the recent past, there is a tendency to revert back to the old system of natural farming or organic farming so as to safeguard the quality of the soil and to protect the atmosphere from pollution, which is a welcome proposition. But we have upset the natural ecosystem to such an extent that it may not be possible to revert to the old system of organic farming in the near future by avoiding the use of pesticides and inorganic fertilizers altogether for crop production. However, it is necessary to take utmost care in the judicial use of such chemicals. Only when absolutely necessary pesticides should be used and when using such chemicals care should be taken to select appropriate chemicals. They should be used at the correct and recommended dosage, and at the proper time when the pests are in the most vulnerable stage to ensure maximum control of the pests.

In the compilation of this book, I have gathered informations from several books, journals and other media and I am very much indebted to the authors concerned. Ackowledgements are due to collegues who have supplied me with informations and also to friends and well wishers who have inspired, helped and encouraged me in my endeavor to complete the book. I wish to place on record my sincere thanks and appreciation to M/S. New India Publishing Agency, 101, Vikas Surya Plaza, Pitam Pura, New Delhi - 110 088 for the excellent manner in which the book has been brought out.

H.Lewin Devasahayam

LIST OF FIGURES

Chapter-1 : Pests of Field Crops

❑❑❑

COLOUR PLATES

Coconut – Rhinoceros beetle and damage

Coconut – Red palm weevil and damage

Cut worm

Cotton boll worm

Sugarcane internode borer

Sugarcane shoot borer

Looper caterpillar

Hawk moth caterpillar

Rose – Leafcutter bee damage

Potato beetle

Lady bird beetle

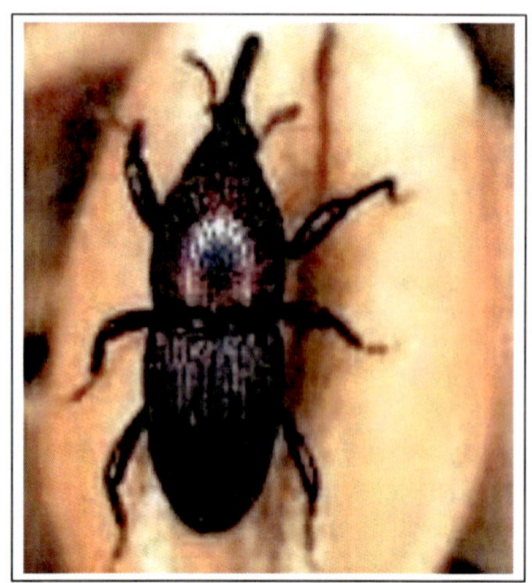

Rice weevil

Mealy bugs and damage

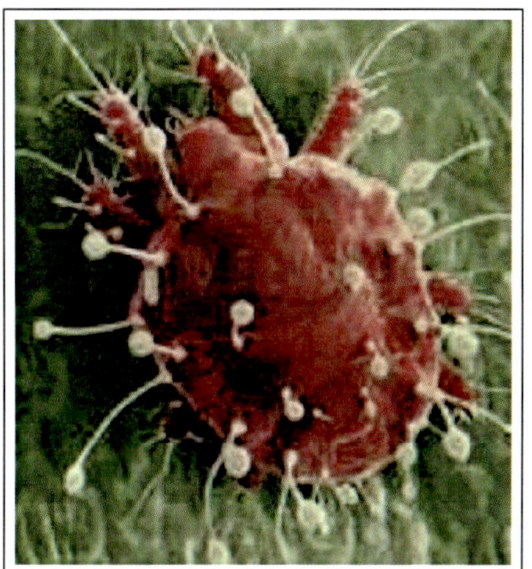

Coconut – Eriophid mite and damage

Leaf miner

Aphids

Mango leaf hopper

Root knot nematode

Grass hopper and damage

Locust

Crow damaging grains

Rat damaging fruit

Squirrel damaging nuts

Parrot damaging fruit

CONTENTS

Chapter-1

Pests of Field Crops

PESTS OF CEREALS

Rice (*Oryza sativa*)

Rice is the staple food of a vast majority of people throughout the world, especially in India. Rice crop is grown in about 75 million acres, which is about one-third of the total world area under rice. The warm and humid environment for the rice crop is best suited for invasion by several insect pests and as many as 85 species of insects are known to infest this crop. The damages caused by different pests extend from the seedling to harvest stages of the crop. Some of these pests are widely distributed and cause extensive damage. The more important pests are described below:

1. Rice stem borer or Rice yellow stem borer

Scirpophaga (Tryporyza) incertulas

Order - Lepidoptera

Family - Pyraustidae

Among the pests, which ravage rice crop, 'rice stem borer' is a major one and infests rice crop at all growth stages, from seedling to flowering. It is a specific pest of rice alone and is more serious during the months of October-January. The pest has a wide distribution in all Asian countries. The damages caused by this pest ranges from 5-20 per cent. Generally, late varieties of rice are more severely infested than the early maturing varieties, because of the possibility of the pest multiplying in larger numbers by the time the late varieties come to maturity. Cold weather with high humidity and low temperature prevalent during October-December is found to be conducive for the multiplication of the insect.

Nature of damage. The external symptom of this pest attack is the gradual fading and ultimate death of the central shoot. The larvae bore into the stem, about 1.0-3.0 cm. above the water level and eat the inner tissues, leading to wilting and eventual death of the central shoot. This is referred to as **'dead heart'**. When the larvae attack seedlings, dead hearts appear invariably, which come off easily when pulled. When the plants are attacked at flowering stage, the inner tissues of the stem are eaten away and the earheads dry up without producing grains, as a result, white-colored, chaffy earheads appear, which are called **'white ears'**. The white ears are clearly visible in the infested fields and they also come off easily when pulled. In the basal portion of the infested stem, near the nodes, minute holes made by the larvae are visible on close examination (Fig.1).

Life cycle of the pest. The adult moths rest during the daytime under the shades of plants and become active after dusk. They fly about and mate after dusk. The female moth lays 200-300 eggs in 2 or 3 compact masses on tender paddy leaves and covers the egg masses with buff-colored anal hairs. Freshly laid eggs are oval, flattened and dirty white in color. Before hatching, the eggs turn brown in color. They hatch in 6-8 days. The young larvae after emerging from the eggs spin fine, silken threads from their saliva and hang loosely from the leaves. In the blowing wind, the larvae swing and reach nearby plants within a radius of 10-20 cm. Afterwards, they crawl towards the base of the plants. Some larvae after hatching just crawl along the leaves towards the base of the plants, while some others fall directly into the water and swim freely, because of the presence of a thin air layer next to the body surface. The larvae worm their way along the leaf sheath, enter the leaf sheath and after feeding for 2-3 days, bore into the stem near the nodal region, 1.0-3.0 cm. above the water level. The newly hatched larvae are pale white, with dark brown head and a prothoracic shield. They cut and eat away the tender inner tissues, causing dead hearts. Similarly before flowering, the larvae bore into the peduncle directly, cut and destroy the inner tissues resulting in white ears. Generally only one larva is found in a stem, but occasionally 2-4 larvae may be found. Sometimes after destroying one plant, the larva migrates to the adjoining plant. In such instances, the larva after coming out of the stem, drifts on water by enclosing itself in a case made of leaf bits till it reaches another plant and attacks it. The larval period extends from 33-41 days. A full-grown larva measures about 2.5 cm. in length, pale yellowish-white in color and smooth, with a well-developed prothoracic shield. Before pupation, the larva covers the exit hole with a web of fine silken thread and pupates within a white, silken cocoon. The pupa is dark brown in color and about 1.2 cm. long. The pupal period lasts for 9-12 days. The adult moth emerging from the pupa comes out through the exit hole already made by the larva before pupation. The female moth has bright, yellowish-brown forewings with a minute, distinct, black spot on each and an anal tuft of buff-colored or yellowish hairs. It is about 1.5 cm. in size across the wings and slightly bigger than the male moth. The male moth has pale yellow forewings and without the black spots. The moths are positive phototrophic in nature and their flight range is about 8.0 km. The moths live for 3-5 days. There may be 3-5 overlapping generations in a season. The entire life cycle is completed in 50-70 days. The larvae are reported to undergo hibernation from November-January in the stubble, 2.0-4.0 cm. below the soil level (Fig.1).

Control Measures

Physical and cultural methods (i) The egg masses on the leaf surface are clearly visible and these can be collected and destroyed (ii) Before planting, leaf tips of seedlings are clipped, thereby the egg masses on the leaf tips can be destroyed (iii) 'Dead hearts' can be removed and destroyed (iv) Light traps can be set up in the fields to attract and destroy the moths (v) After harvest, the fields are ploughed to destroy the larvae and pupae present in the stubble.

Chemical control (i) Chemical control measures should be taken up when the pest level reaches the economic threshold level. Spray application of phosphamidon-110 ml. or fenitrothion-400 ml. or fenthion-200 ml. or endosulfan-400 ml. or monocrotophos-200 ml. or decamethrin-100 ml. mixed in 200 litres of water per acre, using a high volume sprayer affords effective control of the pest. The spray should be directed towards the base of the plants to give a thorough covering of the stem region (ii) Application of granular insecticides such as carbofuran 3G-13 kg. or lindane 10G-5.0 kg. or quinalphos 5G-10 kg., mixed with 10 kg. of sand and spread uniformly on the field after impounding a thin film of water to a height of about 2.5 cm. is also effective in controlling the pest. But the cost of treatment with granular formulations in the main field is comparatively very high.

Biological control (i) Eggs of the pest are parasitized and destroyed by the egg parasites-*Trichogramma australicum, Tetrastichus schoenobii, Telenomus manolus* etc. (ii) The larval parasite-*Isotima javensis* is found to parasitize the larvae (iii) The fungal parasite-*Beauveria bassiana* affects the larvae.

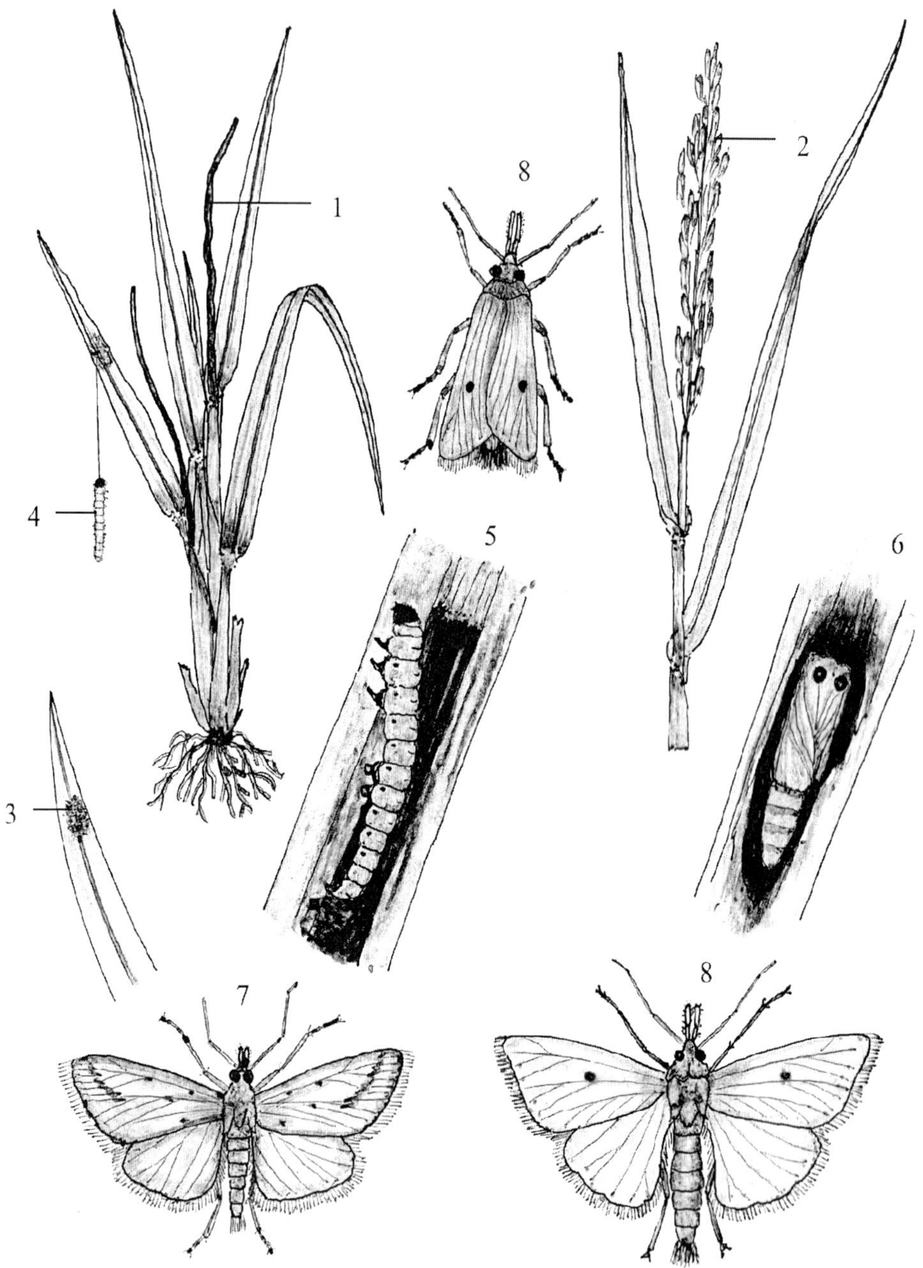

Fig. 1 : Rice Yellow stem borer-*Tryporyza incertulas*

1. Dead heart 2. White ear 3. Egg mass 4. Young caterpillar 5. Larva inside the stem 6. Pupa inside the stem 7. Male moth 8. Female moth

2. Rice leaf folder

Gnaphalocrocis medinalis

Order - Lepidoptera

Family - Pyraustidae

'Rice leaf folder' or 'Rice leaf roller', which was considered as a minor pest has assumed serious proportions and causes severe losses to rice during some seasons. In Tamil Nadu, it is an endemic pest and appears in all the seasons and all cultivated varieties of rice have been found to be susceptible. The infestation is more in fields, which receive excess amount of nitrogenous fertilizer and in shaded places. The pest infestation is more severe during the months October-January. Besides rice, it attacks several graminaceous weeds.

Nature of damage. The newly hatched larva produces fine, silken threads from secretion of its salivary glands and folds or rolls the leaves and webs them. The larva stays inside the folds of the leaves securely, scrapes and feeds on the green portion of the leaves, as a result the scrapped portions become white and papery in patches. Severely affected leaves have many such patches, sometimes extending to the entire length of the leaves. Growth of such infested plant is retarded and grain formation is adversely affected. The pest infests the crop at all growth stages. If the infestation occurs at the flowering stage and if the boot leaf is affected, the yield is reduced to a considerable extent (Fig. 2).

Life cycle of the pest. The female moth lays eggs singly, mostly in lines of 10-15 on the under surface of tender leaves. The eggs are spherical, white in color and smooth. They hatch in 6-7 days. The young caterpillar emerging from the egg starts feeding by scraping the chlorophyll portion of the leaves. Simultaneously the larva rolls the leaf blade and fastens the edges with silky thread. Sometimes the leaf tip is folded and fastened to the basal part of the leaf blade. The larva remains securely inside the leaf roll or fold, scrapes, feeds on the green portion of the leaf blade and grows. The larva is pale yellowish-green in color and smooth. It becomes full-grown in 25-30 days. The grown-up larva pupates inside the leaf roll or inside the leaf fold and in 6-8 days, the adult moth emerges from the pupa. The entire life cycle is completed in 30-42 days. The moths are small in size and brownish-orange in color. The forewings are light brown in color, with two, distinct, dark, wavy lines on each wing. The hindwings are also light brown in color with one dark, wavy line and a dark brown band along the outer margin of each of the wings. There are 3-5 overlapping generations in each season (Fig. 2).

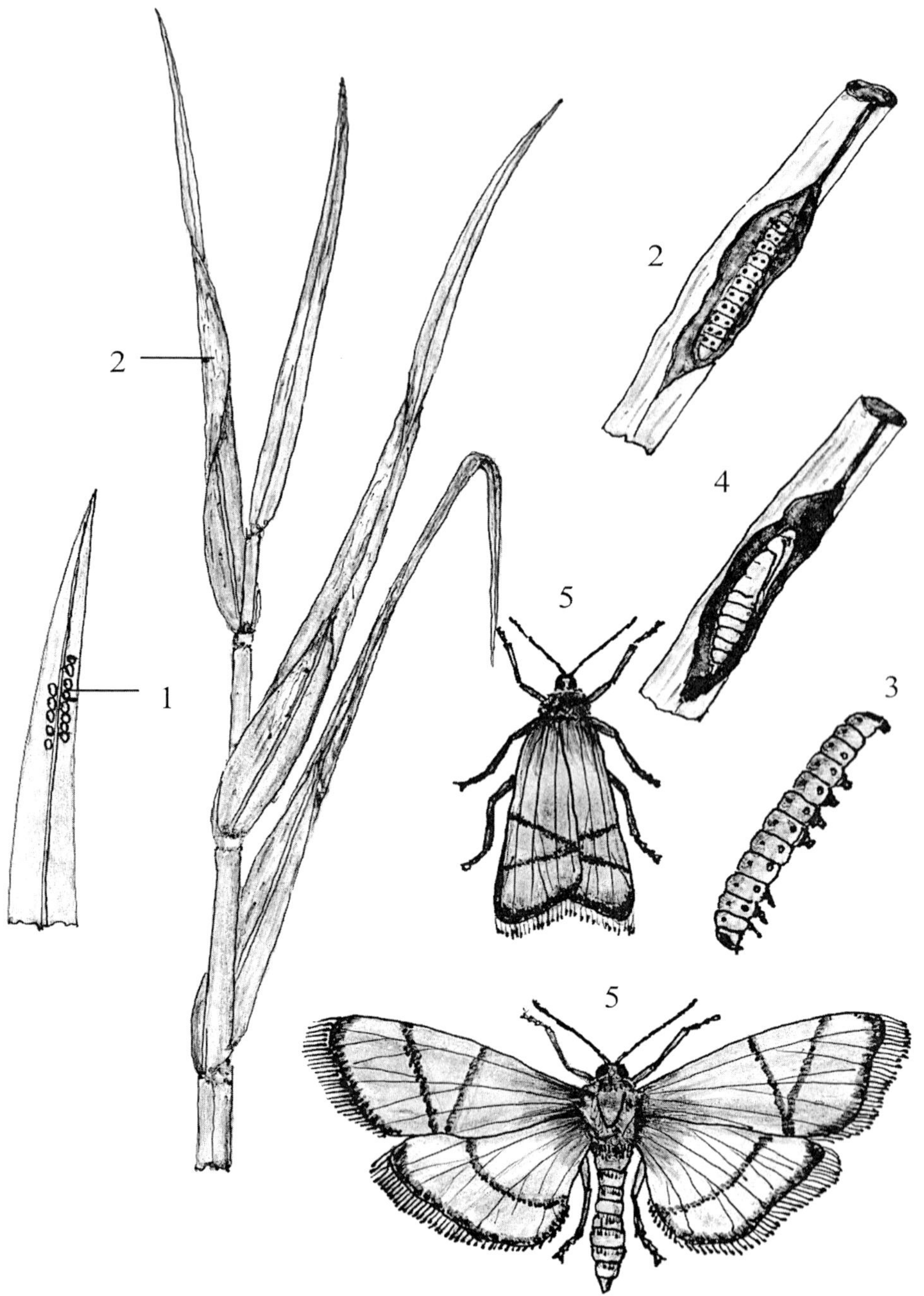

Fig. 2. Rice leaf folder-*Gnaphalocrocis medinalis*

1. Egg mass 2. Larva inside the leaf fold 3. Larva 4. Pupa inside the leaf fold 5. Moth

Control measures

Cultural methods (i) Graminaceous weeds, which serve as collateral hosts of the pest in and around rice fields should be eradicated (ii) More than the required quantity of nitrogenous fertilizer should not be applied. Further, instead of applying nitrogen in a single dose, split application should be followed. Application of large dose of nitrogen at the flowering time should be avoided (iii) By using light traps, the moths can be attracted and destroyed.

Chemical control (i) Foliar spraying with carbaryl WP-1000 gm. or quinalphos-400 ml. or monocrotophos-200 ml. or phosalone-400 ml. or fenitrothion-400 ml. or chlorpyrifos-400 ml. in 200 litres of water per acre, using a high volume sprayer will control the pest effectively (ii) Spraying with neem seed kernel extract-5% also controls the pest effectively.

Biological control. A few egg parasites belonging to *Trichogramma* species parasitize the eggs of the pest. *Elasmus brevicornis* and a few *Bracon* species destroy the larvae. *Tetrastichus israli* parasitizes the pupae.

3. Rice case worm

Nymphula depunctalis

Order - Lepidoptera
Family - Pyraustidae

'*Rice case worm*' is a sporadic pest but may assume very serious proportions in some seasons and during some specific weather conditions, especially when there is continuous water inundation during the young stages of the crop. Besides rice, the pest breeds on a number of grass hosts.

Nature of damage. The pest is more serious in young transplanted crop. The larva cuts the leaf blade into small bits, constructs a tubular case and lives inside the case securely. The case is attached to the stem or leaf at or above the water level or the larva may float on the water along with the case. When about to feed or while swimming, the head and a portion of the body protrudes out of the case. With any slight disturbance, the larva retracts its body and head into the case. The larva feeds by scraping the under surface of the leaf blades, leaving the upper epidermis intact and as a result of such scraping, characteristic whitish, papery patches are seen on the leaf blades. The area near about the cut ends of leaves also appears whitish. Severely infested crop appears as if grazed by cattle. The affected tillers become stunted and lose their vigor and often the plants are killed. Even if the plants are not killed, earhead formation is adversely affected. If the plants are disturbed, the larvae attached to the stem and leaves detach, fall into the water, start swimming and go to some other plant. The larvae are semi-aquatic in nature.

They are capable of swimming in water to some extent and can breathe under water with the help of tracheal gills found on the sides of the body. If the water is drained off, the larvae crawl and remain inside the base of the hills (Fig.3)

Life cycle of the pest. The female moth lays about 50 eggs, singly on the under surface of leaves. The eggs hatch in 2-6 days and the young caterpillars emerging out of the eggs start feeding by scraping the chlorophyll portion from the under surface of the leaf blades for a few days. Then they cut the leaf blades into small bits and construct tubular cases, live inside the cases and feed on the under surface of the leaf blades. With every moulting, the case is replenished. The larva is pale green in color, with a light brownish-orange head and thin hairs on the body and measures about 2.5 cm. in length. The larval period extends over 14-20 days. The grown-up larva pupates inside the case, which is attached to the base of a tiller and the adult moth emerges from the pupa after a period of 4-7 days. The entire life cycle is completed in 20-37 days. The moths are small, whitish in color and on the white wings, light brown, wavy markings are seen and they live for about 4 days (Fig. 3).

Control measures

Physical and cultural methods (i) A long rope is held across the field from the two opposite bunds by two persons and the rope is slowly dragged along the top canopy of the crop, so as to disturb the crop. Because of the disturbance, the larvae attached to the leaf and stem portions fall into the water. A small quantity of kerosene oil is poured over the standing water, which spreads to the entire area as a thin film. Under this condition, the larvae falling into the water cannot breathe and so they die of suffocation (ii) Too much of water should not be allowed to stagnate in the fields.

Chemical control (i) Dusting with carbaryl 10 D at 10 kg. per acre may be resorted to in case of mild infestation (ii) Foliar application of monocrotophos-200 ml. or endosulfan-400 ml. or fenitrothion-400 ml. or phosphamidon-110 ml. in 200 litres of water per acre, with a high volume sprayer effectively controls the pest.

4. Rice thrips

Stenchaetothrips biformis

Order - Thysanoptera

Family - Thripidae

'Rice thrips' is a specific pest of rice and has a wide distribution. It was first reported in Tamil Nadu in 1915 and occurs sporadically in different parts

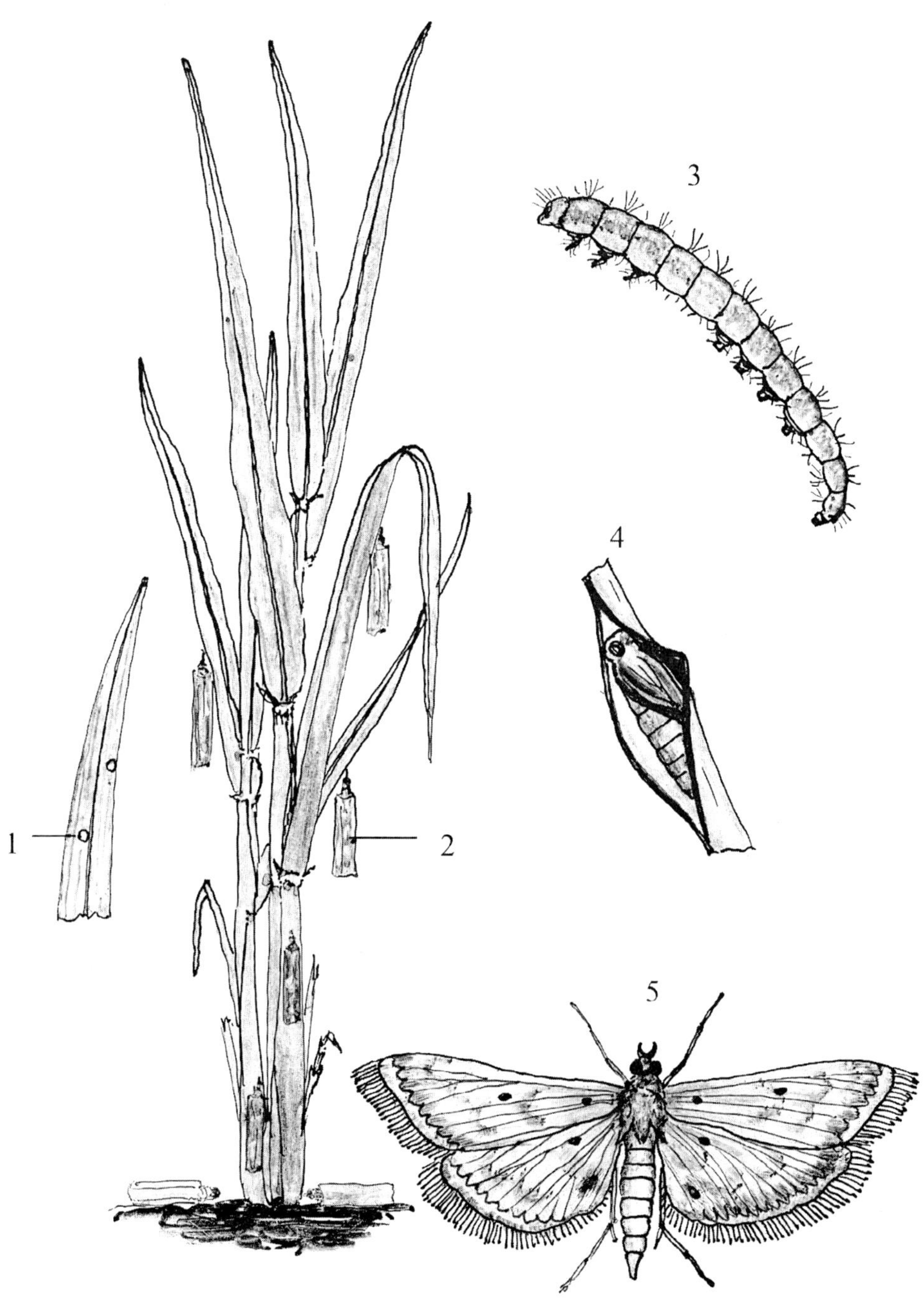

Fig. 3. Rice case worm-*Nymphula depunctalis*

1. Egg 2. Larva inside the case 3. Larva 4. Pupa inside the case 5. Moth

of Tamil Nadu. The pest may assume serious proportion in some seasons and cause extensive damage to the crop and yield loss. It also infests *Echinochloa crusgalli.*

Nature of damage. The pest infests the nursery crop and the young transplanted crop to a large extent. The nymphs and adults have rasping and sucking type of mouthparts and they attack the terminal shoots and tender leaves, lacerate the leaves and suck the sap, as a result the leaf tips become rolled and twisted and eventually the leaf tips dry out. The nymphs and adults are seen in large numbers in the rolled leaf tips. In ill-maintained nurseries and when there is lack of sufficient irrigation in the nurseries and transplanted crops, the infestation is severe and tips of the plants present a burnt appearance. In severe cases, the seedlings may completely dry and may even die. However, after one or two heavy showers, the crop becomes free of the pest.

Life cycle of the pest. The adult female insect inserts the eggs individually into the leaf tissues. The eggs hatch in 3-5 days and the newly hatched young ones are very active. They feed by lacerating the leaves and sucking the sap from young leaves. The nymphal period lasts for 7-14 days, after which they shed their antennae and legs, and attain the pupal stage. After a few days the adult insects emerge from the pupae. The life cycle is completed in 14-20 days. The adults are very minute, dark brown or blackish in color and have fringed wings (Fig. 94).

Control measures

Cultural methods. The nursery and young transplanted crop should be properly irrigated and maintained

Chemical control. (i) For 8 cent nursery area, dusting with carbaryl 10 D at 0.5 kg. gives effective control (ii) Instead of dusting, the nursery can be sprayed with endosulfan-60 ml. in 20 litres of water (iii) In the transplanted crop, foliar spraying of endosulfan-300 ml. or phosalone-400 ml. or phosphamidon-110 ml. or fenitrothion-300 ml. or fenthion-200 ml. in 200 litres of water per acre with a high volume sprayer gives appropriate control.

5. Dark-headed caterpillar

Chilo polychrysa

Order - Lepidoptera

Family - Crambidae

The pest was first noticed in Kerala in 1956 and since then has been reported to occur in Tamil Nadu, Orissa, West Bengal and Assam. In case of severe infestation, the loss in yield may go up to 60 per cent. In South India, the pest occurs from July-January. It also attacks the weed-*Panicum crusgalli.*

Nature of damage. The larva feeds on the inner tissues of the central shoot by boring into it. As many as 7 larvae may be found in a single shoot. Later on, the adjacent tillers are also damaged by the larvae, which migrate and infest them. The infestation may result in the ultimate death of the plants.

Life cycle of the pest. The female moth lays 20-150 eggs on the lower or upper surface of the leaf. The eggs are flat, broadly oval and are laid in two or more longitudinal rows, overlapping each other. The egg period lasts for about 6 days. The full-grown larva is characterized by the presence of three dorsal and two lateral, purplish-brown stripes on the body, a brownish-black head and prothoracic shield. It measures about 2.1 cm. in length. The larval period extends from 23-36 days. The full-grown larva pupates in a thin, silken web inside the stem. After a period of about 4 days, the adult moth emerges from the pupa. The moth has brownish forewings, with 6-7 black spots on each of the forewings.

Control measures. The control measures suggested for the control of rice stem borer may be adopted.

6. **Rice gall fly** or **Rice gall midge**

Orseolia (Pachydiplosis) oryzae

Order - Diptera

Family - Cecidomyiidae

The pest is found in all rice growing tracts of India, including Tamil Nadu, Kerala, Orissa, Telungu Desam, Madhya Pradesh, and Bihar and in some seasons, when the weather conditions are favorable for the pest, it appears in a serious form. The infestation is commonly referred to as **'silver shoot'** or **'onion shoot'** or **'anaikomban'** in Tamil, because of the formation of hollow, pink, purple, dirty white or pale green, cylindrical, tubular structures bearing at their tips a green, reduced and malformed leaf blade with ligules and auricles. In case of severe infestation, the yield loss may go up to 50 per cent.

The pest also attacks a number of grass hosts such as, *Paspalum scrobiculatum, Paspaladium geminatum, Panicum* sp., *Ischaemum ciliare, Cynodon dactylon* and *Eleusine indica.*

Nature of damage. The pest infests rice crop even in the nursery, but 5-6 weeks old, actively tillering transplanted plants are more vulnerable to attack. The maggot bores into the stem, feeds on the shoot apex and the growth of the apical meristem is suppressed. The first instar larva produces a hormone, 'cecidogen', which causes some type of stimulation in the plants and instead of the normal growth of the shoot, the leaf sheath of the innermost leaf primordium elongates into a long, hollow, tubular structure like a long tusk of an elephant, commonly called as 'silver shoot' or 'onion shoot'. This elongation of the leaf sheath is attributed to cell proliferation caused as the result of the hormone secreted by the larva. The silver shoot is actually the modified leaf sheath and its color resembles the color of the normal leaf sheath of the particular rice variety. The infested plants fail to produce earheads (Fig. 4).

Life cycle of the pest. The female midge lays 100-300 eggs, singly or in batches of 2-6, just below or above the ligules of the leaf blade. Occasionally, the eggs may be deposited on the surface of standing water. The eggs are reddish in color, elongate and tubular, with rounded ends. The eggs hatch in 3-4 days. The young maggot, which emerges out from the egg, crawls along the leaf sheath, reaches the stem region, bores into it and finally reaches the shoot apex in 6-12 hours. Only one larva is present in a shoot apex and after the formation of the gall due to its feeding, remains inside the gall throughout its development. The young grubs are whitish-pink in color, while the grown-up grubs are pink in color. They lack legs. The larval stage lasts for 15-20 days. The full-grown grub pupates inside the gall and after a period of 2-8 days, the fly emerges from the pupa. Before emergence, the pupa wriggles up to the tip of the gall and projects half way up the gall. The pupal case is seen protruding out of the mouth of the exit hole after the emergence of the fly. The entire life cycle is covered in 19-21 days, but during winter it may take 32-39 days. After the exit of the fly, the gall dries up. The development of the pest may be staggered due to larval dormancy and the dormant maggots remaining in the axillary shoot apices help in the carry over of the pest. The adult flies are small, delicate, mosquito-like, with long, thin legs. The females are yellowish-red in color, with swollen abdomen, while the males are thin and dark red in color. The wings are almost transparent. The adults are active during the nighttime (Fig. 4).

Control measures

Cultural methods (i) Infested plants should be removed and destroyed (ii) After the harvest of the crop, the field should be ploughed to destroy the larvae and pupae that may be present in the inactive axillary shoots

Chemical control (i) Spray application with endosulfan-400 ml. or phosalone-400 ml. or phosphamidon-110 ml. in 200 litres of water per acre with a high volume sprayer controls the pest. The spray should be directed towards the basal parts of the plants (ii) Seed treatment with chlorpyrifos-0.2% emulsion for 3 hours or seed mixing with chlorpyrifos

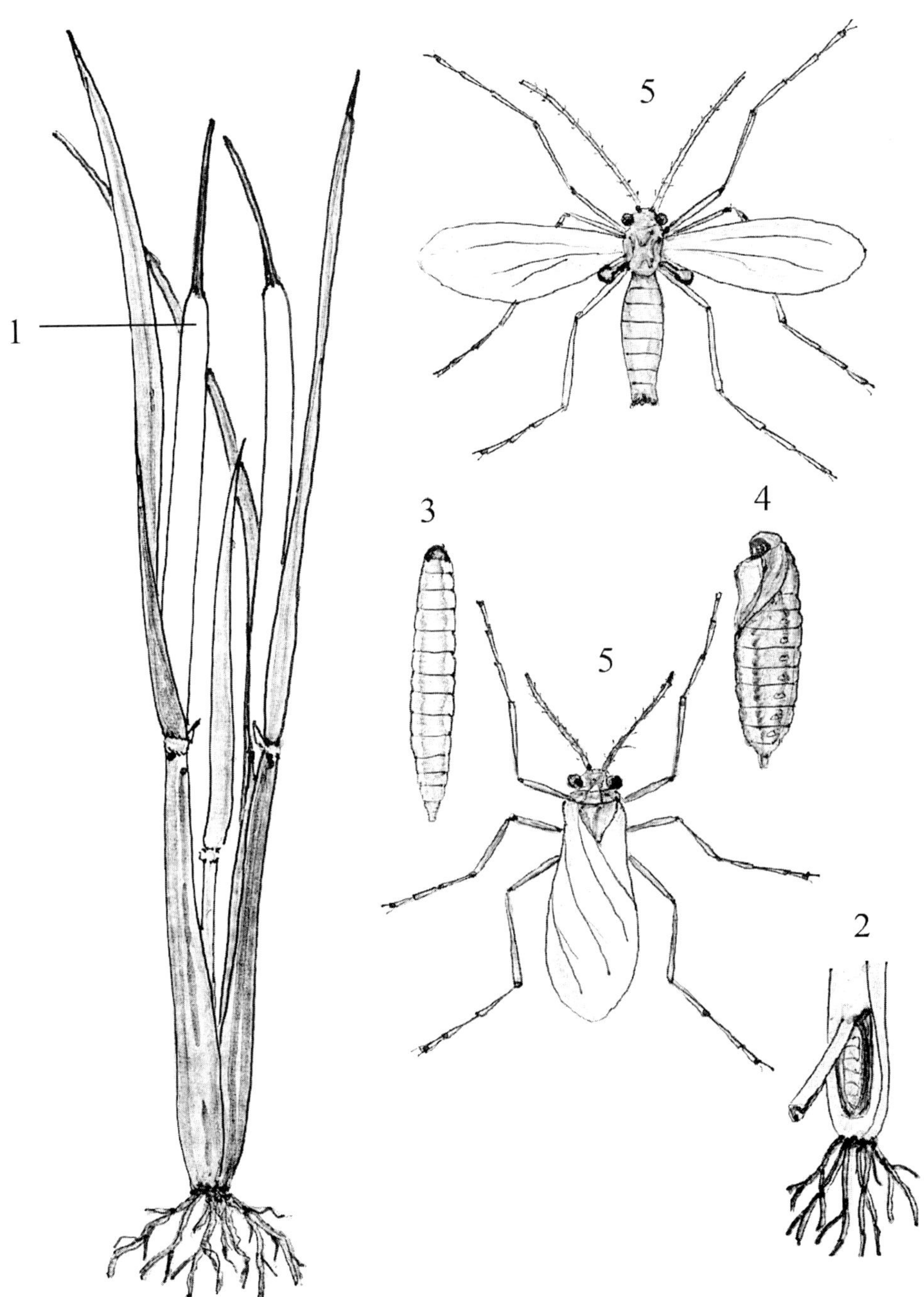

Fig. 4. Rice gall fly-*Pachydiplosis oryzae*

1. Galls in the affected plant 2. Larva inside the stem 3. Larva 4. Pupa 5. Adult fly

(0.75 kg. a.i./100 kg seeds) provides protection of seedling for 30 days in the nursery (iii) Seedling root dip in chlorpyrifos-0.02% emulsion for 12-14 hours before transplanting gives protection for 30 days (iv) Placement of carbofuran 3G at 0.8 kg. a.i./acre near the root zone after planting reduces gall midge infestation.

Biological control. The larval parasites-*Telenomus israli* and *Platygaster oryzae* parasitize the larvae.

Resistant varieties. The varieties Jeya, Vikram, Sureka, Pothana, IET.6187, PTB.18 and PTB.21 are less susceptible.

7. Rice hispa

Dicladispa armigera

Order - Coleoptera

Family - Hispidae

It is a sporadic pest and is found occasionally in all rice growing states of India and is often serious on young rice crop in parts of Tamil Nadu, Telungu Desam, and West Bengal. In case of severe infestation, the yield loss may go up to 25-65 per cent. The pest prefers young plants, which are partly submerged in water. Hot and humid weather conditions favor large-scale pest infestation. It also attacks a number of grasses and sugarcane.

Nature of damage. Both the grubs and adult beetles cause damage to the crop. The beetles scrape and feed on the chlorophyll portion of leaves of plants, especially young plants. The grubs mine into the leaves and eat the inner tissues in-between the veins, but the epidermis is not damaged. The mined portions of the leaves appear as white, straight, discontinuous streaks. When the infestation is severe, many long, parallel streaks appear and the whole leaf turns white and membranous and eventually dry. The affected crop becomes stunted and the leaves may show tip-drying symptoms also (Fig.5)

Life cycle of the pest. The female beetle makes minute slits near the tip of tender leaf blade and lays eggs singly in the slit. One beetle lays 50-60 eggs. The eggs hatch in 3-5 days and the flat, pale yellow grubs emerge. The grub mines into the leaf, feeds on the inner tissues and becomes full-grown in 9-12 days. The grown-up larva pupates inside the mine and emerges out as a adult beetle in 4-5 days. The life cycle is completed in 14-22 days. The beetles are small, steel blue or black in color, with small spines all over the body. There may be about 6 generations in a year (Fig. 5).

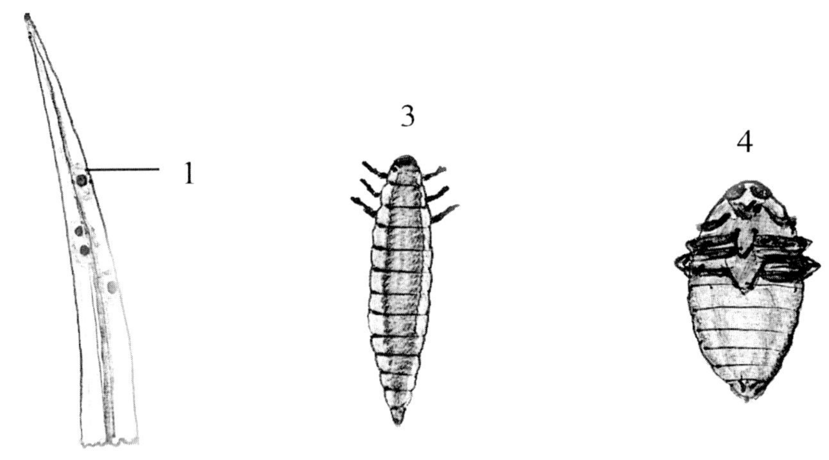

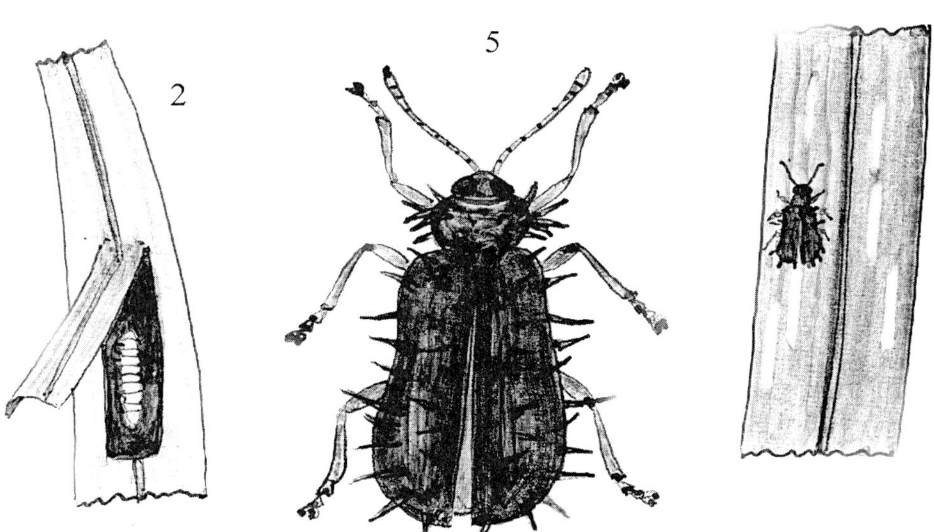

Fig. 5. Rice hispa-*Dicladispa armigera*

1. Egg 2. Larva inside the leaf 3. Larva 4. Pupa 5. Beetle feeding on the green matter of leaf 6. Adult beetle

Control measures

Physical method. The beetles can be collected by using hand nets and destroyed.

Chemical control (i) Dusting with carbaryl 10 D or methyl parathion 2 D at 10 kg. per acre controls the pest (ii) Foliar spraying with cabaryl 50WP-1000 gm. or endosulfan-400 ml. or fenitrothion-400 ml. or chlorpyrifos-400 ml. or quinalphos-400 ml. or fenthion-400 ml. or diazinon 20 EC-400 ml. in 200 litres of water per acre with a high volume sprayer has been recommended for the control of the pest.

8. Rice mealy bug

Ripersia oryzae

Order	-	Hemiptera
Suborder	-	Homoptera
Family	-	Pseudococcoidae

Nature of damage. This is a sporadic pest and occurs in some seasons in a severe form and causes much damage to the crop. The infested plants are stunted, yellowish and appear sickly. In an infested field such infested plants are seen in isolated patches and such plants dry completely within a short period and present a burnt appearance. The damage is caused by hundreds of young and adult mealy bugs attaching themselves to the succulent parts of the stem, protected by the leaf sheaths and sucking the plant sap continuously. This constant drainage of the plant sap affects the growth of the plants drastically and makes them stunted and sickly. The affected plants fail to produce earheads. The insects, which are covered by the leaf sheaths, are not visible from the outside. When the leaf sheath of an infested plant is pulled apart, numerous mealy bugs at all stages of growth are seen adhering to the stem (Fig.6)

Life cycle of the pest. The female mealy bugs are generally sedentary and do not move. The female bug lays spherical, whitish-yellow eggs in rows on the inside of leaf sheath and stem. One insect lays 50-300 eggs. The eggs hatch in 1 or 2 days. The pale yellowish-white young ones, which hatch out of the eggs are quite active and crawl about on the plant for a while and then settle down on the stem, start feeding on the sap from the stem and grow. They are pink-colored, stout and their bodies covered by a white, mealy or waxy coating and are wingless. The young ones become full-grown in 17-37 days. The young and the adults have the same feeding habits and the adult females are generally stationary and wingless. The adults are reddish-white, soft-bodied and are covered by a white, powdery, waxy coating (Fig. 6).

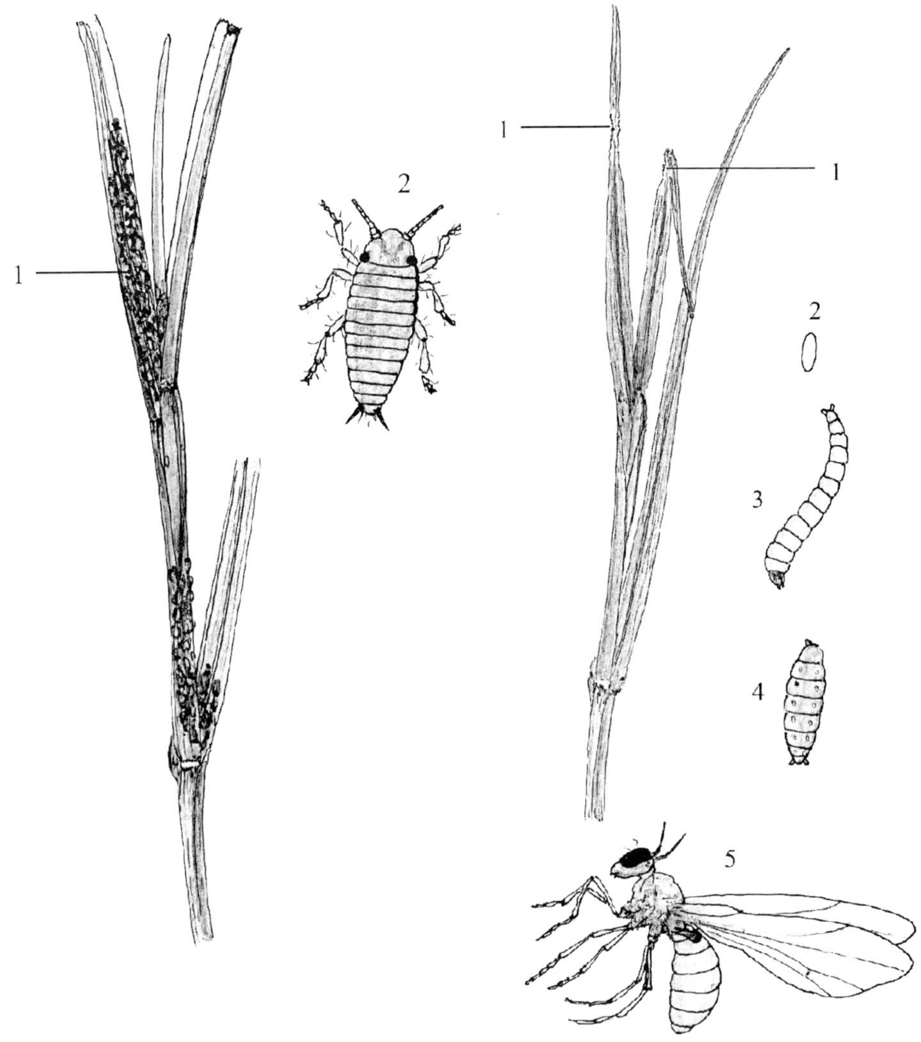

Fig. 6. Rice mealy bug-*Ripersia oryzae*

1. Mealy bugs inbetween stem and leaf sheath 2. Adult insect

Fig. 7. Rice whorl maggot-*Hydrellia sasakii*

1. Infested young leaves 2. Egg 3. Larva 4. Pupa 5. Adult fly

Control measures

Physical and **cultural methods** (i) The infested plants should be removed and destroyed (ii) The mealy bugs have many alternate weed hosts on which they live and multiply. So, the fields, as well as the surrounding areas should be kept free from weeds.

Chemical control. Because a waxy coating covers the body of the insects and as they live concealed within the leaf sheaths, it is difficult to control them with contact and ordinary stomach poisons. Spray application of monocrotophos-400 ml. in 200 litres of water per acre, using a high volume sprayer controls the pest to some extent. The spray should be directed towards the stem region to achieve better results

Biological control. Natural enemies such as ladybird beetles and a few parasitic flies and wasps sometimes attack the colonies of the pest.

9. Rice earhead bug

Leptocorisa varicornis (acuta)

Order	-	Hemiptera
Suborder	-	Heteroptera
Family	-	Coreidae

'The rice earhead bug', known as 'Gundhi bug' in Hindi, is a sporadic pest and occurs throughout India. The pest generally appears during the flowering stage of rice crop and continues up to the milky stage. The pest, which is found throughout the year, is more abundant during the months of August-November. In some of the rice growing tracts and specifically in some of the fields, the pest appears season after season in a severe form. A heavily infested field can be recognized even from a distance by a strong, pungent, bad odor, which emanates from the field because of the presence of the insects. Besides rice, the pest infests Jowar, pearl millet, maize, sugarcane, lesser millets and some grasses.

Nature of damage. Both the nymphs and adults, which appear in large numbers during the flowering stage, feed on the sap from the peduncles, tender stem and milky grains. Mostly the nymphs and adults pierce and suck the milky juice from the milky grains, as a result the grains become chaffy. A distinct hole is seen on the grains, where the bugs have punctured the grains with their proboscis and around the hole a brown spot develops. All the grains in an earhead are not attacked by the pest and become chaffy. However, in a badly infested field, numerous grains are attacked and they turn brown and chaffy (Fig. 8).

Life cycle of the pest. The female bug lays seed-like, dark red eggs in rows of 10-25 on the leaves. One female lays 200-300 eggs. The eggs hatch in about 7 days. The newly hatched nymphs are thin, long, stick- like, green in color, wingless and are very active. They become adults in 12-15 days after 5 moults. The adult is also long, stick-like, about 2.0 cm in length, and yellowish-green in color, with a dark colored, triangular spot on the dorsal

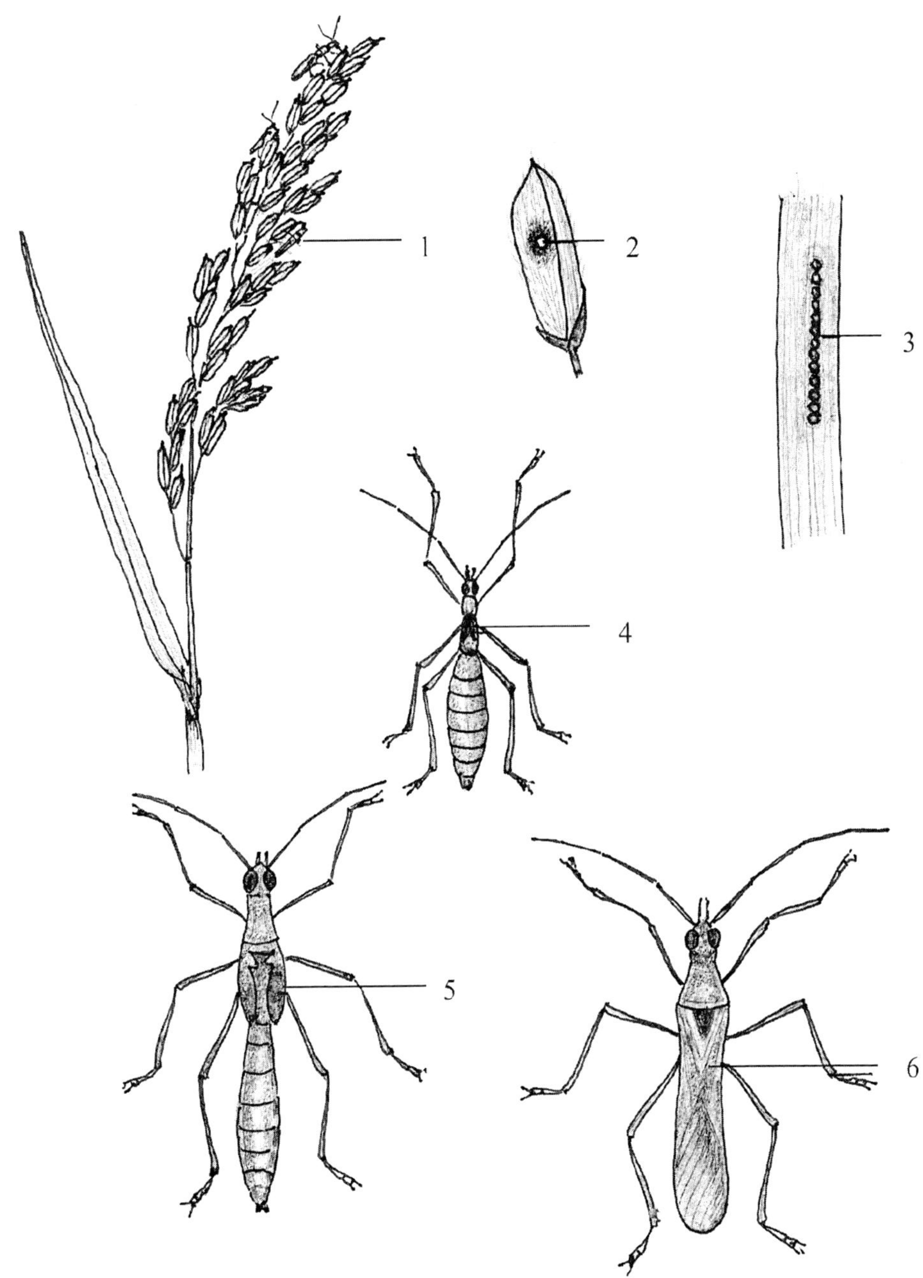

Fig. 8. Rice earhead bug-*Leptocorisa varicornis*

1. Nymphs and adults sucking sap from the grains 2. Hole in the grain at the point of sucking 3. Eggs 4. 5. Nymphs 6. Full grown bug

side of the thorax and has two pairs of wings. The life cycle is completed in 20-30 days. The insects are usually very active from morning to evening and during the hotter part of the day, they shelter under shades of plants. The adults may live for 3-4 months. There are several overlapping generations in a year (Fig. 8).

Control measures

Physical and cultural methods (i) The nymphs and adults can be collected by using hand nets and destroyed (ii) The adults can be attracted and destroyed by setting up light traps or bonfires (iii) The rice field and surrounding areas may be kept free from weeds and other alternate hosts (iv) The egg masses, which are clearly visible on the leaves can be collected and destroyed.

Chemical control (i) Dusting with carbaryl 10 D at 10 kg/per acre, so as to give a thorough covering of the earheads, using a duster controls the pest. Two applications have to be given at 10 days interval from the time of flowering. If the pest incidence is rather severe, dusting with mythyl parathion 2 D at 10 kg/acre may be adopted (ii) Spraying with malathion-400 ml. in 200 litres of water per acre, so as to give a thorough covering of the earheads is more effective.

10. Rice green leaf hopper (GLH)

Nephotettix virescens (impicticeps) and *N. nigropictus (apicalis)*

Order	-	Hemiptera
Suborder	-	Homoptera
Family	-	Jassidae

The pests are found in all rice growing regions of India and many other countries. The pests are more prevalent during the months of July-September, when the temperature is moderate and humidity is high. The insects are known to thrive in a few graminaceous weed hosts also. Besides causing direct damage to rice crop by sucking the sap from the tissues, the pests are responsible for transmitting the virus causing **'rice tungro virus disease'** and the mycoplasma causing **'rice yellow dwarf disease'**.

Nature of damage. The pests are seen mostly on the foliage of the plants. Both the nymphs and adults suck the sap from the leaves and succulent stem portions and cause damage. The leaves of infested plants show yellowing, with drying of leaf tips. In case of severe infestation, the whole plant turns yellow, becomes stunted and eventually dies. Young plants are more prone to attack by the pests than the older plants. The damage caused by *Nephotettix virescens* is more than the damage caused by *N. nigropictus* (Fig. 9).

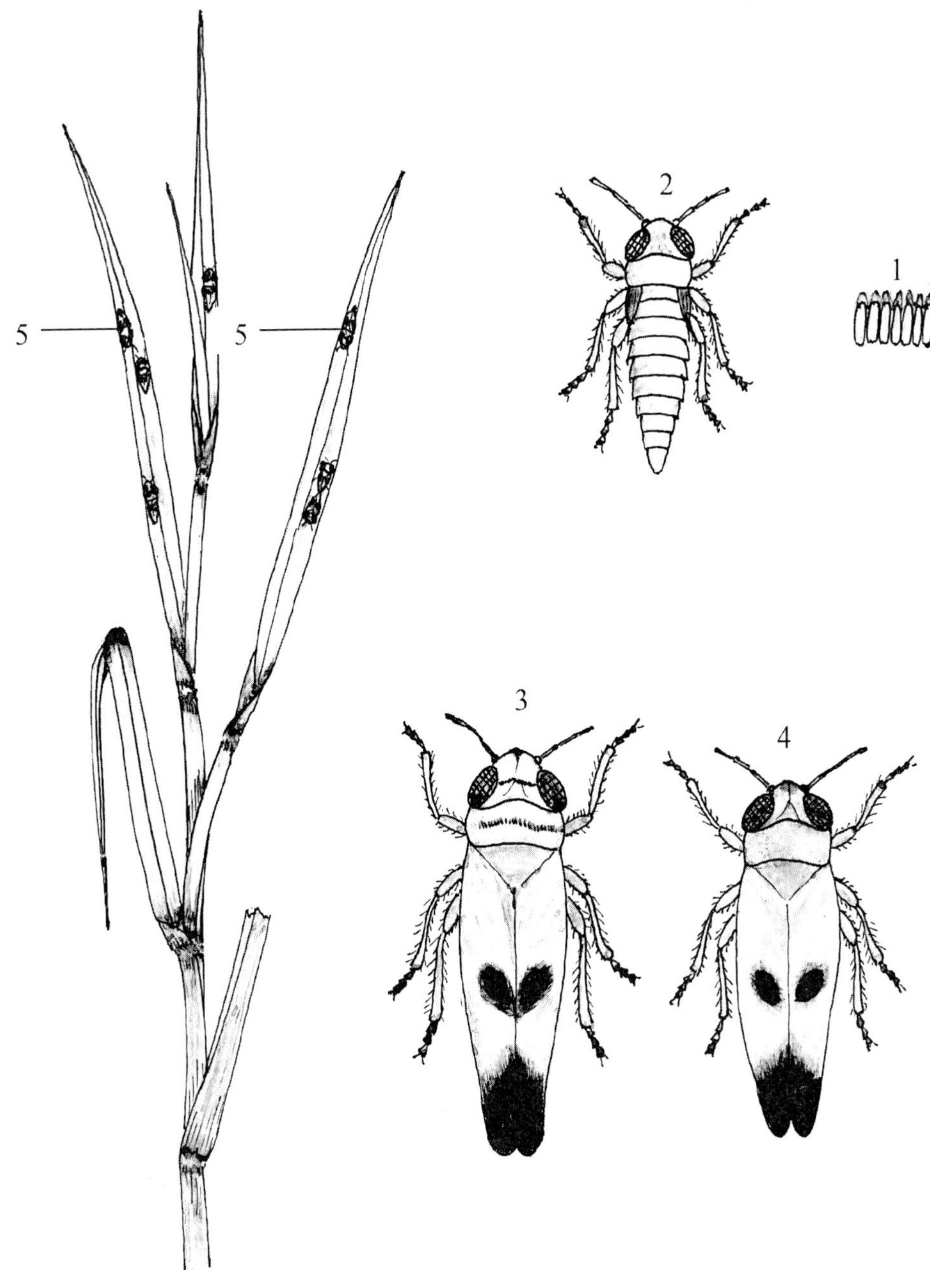

Fig. 9. Rice green leaf hopper-*Nephotettix* species

1. Eggs 2. Nymph 3. *Nephotettix nigropictus* or *N. apicalis* 4. *Nephotettix virescens* or *N. impicticeps* 5. Nymphs and adults sucking sap from leaves

Life cycle of the pests. The female hopper inserts the eggs inside the epidermal layer of the leaf sheath of plants, preferably young plants in rows of 1-20 in each row. A single female hopper can lay up to 420 eggs in about 44 egg masses. The eggs are cylindrical in shape and translucent in the beginning, which turn brown before hatching. They hatch in 6-7 days. The nymphs, which emerge out of the eggs are pale green in color, wingless, but are very active. They hop from leaf to leaf and from plant to plant and suck the plant sap from leaves and tender stem portions and grow. They moult their skin 5 times during their growth and become adults in 18-20 days. The life cycle is completed in about 25 days. The adults live for 50-60 days. They are small, 4.0-6.0 mm. in length with the front portion broader and the hind portion slightly narrower and green in color. The females are slightly bigger than the males. The life cycle of both the species is almost similar.

The males of *Nephotettix virescens,* have two, black spots on the forewings, which do not extend up to the black distal portion and there is no black, sub-marginal band on the crown of the head. The males of *Nephotettix nigropictus* have two, black spots, extending up to the black distal portion on the forewings, a black tinge along the anterior margin of the pronotum and a sub-marginal band on the crown of the head. The females are green and a black tinge on the pronotum is absent. Only a few females have black spots on the wings (Fig. 9).

Control measures

Physical and cultural methods (i) The fields, as well as the surrounding areas should be kept free from weeds, which may harbor the hoppers in the absence of rice crop (ii) Light traps should be set up to monitor the occurrence of the pest and to destroy them.

Chemical control

Nursery treatment (i) Foliar spraying with monocrotophos-40 ml. or endosulfan-40 ml. or phosalone-40 ml. or quinalphos-40 ml. or phosphamidon-10 ml. in 20 litres of water for 8 cent nursery area with a hand operated sprayer is effective (ii) Application of phorate 10 G-400 gm or carbofuran 3 G-1400 gm or quinalphos 5 G-800 gm mixed with 1,0 kg of fine sand uniformly over a thin film of water and impounding the water in the nursery for a few days is effective in controlling the pest.

Main field treatment (i) If the pest appears in the main field, foliar spraying with monocrotophos-400 ml. or fenthion-200 ml. or phosphamidon-110 ml. in 200 litres water per acre may be given with a high volume sprayer on the fifteenth day and thirtieth day after transplanting (ii) Foliar application of neem oil-2.0% (neem oil-4.0 litres and liquid soap-200 ml) is also effective.

Resistant varieties. TKM.6, ASD.5, Co.25, ADT.37 and PY.3 have been found to be resistant to the pests.

11. Rice brown plant hopper (BPH) or Fulgorid bug

Nilaparvata lugens

Order - Hemiptera
Suborder - Homoptera
Family - Fulgoridae

The pest was considered as an occasional pest on kuruvai crop at the short blade stage in South India. But it occurs in some seasons in epidemic proportions and causes severe damage to the crop and results in heavy yield loss. The pest appeared in a very serious form in Kerala in 1973 and the yield loss recorded ranged from 10-70%. It appears in a severe form during some seasons especially on high yielding varieties of rice in Tamil Nadu, Telungu Desam, Karnataka, West Bengal, Maharashtra Madhya Pradesh, Uttar Pradesh, Haryana and Punjab. Besides causing direct damage to rice crop, it is responsible for transmitting the virus, which causes **'grassy stunt disease'**. It also attacks cholam, lesser millets, sugarcane and some weeds found commonly in rice fields.

Nature of damage. The pest occurrence is severe during the flag leaf stage of the crop. Both the nymphs and adults congregate in large numbers on the stem portion just above the water level and suck the sap directly from the phloem vessels. Due to continuous draining of sap from the plants, they turn brown and dry within a short period of time. In the initial stages, this type of drying is seen in a few patches in the field. Gradually the patches enlarge and the plants show a burnt-up appearance and is known as **'hopper burn'**. Within a few days, the entire field is infested and large scale drying occurs. While feeding, the insects inject some toxic substances along with the saliva, which is also responsible for producing hopper burn. Due to the attack by the pest, the root growth of the plants is also affected and the protein and water content of the leaves are also reduced to a considerable extent and the plants die before the milky stage. Even if earheads are formed, the grains become completely chaffy. The castings of the insects are seen floating on the surface of water or sticking on to the stem portions in large numbers (Fig. 10).

Life cycle of the pest. The female hopper generally lays eggs in two rows, at 1-16 eggs per row. One insect can lay up to 650 eggs. The eggs are inserted inside the tissues of leaf sheaths or on both the sides of the mid rib and the tips of the eggs are joined by a gummy substance secreted by the mother and the rows of eggs appear as black lines. The eggs are small, about 1.0 mm. in length and look like banana fruits. They hatch in 6-7 days and the young nymphs emerging out of the eggs look whitish in color. They moult their skin 4 times and finally turn light brown in color. The nymphal period lasts for 14-28 days. They remain on the lower part of the stem just above the water level most often. The adult hoppers with wings are very active, especially during the morning hours. The adults live for 10-15 days. The entire life cycle occupies 20-35 days. The adult hoppers are yellowish-brown in color, with dull white streaks on the thorax and black lines on the lateral sides of the head. The females are slightly bigger than the males and measure 3.5-4.5 mm long. Two forms namely, 'macropterous' (long-winged) and 'brachypterous' (short-winged) are found, of which the former is responsible for migratory movement and colonization in other neighbouring rice crops (Fig. 10).

Control measures

Physical and cultural methods (i) The rice fields and surrounding areas should be kept free of weeds and other alternate hosts (ii) After the harvest, the field should be ploughed and the stubble buried under the soil to prevent the growth of self-sown and ratoon crops, so as to avoid multiplication of the pest during the off-seasons (iii) Impounding too much of water in the fields should be avoided and excess water should be drained off (iv) Light traps should be set up to monitor the occurrence of the pest and to destroy them (v) Delta sticky traps may be set up just above the crop canopy to trap and destroy the flying adults and hopping nymphs. Otherwise improvised sticky traps with some adhesive material on both the surfaces of cardboards or wooden boards may be set up (vi) In tracts where the pest appears every year, strip planting should be adopted to facilitate pest monitoring and plant protection operations. Six feet wide planting strips alternated with an interspace of one foot is ideal.

Chemical control

Nursery treatment. If the pest appears in the nursery, 10 days prior to pulling out the seedling, water is impounded in the nursery to a height of 2.5 cm. and carbofuran 3 G-1.4 kg., mixed with 1.0 kg. of sand is broadcasted uniformly over 8 cent nursery area and the water is retained for a few days.

Main field treatment (i) If the pest appears before the flowering stage, the crop may be sprayed with monocrotophos-200 ml. or phosalone-400 ml. or phosphamidon-110 ml. or chlorpyrifos-400 ml. or lindane-400 ml. or methyl demeton-400 ml. or dichlorvos-200 ml. in 200 litres of water with a high volume sprayer. The spray should be directed to the basal parts, so that the stem portions are completely covered (ii) If the pest occurs during the milky stage, dusting with carbaryl 10 D at 10 kg. per acre may be taken up, so as to cover

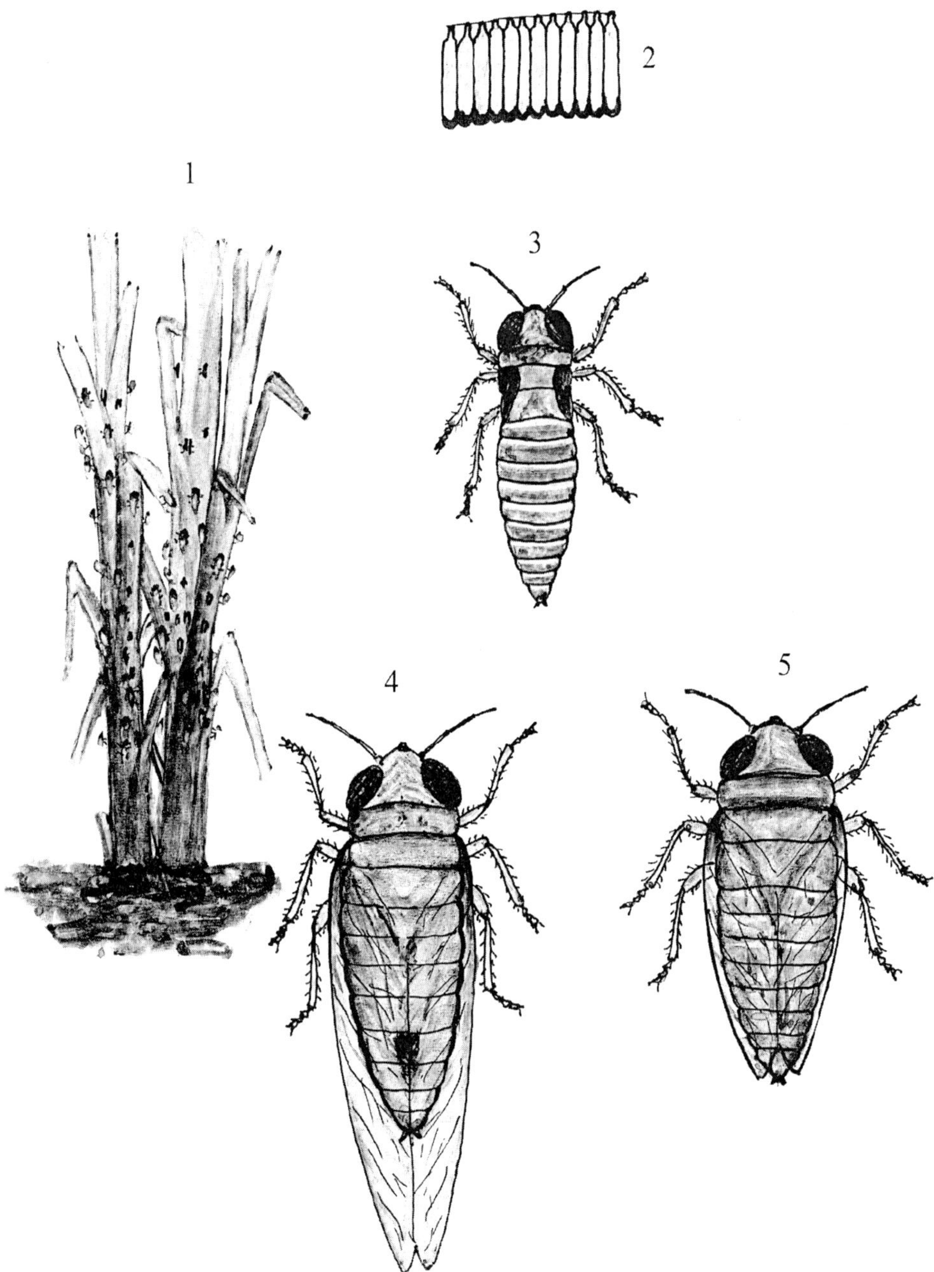

Fig. 10. Rice brown plant hopper-*Nilaparvata lugens*

1. Nymphs and adults sucking sap from the stem region 2. Eggs 3. Nymph 4. Well developed long-winged adult 5. Short-winged adult

the basal parts of the plants thoroughly (iii) Spraying with neem seed kernel extract-5% or neem oil-3% is also effective in controlling the pest.

Some insecticides may control sucking pests, such as brown plant hopper, white flies, aphids etc. effectively in the initial stages. However, repeated use of these insecticides may lead to 'resurgence'. Insecticides, such as synthetic pyrethroids, fenthion, quinalphos and methyl parathion may have such qualities. So, use of these chemicals more than once or twice continuously should be avoided.

Biological control (i) The predatory spider-*Lycosa pseudoannulata* commonly found in rice fields feeds on the adults and nymphs and can destroy several hoppers in a day (ii) Another predator-*Cyrtorhinus lividipennis* also feeds on all the stages of the pest, primarily on eggs.

Resistant varieties. The varieties, Co.25, Co.42, ASD.5, ADT.36, ADT.37 and PY.3 are found to be resistant to the pest.

12. Rice grasshopper

Hieroglyphus banian

Order - Orthoptera

Family - Acrididae

The pest is commonly found in all rice growing areas in India. Besides rice, the pest attacks sugarcane, lesser millets and some other grass weeds.

Nature of damage. The nymphs and adults cause damage to the leaves of the crop by notching the leaf blades from the leaf edges. Various types of irregular, angular-shaped cut edges are seen in the leaves of the infested crop. Sometimes the leaves are cut deep up to the midribs. However, the leaf midribs and stem portions are not attacked. At the flowering stages, the insects knaw the peduncles of the earheads, as a result the earheads break and hang down or white ears are formed leading to severe crop losses. The insects also nibble and eat the floral parts.

Life cycle of the pest. The female grasshopper digs a deep hole in soft soil, mostly by the side of a bund with its ovipositor up to a depth of 5.0-8.0 cm. and lays 30-40 eggs in a spiral arrangement, which is called an 'egg pod'. The eggs are cylindrical, with rounded ends, whitish in color and look like rice grains. After laying eggs, the female seals the hole with a gummy secretion and then covers it with soil with its hind legs. The eggs laid during the months of October-November, hatch during June-July next year after the summer showers and the nymphs emerge out. The nymphs are pale green or pink in color and lack wings. In all other respects, they resemble the adult grasshoppers. They moult their skins 5-6 times at 10-15 days interval and become adults with wings in 75-105 days. The adults are big, about 3.75 cm. in length, green in color, with three, black, transverse lines on the lateral sides of the prothorax. There is only one generation in a year (Fig. 58).

Control measures

Physical and cultural methods (i) Before planting, the fields should be well ploughed and the bunds trimmed properly, so that the egg pods are brought out and destroyed (ii) The fields and surrounding areas should be kept free from weeds, which may serve as alternate hosts (iii) The nymphs and adults may be caught using hand nets and destroyed.

Chemical control (i) When the pest occurs in large numbers, dusting with carbaryl 10 D at 10 kg./ acre may be adopted (ii) Foliar spraying with endosulfan-400 ml. or chlorpyrifos-400 ml. or fenitrothion-400 ml. affords good control of the pest.

13. Rice whorl maggot

Hydrellia sasakii (philippina)

Order - Diptera

Family - Ephydridae

The pest was first recorded in 1963 and has gained importance, especially in some of the high yielding varieties in Telungu Desam, Tamil Nadu and Orissa. The pest also thrives on *Cynodon dactylon, Echinochloa crusgalli,* and a few other grass hosts.

Nature of damage. The grubs bore into the shoots of young plants and feed on the inner tissues. When the young leaves unfurl, narrow sripes of whitish area are seen in the margins, near the leaf tips at the point of entry of the grubs. Often the leaves shrink and break at the white patches. Severely affected plants show stunted growth (Fig. 7).

Life cycle of the pest. The pest occurs in a severe form in water inundated fields, when the plants are young. Stagnant water attracts the flies. They are active during the daytime and afterwards rest on the leaves just above the water level. The female fly lays about 100 eggs, singly within 3 or 4 days on tender leaves of transplanted crop, which are about 30 days old. Very often, the eggs are laid on plants near the bunds. The eggs are long, cylindrical and are whitish in color and are pasted to the leaves with a gummy substance. The eggs hatch in 2-6 days. The young grubs, which emerge out from the eggs are translucent or pale white in color. They wriggle in the water droplets found on the surface of the leaves and reach the terminal shoot, bore into it, feed on the inner tissues and become full-grown in 8-17 days. They lack legs and are yellow in color. They pupate inside the shoot. The pupae are deep brown in color. The adult flies emerge from the pupae in 5-9 days. The flies are very small, appear as mosquitoes, gray in color with long legs and transparent wings. They are negatively phototrophic (Fig. 7).

Control measures

Cultural methods (i) Stagnating excess of water in the fields should be avoided (ii) The fields and surrounding areas should be kept free from weeds, which may harbor the pest.

Chemical control (i) Foliar spraying with endosulfan-400 ml. or quinalphos-400 ml or chlorpyrifos-400 ml. is effective in controlling the pest (ii) Application of carbofuran 3 G-13 kg. or phorate 10 G-5.0 kg. at the time of transplanting controls the pest. But the cost of this treatment is very high.

14. Rice yellow hairy caterpillar

Psalis pennatula

Order - Lepidoptera

Family - Lymantidae

The pest is often found on many grass hosts and sometimes infests rice crop and causes severe damage

Nature of damage. The pest is a leaf feeder and the caterpillars cut and feed on the leaves (Fig.12)

Life cycle of the pest. The female moth lays spherical eggs in clusters on the under surface of leaves. The young caterpillars emerging from the eggs scrape and eat the green matter of the leaves in the beginning. Afterwards, they start eating the leaves and grow. The full-grown caterpillars are yellowish-brown in color, with orange-colored head and have lateral red stripes on the body. They have hairs all over the body surface, with two prominent, longer tufts of hairs on the anterior part and one tuft on the posterior part of the body. The full-grown larvae measure about 3.75 cm. in length. The larva makes a pale white cocoon with silken thread and frass and pupates inside it. The pupae are found sticking on to the leaf surface. The moths are stout, with their forewings straw-colored and hindwings dirty white in color (Fig.12).

Control measures

Chemical control. When the pest occurs in large numbers, foliar spraying with endosulfan-400 ml. or quinalphos-400 ml. or fenitrothion-200 ml. or methyl parathion-200 ml. in 200 litres of water per acre, with a high volume sprayer may be taken up.

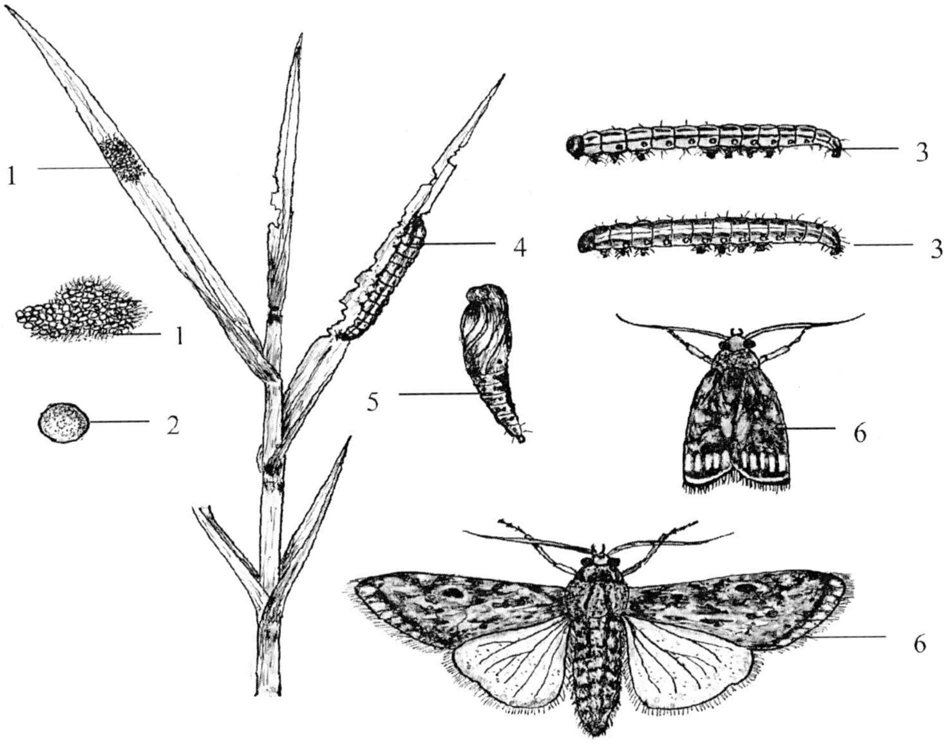

Fig. 11. Rice swarming cater pillar-*Spodoptera mauritia*

1. Egg mass 2. Single egg 3. Larva 4. Infested plant 5. Pupa 6. Adult moth

15. **Swarming caterpillar** or **Army worm**

Spodoptera mauritia

Order - Lepidoptera

Family - Noctuidae

The pest is a sporadic one and appears in a severe form during some seasons and causes severe economic losses to rice crop. Mostly it appears during July-September. Severely infested crop appears as if grazed by cattle. Besides rice, the pest is known to attack and breed on maize, sorghum, lesser millets, sugarcane, wheat, barley and many other grasses.

Nature of damage. The pest appears suddenly in swarms of thousands, completely eat the foliage of the entire crop and marches on to another field like an army and so, it is known as **'army worm'**. The caterpillars feed by night and hide in cracks, crevices, under mud clods etc. during the daytime (Fig. 11).

Life cycle of the pest. The female moth lays 200-400 eggs in 5-6 clusters on the under surface of leaves of rice and various kinds of wild grasses and covers the egg masses with buff-colored or grayish hairs. The eggs are small, spherical and light yellow in color. The eggs hatch in 6-8 days and the young caterpillars, which emerge out are green in color. They start feeding immediately on young leaves. The caterpillars caste their skin 4-5 times and become full-grown in about 20 days. Grown-up larva is about 3.5 cm. long, cylindrical, smooth, and dark to pale green in color, with dull reddish and yellowish dorsal and sub-dorsal stripes. There is a great deal of color variations in the caterpillars. The full-grown caterpillar pupates in an earthen cocoon under the soil. The pupae are dark brown in color. The moth emerges from the pupa in 10-15 days. The life cycle is completed in 30-40 days. There may be 2-4 generations in a year. The moth is of medium size, grayish-brown in color, with a prominent black spot near the anterior margin and irregular, wavy, black band along the lateral margin of the forewings. The hindwings are white in color (Fig.11).

Control measures

Physical and cultural methods (i) The caterpillars, which are big and easily visible can be collected during the evening hours and destroyed (ii) Flooding the nursery forces the larvae to come out from their hiding places and are picked up by insectivorous birds (iii) Ducks allowed into the fields also feed on the caterpillars

Chemical control (i) Foliar spraying with endosulfan-400 ml. or quinalphos-400 ml. or chlorpyrifos-400 ml. or fenitrothion-200 ml. or fenthion-200 ml. or methyl parathion-200 ml. in 200 litres of water per acre with a hand operated sprayer during the evening hours is effective in controlling the pest. (ii) Poison bait consisting of a mixture of rice bran-5.0 kg., molasses-1.0 kg. and carbaryl 50 WP-500 gm. may be used to attract and destroy the pest (Refer pests of cotton, page 162).

16. Rice black bug

Scotinophara coarctata and *S. lurida*

Order - Hemiptera

Suborder - Heteroptera

Family - Podopidae

It is a sporadic pest of rice and appears in some parts of Tamil Nadu during some seasons in a serious form. Besides rice, the pest is known to infest wheat, maize, sorghum, sugarcane, and species of *Echinochloa, Cyperus* etc. and wild rice varieties.

Nature of damage. Both the nymphs and adults of the pest harbor at the base of the plants, suck and feed the sap from the stems and also from leaf sheaths, leaves and panicles. The infestation may occur from tillering to flowering stages of the crop. As a result of continued sucking of sap by large number of the insects, the plants become reddish-brown or yellow. The bugs prefer feeding on the nodal regions, because of the availability of more sap in these parts. When the pest occurs in large numbers during the tillering phase, dead hearts may appear, besides stunted growth and reduction in the number of tillers. Infestation during the flowering stage results in the production of white ears and chaffiness. Severely infested plants wilt, dry and present a burnt appearance.

Life cycle of the pest. The adult female lays 25-50 eggs in 3-5 linear rows on the surface of old leaves or on the stem portion near the soil level. The nymphs emerge out of the eggs in 4-7 days. They moult 4-5 times during their growth period and become full-grown in 30-50 days. The adult bugs live for 60 days or more. The males have a longer longevity than the females. Full-grown bugs are deep brown to brownish-black or shiny black in color, 7.0-9.0 mm. in length, with black head and neck region and yellowish-brown antennae. There may be a few pale yellowish dots on the thoracic region (Fig. 13).

Control measures

Cultural methods (i) Rice fields and surrounding areas should be kept free from weed hosts (ii) Stagnating excess quantity of water in the fields should be avoided.

Chemical control (i) In case of mild infestation, dusting with carbaryl 10 D at 10 kg/acre is effective in controlling the pest (ii) Spraying with monocrotophos-400 ml. or neem seed kernel extract-5% controls the pest. Before dusting or spraying, the water in the field should be drained off and the dust or spray should be directed towards the basal part of the plants.

Biological control. The egg parasite-*Telenomus triptus* parasitizes and destroys the eggs.

Resistant varieties. Varieties such as TKM.9, IR.50 and ADT.36 are prone to attack by this pest. So, in places where this pest is prevalent, growing these varieties should be avoided.

17. Climbing cutworm

Mythimna albistigma

Order - Lepidoptera

Family - Noctuidae

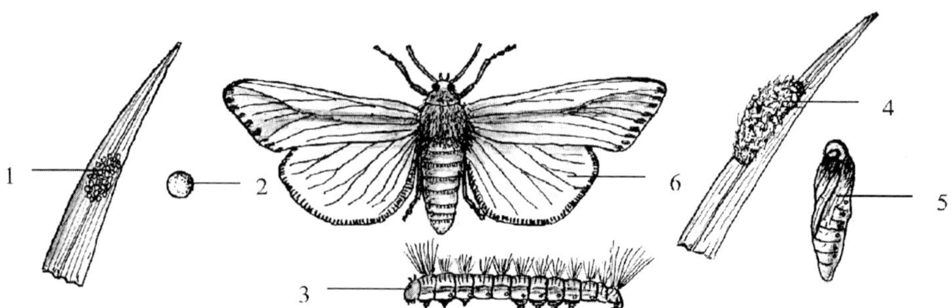

Fig. 12. Rice Yellow hairy caterpillar-*Psalis pennatula*

1. Egg mass 2. Single egg 3. Caterpillar 4. Pupa inside cocoon 5. Pupa 6. Adult moth

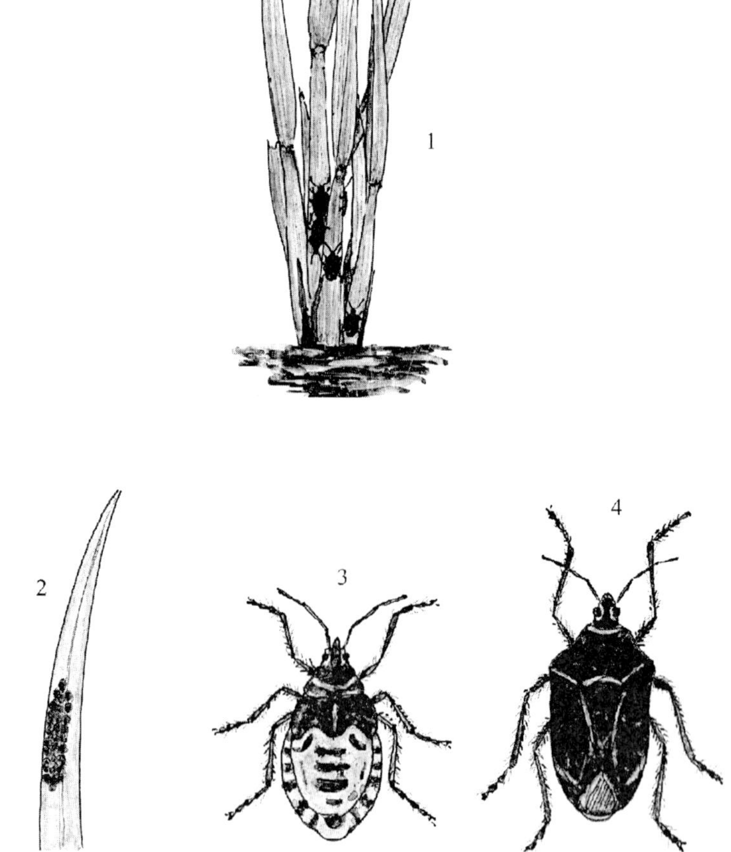

Fig. 13 Rice black bug-*Scotinophara coarctata*

1. Nymphs and adults sucking sap from the stem region 2. Eggs 3. Nymph 4. Full grown adult bug

It is a sporadic pest and occurs only during certain years, especially when there is heavy rainfall and the fields are flooded. It is a very destructive, polyphagous pest. Besides rice, it attacks wheat, maize, barley, sugarcane, lesser millets and several grasses. The pest generally appears during the later stages of crop growth and causes severe yield loss.

Nature of damage. The caterpillars, which appear in swarms, are nocturnal in habit. They hide themselves in folds of leaves or cracks and crevices in the soil during the daytime. During the nighttime they climb up the plants, bite and cut off the half ripe earheads.

Life cycle of the pest. The life cycle of the pest is very much similar to that of the swarming caterpillar. The female moth lays 500-700 eggs in batches or clusters, mostly on the lower leaves and covers the egg masses with grayish hairs. The eggs are small, spherical and greenish-white in color. They hatch in 4-5 days and the newly hatched young caterpillars are pale green in color and have looping habit of crawling till they are half grown. The caterpillars hide during the daytime at the base of the hills or cracks in the soil. They climb up the plants at dusk and cut and feed on small and big raches and earheads, which are nearing maturity. When the food supply in a field is exhausted, they march in swarms like an army and attack the neighbouring fields. Full-grown caterpillar is about 3.5 cm. in length, smooth and cylindrical, with a dark narrow, broken stripe along the center of the back, bordered by a wide, darker, mottled stripe. There are three dark stripes on the lateral sides also. The head is marked with dark lines. There are much color variations in the larvae. The caterpillars moult four times before they become full-grown. The larval period lasts for 21-28 days. The grown-up caterpillar constructs an earthen chamber in the soil and pupates inside it. If there is standing water in the field, the caterpillar pupates naked in the frass at the base of the plant. The pupa is dark brown in color and about 1.25 cm. long. The adult moth emerges from the pupa after 10-12 days. The moths are brown or brownish-gray in color, with a wingspan of 3.0-5.0 cm. There is a single, white spot in the center of each forewing. The moths are strong fliers. They remain hidden during the daytime and become active after dusk. They are positive phototrophic. There may be up to five generations in a year.

Control measures

Physical and cultural methods (i) Rice fields and surrounding areas should be kept free from weeds and other alternate hosts (ii) Flooding the fields for a few days at the time of maturity, forces the larvae to come out from their hiding places and are picked up by insectivorous birds (iii) Ducks allowed into the fields also feed on the caterpillars (iv) The caterpillars can be hand picked and destroyed.

Chemical control (i) Foliar spraying with endosulfan-400 ml. or quinalphos-400 ml. or chlorpyrifos-400 ml. or fenitrothion-200 ml. or fenthion-200 ml. or methyl parathion-200 ml. in 200 litres of water per acre with a hand operated sprayer during the evening hours is effective in controlling the pest. (ii) Poison bait consisting of a mixture of rice bran-5.0 kg., molasses-1.0 kg. and carbaryl 50 WP-500 gm. may be used to attract and destroy the pest (Refer pests of cotton, page 162).

18. Rice root weevil

Echinoenemus oryzae

Order - Coleoptera

Family - Curculionidae

The pest was first recorded in Haryana state in 1953 and in Telungu Desam in 1954. Since then, it has been reported from parts of Tamil Nadu, Madhya Pradesh and Uttar Pradesh in a sporadic form. The pest is known to attack a number of grass hosts, besides rice.

Nature of damage. The pest is active only during the months of July-September. The grubs feed on the roots and root hairs of the transplanted rice crop and affect the plant growth. The infested crop becomes stunted and tillering is adversely affected. Severely affected plants may be killed.

Life cycle of the pest. The female weevil lays eggs on the basal part of the plants. The grubs, which emerge from the eggs in a few days, are aquatic. They feed on the root hairs and roots and grow. The grubs are small, whitish, translucent, and apodous and have 6 pairs of prominent tubercles on the dorsal side of the abdomen, connected to the tracheal system, which aid them in the respiration process. The aquatic grubs obtain oxygen from the air spaces inside the roots of the host plants. The grubs become full-grown by the middle of September. The grown-up grub is about 6.0 mm. in length. They burrow deep into the soil to a depth of 8.0-20.0 cm. and pupate inside earthen cocoons and pass the rest of the period as pupae. The weevils emerge from the pupae during June-July of the next year, after the receipt of the monsoon rains and the weevils are seen sitting in large numbers on rice plants and grass hosts at the time. The weevils are small and brownish in color. There is only one generation in a year.

Control measures

Physical and cultural methods (i) Rice fields and surrounding areas should be kept free from weeds, which may harbor the pest (ii) The fields should be ploughed during the summer months before the onset of rains to expose and destroy the pupae (iii) Adult weevils resting on rice plants and grasses may be collected by using hand nets and destroyed (iv) Application of superphosphate at 40 kg./ acre at the time of transplanting has been found to deter active feeding of the roots by the grubs.

Chemical control (i) Application of carbaryl 10 D at 20 kg/acre at the time of planting controls the pest (ii) Spraying with malathion-400 ml. or endosulfan-400 ml. or methyl parathion-200 ml. is effective in destroying the adult weevils resting on rice plants.

19. Rice horned caterpillar

Melanitis leda

Order - Lepidoptera

Family - Satynidae

The leaf feeding caterpillar pest is found all over South India. It attacks rice crop at all growth stages.

Nature of damage. The caterpillars of this butterfly pest feed on the leaf blades of rice plants.

Life cycle of the pest. The female butterfly lays spherical, white eggs singly on the leaves. The caterpillars are elongated and green in color, with a rough body surface. They have two, red, horn-like processes on the head and two yellow processes at the anal end. The full-grown larva pupates in a chrysalis, which is suspended from the leaf. The adult butterfly is dark brown in color, with large wings having a black and yellow, eye-like spot on each of the forewings.

Control measures. The measures suggested for the control of other leaf feeding caterpillar pests of rice may be followed.

20. Rice skipper

Pelopidas mathias

Order - Lepidoptera

Family - Hesperiidae

It is another leaf feeding caterpillar pest found commonly all over India. The pest attacks the crop at all growth stages.

Nature of damage. The caterpillar folds the leaves longitudinally, fastening the edges together, remains inside the fold and feeds on the green matter resulting in stunted growth of the plant.

Life cycle of the pest. The female skipper butterfly lays eggs singly on the leaf blades. The caterpillar is elongated, green in color with a smooth body. The neck is constricted and the head is prominent with a red 'V' mark on it. The pupa is held by a girdle and is attached to the leaf blade inside the leaf roll at its anal end. The adult butterfly is dark brown, with two white spots on the forewings.

Control measures. The measures suggested for the control of other leaf feeding caterpillar pests of rice may be followed.

Pests of minor importance. Besides the pests described above in detail, rice crop is subjected to attack by several other pests, which are considered to be of minor importance. However, when conditions are favorable for their occurrence and spread, they may become destructive and may cause extensive damage to the crop and may result in appreciable yield loss; The rice white leafhopper-*Tettigella spectra* is often found along with the green leaf hoppers. Both the nymphs and adults suck the sap from the leaves, causing yellowing of leaves and stunting of plants. But the pest is not a vector of any virus diseases of rice; the white-backed rice plant hopper-*Sogatella furcifera* also causes similar damage to rice crop; the rice striped bug-*Tetroda histeroides* is another sucking pest. The nymphs and adults of this bug pest suck the sap from the stem, resulting in stunting and yellowing of the plants; another grass hopper pest-*Oxya nitidula* also causes damage to rice crop by cutting and feeding on the leaves; the borer pests viz., the striped borer-*Chilo suppressalis (infuscatellus), Scirpophaga (Tryporyza) innotata* and the pink borer-*Sesamia inferens* cause damage to the plants by boring into the stem region and feeding on the inner tissues; the blister beetles-*Psalydolytta rouxi* and *Cylindrothorax tenuicollis* and the coccinellid beetle-*Alesia discolor* feed on the pollen of rice flowers; the panicle thrips-*Haplothrips ganglbaueri* lacerate and feed on the rice inflorescence, causing deterioration of rice grains.

MILLETS

SORGHUM or CHOLAM *(Sorghum vulgare)*

1. Sorghum stem borer

Chilo zonellus (partellus)

Order - Lepidoptera

Family - Crambidae

The pest has a wide distribution and is found in all the sorghum growing tracts in South India. Besides sorghum, the pest is known to attack pearl millet, maize, finger millet, sugarcane etc.

Nature of damage. The pest attacks seedlings from the age of 15-20 days up to the grown up stage of the plants. The young caterpillars emerging from the eggs scrape and feed on the green matter of the leaves. Then the caterpillar burrows through the stem of the young plants, feeds on the inner tissues and severes the central shoot. The borer-affected plants show dead central shoots known as **'dead hearts'**. The infested seedlings may eventually die. Though grown-up plants are also affected, the damage to young plants is far more serious. In the infested grown-up plants, when the shoots are bored, the new leaf, which unfurls shows small holes on either sides of the midrib. The infested plants fail to produce earheads or produce poor quality earheads (Fig. 14).

Life cycle of the pest. The female moth lays batches of scale-like, flattish, oval, overlapping eggs, mostly on the under surface of leaves by the side of the midrib. A single moth may lay 200-300 eggs. The eggs hatch in about 7 days. The young caterpillar gradually bites its way and burrows into the stem. It feeds on the inner tissues, severes and kills the central shoot in the young plant. Sometimes the young larva bores through the midrib and finally enters the stem. Most of the second instar caterpillars migrate to neighbouring plants kill the central shoots, resulting in the formation of dead hearts. The larval period lasts for 25-30 days. The full-grown caterpillar is about 2.5 cm. in length, yellowish-brown in color, with a dark brown head and a hard, chitinous shield over the prothorax and black, warty spots on the body. Full-grown larva makes an exit hole in the stem, covers it with a mesh made of silken thread and pupates inside the stem. The adult moth emerges from the pupa in 7-10 days. The moth is yellowish-brown in color, medium sized and measures about 3.0 cm. across the wings. The forewings are straw-colored, with two rows of black spots near the lateral margins. The hindwings are yellow in color. The life cycle is completed in 40-50 days. There may be 5-6 overlapping generations in a year. The moths are nocturnal in habit and are attracted to light (Fig. 14).

Control measures

Physical and cultural methods. (i) Seedlings showing dead heart symptoms should be removed and destroyed (ii) Light traps should be set up to attract and destroy the moths (iii) After harvest, the stubble should be removed or ploughed deep under the soil (iv) In tracts where the pest occurs regularly, the seed rate may be slightly increased, so that even after the removal of infested seedlings, normal plant population can be maintained.

Chemical control

Irrigated crop (i) Foliar spraying with endosulfan-300 ml. or carbaryl 50 WP-600 gm. in 300 litres of water per acre with a high volume sprayer controls the pest effectively. Young plants of age 15-20 days should also be sprayed with any one of these insecticides (ii) Soil application of carbofuran 3 G-13 kg. or phorate 10 G-4.0 kg. or endosulfan 4 G-10 kg. and irrigating the field

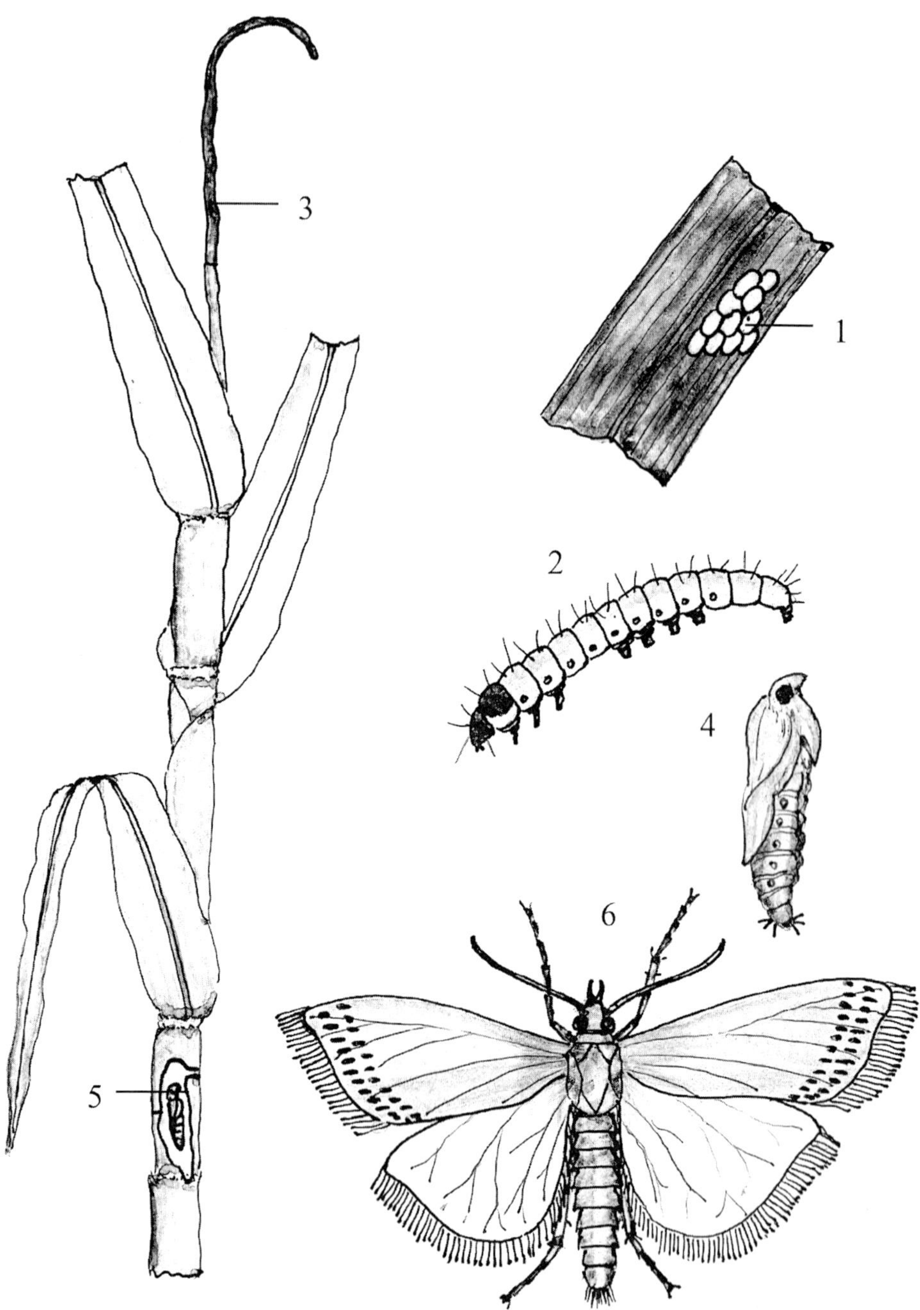

Fig. 14. Sorghum stem borer-*Chilo partellus (or) C. zonellus*

1. Egg mass 2. Larva 3. Dead heart 4. Pupa 5. Pupa inside the stem 6. Moth

immediately after is also effective in controlling the pest. But, the cost of granular application is high.

Rainfed crop (i) In case of mild infestation, dusting with phosalone 4 D or phenthoate 2 D at 10 kg./ acre will control the pest (ii) In case of severe infestation spraying with endosulfan-300 ml. or carbaryl 50 WP-500 gm. controls the pest effectively. While spraying. the spray should be directed towards the stem portion of the plants.

Biological control (i) A few species of *Trichogramma* and *Telenomus* parasitize and destroy the eggs (ii) Some larval parasites viz., *Bracon chinensis* and *Chelonus narayani* destroy the larvae (iii) The pupal parasite-*Tetrastichus ayyari* destroys the pupae (iv) The fungal parasite-*Aspergillus flavus* attacks the larvae, causes diseases and destroys them (v) The coccid beetle-*Coccinella undicempunctata* is predatory on the larvae.

2. Sorghum shoot fly

Atherigona soccata

Order - Diptera

Family - Anthomyiidae

It is one of the most common and serious pests of sorghum and occurs in all sorghum-growing states of India. The infestation is more severe during the months of October-December in South India. The pest attacks only the seedlings and the damage to seedlings may go up to 80 per cent or more under conditions favorable for the occurrence and spread of the pest.

Nature of damage. The young maggot, which hatches out of the egg crawls in-between the leaf sheath and the stem, bores into the shoot, bites and eats the inner tissues, as a result the shoot is severed from inside leading to rotting of the cut portion of the shoot and death of the growing shoot, thus producing 'dead heart'. Infested seedlings may die. The maggots attack seedlings up to the age of about 30 days only (Fig. 15).

Life cycle of the pest. The female fly lays eggs in rows of 3-4 eggs on the under surface of leaves or on the leaf sheath mostly during the morning hours. One female lays 20-25 eggs during its life span of 12-14 days. The eggs are whitish in color, long and cylindrical in shape with the distal end somewhat flattened. The eggs hatch in 1-2 days. The maggot is pale yellow in color, without a well-defined head and without legs. The anterior portion is slightly narrower than the posterior portion. The maggot becomes full-grown in 12-15 days after passing through 4 larval instars. The grown-up larva pupates within the stem and emerges as an adult fly from the seed-like pupa in about 7 days. The fly is very small, mosquito-like and whitish-gray in color (Fig. 15)

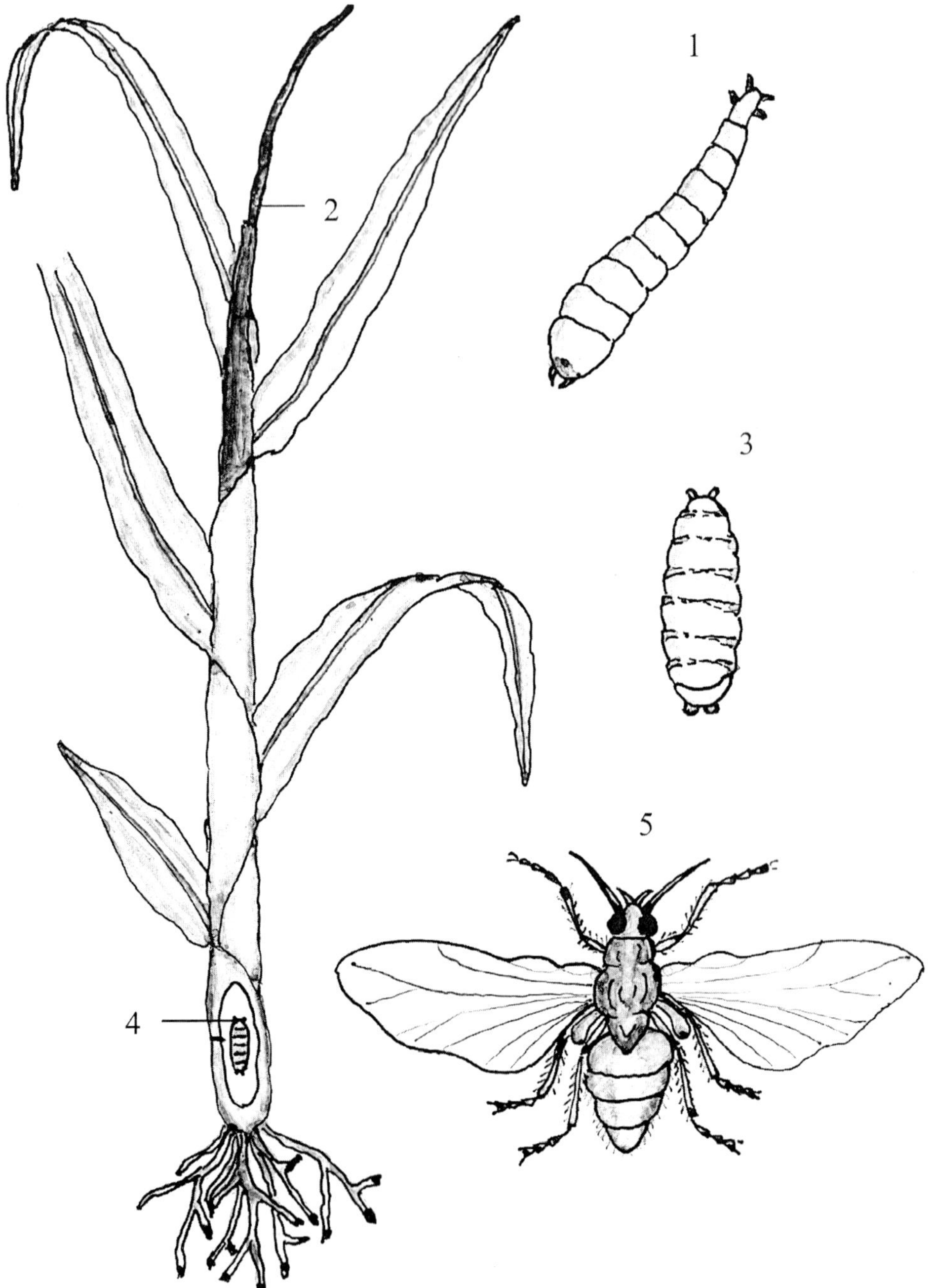

Fig. 15. Sorghum shoot fly-*Atherigona soccata*

1. Larva 2. Dead heart 3. Pupa 4. Pupa inside the stem 5. Fly

Control measures

Physical and cultural methods (i) The seed rate may be slightly increased (5.0 kg./ acre) and the infested seedlings showing dead heart symptoms may be removed and destroyed at the time of thinning out or at the time of first weeding (ii) The crop should be sown early, immediately after the onset of the monsoon rains, so that the seedlings may escape infestation.

Seed treatment. Seed treatment with chlorpyrifos is effective in preventing infestation. Chlorpyrifos-4.0 ml. and gum-0.5 gm. are mixed in 80 ml. of water and 1.0 kg. of seeds are put inside the mixture and mixed thoroughly, shade dried and then sown.

Chemical control (i) Soil application of carbofuran 3 G-13 kg. or phorate 10 G-4.0 kg. or disulfotan 5 G-8.0 kg. or solvirex 5 G-8.0 kg., mixed with 10.0 kg. of sand and applied uniformly along the sowing furrows at the time of sowing, followed by irrigation is effective in preventing infestation (ii) Spraying with endosulfan-200 ml. methyl demeton-200 ml. or dimethoate-200 ml. in 200 litres of water per acre with a high volume sprayer, so as to cover the stem portion of the seedlings thoroughly is also effective in controlling the pest.

3. Sorghum earhead bug

Calocoris angustatus

Order	-	Hemiptera
Suborder	-	Heteroptera
Family	-	Miridae

It is a serious sucking pest of sorghum all over India. In South India, the pest infestation is more severe in the irrigated crop raised during April-June and is much less in the rainfed crop raised during August-January. It is called **'kadir navai poochi'** in Tamil. The loss in grain yield may go up to 15-30 per cent, when the infestation is severe. Varieties producing compact earheads are more prone to attack by this pest. Besides sorghum, the pest infests pearl millet, maize, finger millet, lesser millets, sugarcane and many grasses

Nature of damage. The pest appears at the time of emergence of earhead from the boot leaf. Both the nymphs and adults infest the earheads in large numbers, pierce the developing grains and suck the sap from them and cause damage to the developing grains. The affected grains shrink and become chaffy. In case of severe infestation, hundreds of nymphs and adults are found on a single earhead.

Life cycle of the pest. The adult female bug inserts 2-15 eggs in clusters in the flowers of the earheads One female lays 150-200 eggs. The eggs are light blue in color and cigar-shaped. They hatch in 5-6 days. The nymphs emerging from the eggs are yellowish-green in color, without wings and are very active. In all other characters they resemble the adult bugs. During the growth, the nymphs moult 4-5 times and become full-grown bugs with wings. The adult bugs are pale green in color, long and stick-like, with long legs and measure 0.5-0.7 cm. in length. The life cycle is completed in 20-26 days. There are two generations in a season.

Control measures

Physical method. The nymphs and adults can be caught by using hand nets and destroyed.

Chemical control. Dusting the earheads thoroughly with carbaryl 10 D or malathion 5 D or phosalone 4 D at 10 kg./ acre, once immediately after emergence of the earheads, followed by another dusting 10 days later controls the pest effectively.

Biological control. A reduviid bug-*Reduviolus* sp. is predatory on the bugs.

4. Sorghum earhead midge

Stenadiplosis (Contarinia) sorghicola

Order	-	Diptera
Family	-	Cecidomyiidae

The pest is prevalent in all sorghum-growing countries of the world and is considered to be one of the major pests in India. In case of severe infestation, the loss in yield may go up to 20-50 per cent

Nature of damage. The maggots bore into developing grains at the milky stage and feed on the contents inside the grains. The affected grains, when pressed between the fingers, exude a reddish-colored liquid. Eventually the infested grains shrivel and turn chaffy. In such chaffy grains, the holes made by the maggots are distinctly visible.

Life cycle of the pest. The adult female midge inserts the eggs singly in the flowers of the earhead before the milky stage. One female lays 30-100 eggs. The eggs hatch in 2-3 days. The young maggot on emergence from the egg, bores into the grain, feeds on the contents inside the grain and grows. The larva is reddish in color and lacks legs. The full-grown larva pupates inside the grain itself in a white cocoon. The adult midge emerges from the pupa through a hole between the tip of the glumes, leaving behind the white pupal case attached to the tip of the floret. The life cycle is completed in 17-19 days. The midges are small, fragile, mosquito-like, orange-colored, with dark pinkish abdomen and thin, transparent wings.

Control measures

Cultural methods (i) Sowing should be taken up early in the season, immediately after the receipt of monsoon rains, so that the crop may escape from the pest infestation (ii) The adult flies are attracted to light. So, light traps should be set up to attract and destroy the flies (iii) After threshing, the chaff and other crop debris should be burnt to destroy the pupae that may be present inside the chaffy grains.

Chemical control. Dusting the earheads thoroughly immediately after their emergence from the boot leaf with carbaryl 10 D or malathion 5 D or phosalone 4 D at 10 kg. per acre is effective in controlling the pest.

5. Sorghum earhead webber

Cryptoblabes gnidiella

Order - Lepidoptera

Family - Pyraustidae

The pest occurs commonly in parts of South India and attacks the developing grains. It occurs mostly during the summer months and also during October-December. It infests maize crop also.

Nature of damage. The caterpillar webs together adjacent grains in the ear, remains securely inside the web and feeds on the developing grains and causes direct damage.

Life cycle of the pest. The young larvae hatching out from the eggs laid by the moth on the floral parts, remain near the rachis, web together adjacent grains and feed on the developing grains and grow. The grown-up caterpillar is long, narrow, light brown in color with a dark head and dark lateral lines on the body. The full-grown larva pupates inside the web in a cocoon. The adult moth, which emerges from the pupa, is small, brown in color with brown forewings and light brown hindwings.

Control measures. Dusting the earheads thoroughly after their emergence with carbaryl 10 D or endosulfan 4 D at 10 kg./ acre or spraying with carbaryl 50 WP-400 gm. or endosulfan-400 ml. in 200 litres of water per acre is effective in controlling the pest.

Minor pests. Besides the pests described above in detail, several other pests, which occur less frequently and not of much economic importance are also known to infest sorghum. The semi-looper-*Eublemma (Antoba) silicula* webs the floral parts with silken thread, remains securely inside, eats and destroys the immature grains; the gram caterpillar-Helicoverpa *(Heliothis) armigera* hides in-between the branches of the earheads and feeds on the maturing grains; the cut worm-*Agrotis ipsilon* cuts and feeds on the leaves and shoots; the hairy caterpillars-*Amsacta albistriga* and *A. moorei* cut and feed on the foliage; the ragi pink borer-*Sesamia inferens* bores into the stem of plants and causes damage; the rice grain moth-*Sitotroga cerealella* bores and eats the grains in the field; the nymphs and adults of the plant bugs-*Nezara viridula* and *Dolycoris indicus* infest the ears and suck the sap from the developing grains; both the nymphs and adults of the aphids-*Aphis sacchari* and *Rhopalosiphum maidis* suck the plant sap from the leaves and shoots; the nymphs and adults of the brown aphid-*Hysteroneura setariae* suck the sap from the leaves and immature grains; the nymphs and adults of the ragi root aphid-*Tetraneura nigriabdominalis* suck the sap from the roots; the white grub-*Holotrichia consanguinea* cuts and feeds on the roots; the rice grasshopper-*Hieroglyphus banian,* the short-winged Phadka grasshopper-*Hieroglyphus nigrorepletus* and the Deccan wingless grasshopper-*Colemania sphenaroides* cut and feed on the leaves; colonies of nymphs and adults of the mites-*Oligonychus indicus* and *Schizotetranychus andropogoni* suck the sap from the under surface of the leaves and cause reddish-brown spots and patches.

FINGER MILLET or RAGI *(Eleusine coracana)*

1. Ragi pink borer

Sesamia inferens

Order - Lepidoptera

Family - Noctuidae

The pest occurs in all finger milet-growing countries of the world, including India. Though the pest is more common in wheat in Northern India, it is considered to be a major pest of ragi in South India and sometimes causes considerable damage to the crop. It also attacks maize, barley, sugarcane, sorghum, jute, guinea grass and sometimes rice crop also.

Nature of damage. The caterpillar bores into the stem, feeds on the inner tissues of the central shoot and severes it, leading to **'dead heart'** in the young plant. When the pest occurs during the flowering stage, the inner tissues of the peduncle is eaten and severed, as a result the earhead dries and all the grains become chaffy. If the dead hearts or dried earheads are pulled gently, they come off easily. A borer after destroying one shoot migrates to another plant and can thus infest many plants. In a single plant, up to 5 caterpillars may be found.

Life cycle of the pest. The female moth lays light yellow-colored, spherical eggs in bunches on the inner side of the leaf sheaths. The eggs hatch in about 7 days. The young caterpillar on emergence from the egg, bores into the stem, feeds on the inner tissues of the central shoot and becomes a full-grown caterpillar in 25-54 days. The grown-up larva is cylindrical, smooth, uniformly pinkish-brown in color, with a reddish-brown head and about 2.5 cm. in length. It pupates inside the burrow of the stem and emerges as an adult moth in 8-12 days. The moth is somewhat stout, straw-colored, with straw-colored forewings and whitish hindwings. The moths are nocturnal in habit (Fig. 16).

Control measures

Cultural methods. Seedlings showing dead heart symptoms should be removed and destroyed.

Chemical control. Spraying with endosulfan-400 ml. or carbaryl 50 WP-400 gm. or phosalone-400 ml. in 200 litres of water per acre with a high volume sprayer affords good control. The spray should be directed towards the basal part of the plants.

2. Ragi white borer

Saluria inficita

Order - Lepidoptera

Family - Phycitidae

Nature of damage. The pest appears both in the nursery and in the main field during the flowering stage. The caterpillars bore into the stem, feed on the inner tissues, severing the central shoot, resulting in dead hearts and dried earheads. It is a specific pest of finger millet in South India and rarely occurs on tenai and rice crops (Fig. 17).

Life cycle of the pest. The life cycle is almost similar to that of the pink borer. The female moth lays light yellow-colored eggs in clusters on the inner side of the leaf sheaths. The young caterpillars, which hatch out of the eggs bore into the stem, eat the inner tissues and grow. They are found at the lower portions of the stem close to the soil surface. The larva is yellowish-white in color, with yellow-colored head. The full-grown larva pupates inside the stem itself and emerges as an adult moth. The moth is medium-sized, dark brown in color and has dark brown forewings, with a dull white band along the margin of each forewing. The hindwings are whitish in color (Fig. 17).

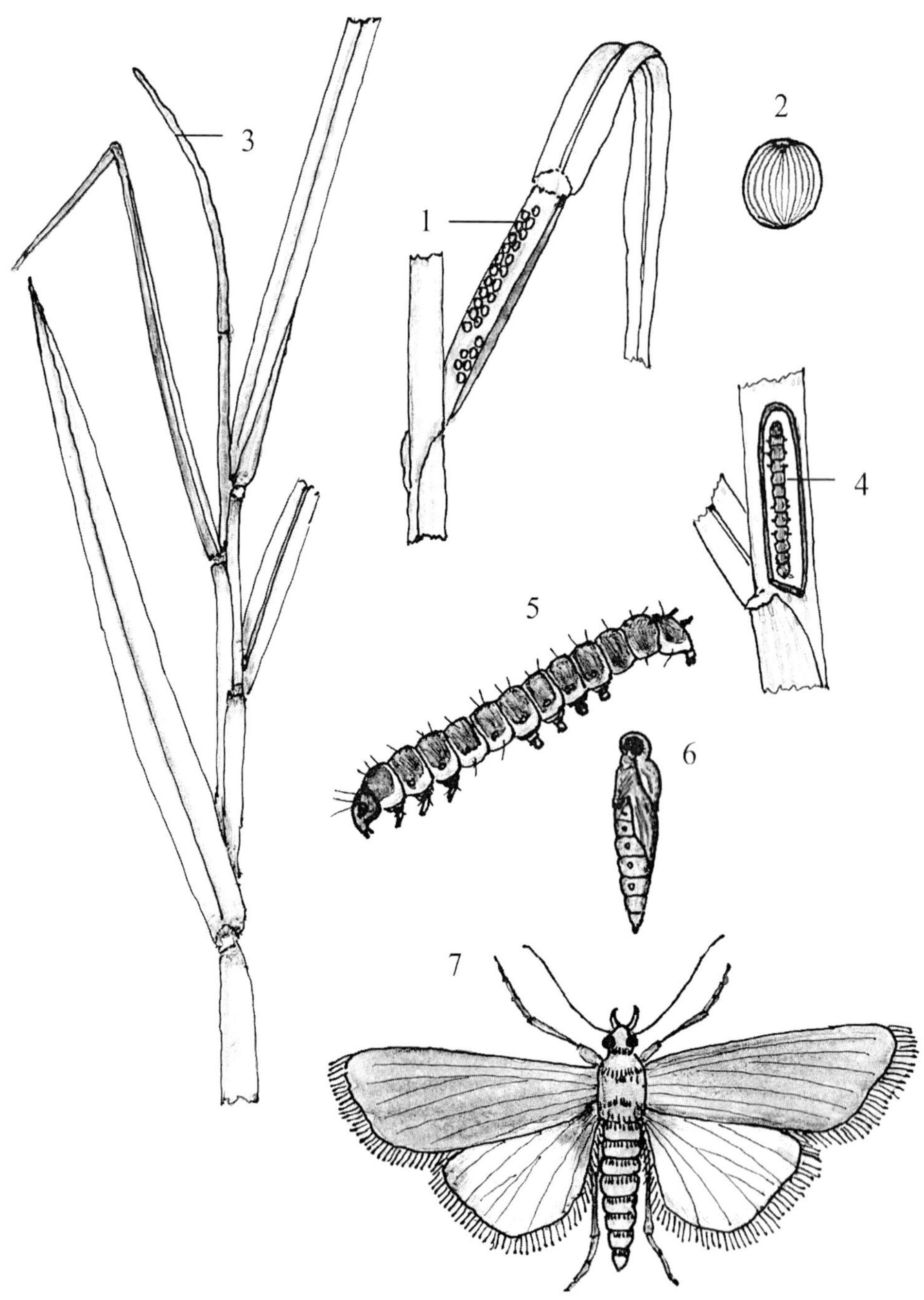

Fig. 16. Ragi pink borer-*Sesamia inferens*

1. Egg mass 2. Egg 3. Dead heart 4. Larva inside the stem 5. Larva 6. Pupa 7. Moth

Control measures. The control measures suggested for the control of ragi pink borer may be followed.

3. Cut worm

Spodoptera exigua

Order - Lepidoptera

Family - Noctuidae

Nature of damage. This cosmopolitan and polyphagous, leaf-feeding caterpillar pest appears suddenly in large numbers and defoliates the plants, especially seedlings in the nursery. The caterpillars are voracious feeders and after destroying the crop in one field they move to adjacent fields in swarms of hundreds to thousands, mostly after dusk and feed on the foliage. Besides ragi, the pest infests several other crops including cotton, eggplant, onion, cowpea, chillies, sunflower, daincha, and castor.

Life cycle of the pest. The life cycle of the pest is very much similar to that of *Spodoptera litura.* The female moth lays clusters of eggs on the lower portions of young plants. The larvae hide in the soil during the daytime and feed on the leaves at night. They become full-grown in 10-16 days. The grown-up larva is brownish-green in color with dark, wavy lines on the dorsal surface and yellow stripes laterally. It pupates in the soil in an earthen cocoon and emerges as a moth in 7-11 days. The moth is brownish in color with brownish forewings and white hindwings.

Control measures. The control measures suggested for the control of *Spodoptera litura* may be followed (Page 162).

4. Black hairy caterpillar

Estigmene lactinea

Order - Lepidoptera

Family - Arctiidae

Nature of damage. It is a sporadic pest and in certain seasons, it is found in some of the districts of Tamil Nadu and in some states of South India. It is often found to occur along with the red hairy caterpillar and has quite similar habits. The larvae feed on the leaves and defoliate the plants. The pest also infests sorghum, pearl millet, pulse crops etc.

Life cycle of the pest. The female moth lays eggs on the leaves of its host plants. The caterpillars hatching from the eggs feed on the leaves and ears and grow to a length of about 5.0 cm. The larva has a black head and hairs all over the body. It is also known as the **'wooly bear caterpillar'**, because of its

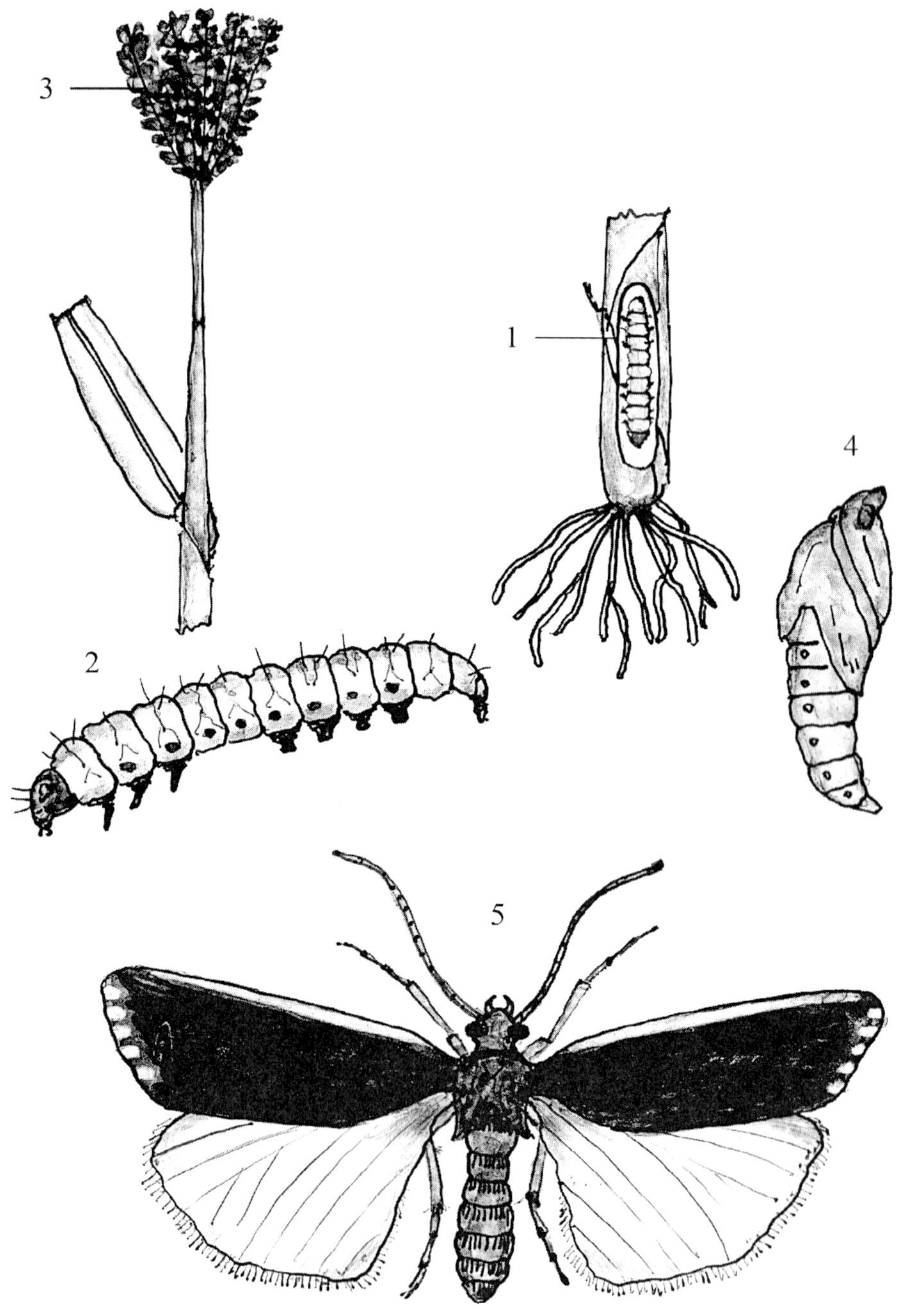

Fig. 17. Ragi white borer-*Saluria inficita*

1. Larva inside the stem 2. Larva 3. Chaffy earhead 4. Pupa 5. Moth

wooly, black body and its bear-like movement. The full-grown larva pupates inside the soil in an earthen chamber. The moth emerging from the pupa is large-sized, beautiful and creamy-white in color, with characteristic crimson markings on the head and body.

Control measures. The physical, cultural and chemical methods of control suggested for the control of groundnut red hairy caterpillar may be followed (Page 66).

5. Ragi root aphid

Tetraneura nigriabdominalis

Order	-	Hemiptera
Suborder	-	Homoptera
Family	-	Aphididae

The pest, which attacks the roots of plants, sometimes causes extensive damage by killing the plants in large numbers. Besides ragi, the pest attacks sorghum, pearl millet, lesser millets and a number of grasses.

Nature of damage. Both the nymphs and adults congregate on the roots of ragi in large numbers under the soil, pierce and suck the sap from the roots. Because of continued drainage of the sap, the infested plants become weak, wilt and die in patches. Many other insects, mostly black ants *(Camponotus compressus)* are found around such infested plants, which come to feed on the honeydew secretion of the insects. The black ants have a symbiotic association with the aphids. They take care of the aphids, at the same time get honeydew from them. If such infested plants are pulled out, a large number of the pest at all growth stages are found adhering to the roots.

Life cycle of the pest. The female aphids are parthenogenetic and viviparous. The pest continues its life in a number of wild grasses and weed hosts throughout the year. There may be two or three wingless generations followed by a winged generation. The winged females fly to other fields and infest the crops. Full-grown aphid is very small, soft-bodied and pink-colored. The females have big abdomen and look almost globular.

Control measures

Chemical control (i) Incorporating malathion 5 D or endosulfan 4 D at the rate of 20 kg. per acre in the soil prior to sowing prevents the occurrence of the pest (ii) Spot drenching to completely wet the root region and the soil around the root region of the infested plants with dimethoate-0.06% (2.0 ml./ litre of water) or carbaryl 50 WP-0.1% (2.0 gm./ litre of water) eradicates the pest (iii) Mixing small quantities of crude oil along with irrigation water also helps to eradicate the pest.

6. White grub

Holotrichia consanguinea

Order - Coleoptera

Family - Melolonthidae

Nature of damage. The pest is found to infest ragi crop in some places in Tamil Nadu during July-September. The grubs of this pest remain under the soil, cut and feed on the roots of ragi plants. Rarely the grubs may bore into the stem. Severely infested plants become weak and may die eventually. It is a potential pest of sugarcane.

Life cycle of the pest and control measures. Refer pests of sugarcane (Page 149, Fig. 59).

Pests of minor importance. Apart from the pests dealt with above, finger millet is subjected to attack by many other pests, which may occur sporadically and are not of much economic importance. Most of the pests, which attack sorghum, invariably attack finger millet and other millets to a more or lesser extent. The earhead webber-*Cryptoblabes gnidiella* and the semilooper-*Antoba silicula* web together adjacent grains in the earhead and feed on the developing grains; the caterpillars of the rice grain moth-*Sitotroga cerealella* feed on the contents of the grains in the field; the weevil-*Myllocerus viridanus* and the wingless grass hopper-*Neorthacris simulans* feed on the foliage; the pentatomid bugs-*Dolycoris indicus* and *Nezara viridula,* the rice earhead bug-*Leptocorisa acuta* and the mirid bug-*Calocoris angustatus* suck the sap from the developing grains; the brown aphid-*Hysteroneura setariae* and the thrips-*Rhopalosiphum maidis* suck the sap from tender leaves and shoots; the nymphs and adults of the ragi jassid-*Cicadulina bipunctella bipunctella,* besides causing direct damage by sucking the sap from the leaves and stem, is the vector of 'ragi mosaic virus disease'; the surface grasshoppers-*Chrotogonus* spp. notch and feed on the leaves.

PEARL MILLET, CUMBU or BAJRA *(Pennisetum typhoides)*

1. Cumbu shoot fly

Atherigona approximata

Order - Diptera

Family - Anthomyiidae

Nature of damage. Among the pests infesting pearl millet, it is considered to be the most serious one and is mostly prevalent during the colder months.

The grubs infest the seedlings up to the age of about 30 days, as well as the peduncles of the inflorescence during the flowering time. The young grubs after hatching from the eggs, bore the stem of seedlings, eat the inner tissues and severe the central shoot, causing dead hearts. Once the grub damages the primary shoot, secondary tillers may appear, which are also infested. When the pest occurs in a severe form, about 25 per cent of the seedlings are infested and they may die prematurely. At the flowering stage, the grub bores the peduncle of the earhead at any point, feeds on the inner tissues, as a result the earhead above the point of attack of the grub, turns into a narrow, white, whip-like structure without any grain formation. Under such conditions, the grain loss may go up to 10-20 per cent

Life cycle of the pest. The female fly lays eggs in batches of 2-3 eggs per batch on the under surface of leaves, when the seedlings are at the two-leaf stage. The eggs are white-colored and elliptical in shape. At the flowering stage, the fly lays eggs on the peduncle or on the inflorescence. The eggs hatch in about 3 days and the grubs become full-grown in 12-15 days. The grown-up maggot is white in color and lacks legs. The maggot pupates inside the stem or peduncle and emerges as an adult fly in about 7 days. The fly is very small, delicate, pale grayish in color, with thin, transparent wings.

Control measures

Cultural methods (i) The sowings should be completed early, soon after the onset of monsoon rains (ii) A higher seed rate of 4.0 kg./acre may be adopted and at the time of thinning out or first weeding, the affected seedlings are removed and destroyed, at the same time the normal plant population in the field can also be maintained (iii) Plants showing dead heart symptoms and severely infested earheads should be removed and destroyed.

Chemical control

Nursery treatment (i) Carbofuran 3 G-1500 gm. or phorate 10 G-500 gm. is mixed with sand-1.0 kg. and broadcasted uniformly over the soil for 8 cent nursery area and irrigated (ii) Instead of granular application, the nursery crop may be sprayed with endosulfan-45 ml. or methyl demeton-30 ml. or dimethoate-30 ml. in 20 litres of water for 8 cent nursery area with a high volume sprayer at the two-leaf stage of the seedlings.

Direct sown, irrigated crop. The crop may be sprayed with methyl demeton-200 ml. or dimethoate-200 ml. or endosulfan-200 ml. in 200 litres of water per acre on the 14th and 21st day after sowing. The spray should be directed towards the stem portion.

Rainfed crop (i) The seedlings may be dusted with carbaryl 10 D or endosulfan 4 D or phosalone 4 D at 10 kg. / acre one week after germination.

If necessary, a second dusting is given 10 days later (ii) To prevent pest occurrence at the flowering stage dusting with carbaryl 10 D or malathion 5 D or phosalone 4 D at 10 kg./ acre may be given, so as to cover the earheads thoroughly.

Other pests of pearl millet. Most of the pests noticed on sorghum and finger millet infest pearl millet also and occasionally cause severe damage to the crop resulting in appreciable yield loss. The leaf feeding red hairy caterpillar pests-*Amsacta albistriga* and *A. moorei* and the black hairy caterpillar pest-*Estigmene lactinea* defoliate cumbu plants; the caterpillars of ragi pink borer-*Sesamia inferens* and sorghum stem borer-*Chilo partellus* bore into the stem of plants and cause damage; the larvae of the semilooper pest-*Eublemma (Antoba) silicula* web adjacent grains in the inflorescence and feed on the developing grains; the slender, yellowish-green caterpillars of the leaf roller pest-*Marasmia trapezalis* folds and rolls the leaves at the top portion, remains inside, scrapes and feeds on the green matter; the wingless grasshopper pest-*Neorthacris simulans* cuts and feeds on the leaves; the adults of the weevil pests-*Myllocerus discolor* and *M. viridanus* notch and feed on the leaves; the nymphs and adults of the pentatomid bugs-*Nezara viridula* and *Dolycoris indicus* suck the sap from the leaves; the adults of the blister beetle-*Lytta tenuicollis* feed on the pollen from flowers and leads to poor development of grains; the grubs of the cumbu midge-*Geromyia penniseti* feed on the ovaries of flowers and affect grain formation.

MAIZE or CORN *(Zea mays)*

Most of the pests infesting other millets attack maize also. The cholam stem borer-*Chilo zonellus (partellus)* and the ragi pink borer-*Sesamia inferens* feed on the leaves during the early stages of the crop and later bore into the stem and cobs; the grubs of the shoot fly-*Atherigona orientalis* mine into the stem of seedlings, killing the shoots and produce dead hearts; the gram caterpillar-*Helicoverpa (Heliothis) armigera* damages the developing grains in the cobs to a large extent; the army worm-*Mythimna separata* feeds on the foliage; maize cobs are occasionally infested by the polyphagous cut worms-*Agrotis ipsilon, Spodoptera litura* and *S. exigua*; the earhead web worm-*Cryptoblabes gnidiella* feeds on the developing grains; the red hairy caterpillars-*Amsacta moorei* and *A. albistriga* feed on the foliage; the ash weevil-*Myllocerus discolor* notches and feeds on the leaves; the Phadka grass hopper-*Hieroglyphus nigrorepletus* bites and feeds on the foliage; the sugarcane leaf hopper-*Pyrilla perpusilla* and the aphid-*Rhopalosiphum maidis* suck the sap from the leaves; the termites-*Odontotermes obesus* and *Microtermes obesi* live underground and feed on the roots and stem portion of seedlings, as well as grown-up plants.

ITALIAN MILLET or TENAI *(Setaria italica)*

1. Tenai stem boring beetle

Anadastus parvulus

Order	-	Coleoptera
Family	-	Languriidae

Nature of damage. It is a specific pest of tenai and occurs regularly in some parts of South India, including Tamil Nadu. The grubs of this beetle bore into the stem and cause withering and drying of the earheads.

Life cycle of the pest. The female beetle inserts eggs singly into the tissues of the stem of plants. The eggs are small and cigar-shaped. The grubs that hatch out of the eggs are pale yellow in color. The young grub bores into the stem and makes a ring-like hollow inside the stem, feeds on the inner tissues and grows. The full-grown grub is yellowish-white in color, 0.8-1.0 cm. long and has a pair of short, chitinous spines on the anal segment. The grown-up larva pupates inside the hollow of the stem. The pupa is yellowish in color in the beginning and gradually turns brown, mottled with dark brown patches. The adult beetle emerging from the pupa comes out through an exit hole already made by the grub in the stem. The beetle is small and smooth, with a red head and thorax. The wing covers are deep blue in color. The life cycle of the pest is completed in 40-50 days.

Control measures

Physical methods. The adult beetles may be collected by using a hand net and destroyed.

Chemical control. Spraying the crop with endosufan-400 ml. or fenitrothion-400 ml. or fenthion-200 ml. or phosphamidon-110 ml. or monocrotophos-200 ml. or decamethrin-100 ml. in 200 litres of water per acre with a high volume sprayer controls the pest effectively.

Other pests of tenai. The grubs of the shoot flies-*Atherigona destructor* and *A. artipalpis* bore into the stem of young plants and cause death of the central shoot; the cut worm-*Mythimna separata* and the weevils-*Myllocerus dentifer* and *M. transmarinus* feed on the foliage; the sucking pests-*Leptocorisa acuta* and *Scotinophara coarctata* suck the sap from the developing grains and leaves.

WHEAT *(Triticum vulgare)*

Many of the pests that infest millets, as well as a few other pests are found to attack wheat. The grubs of the wheat shoot flies-*Atherigona bituberculata* and *A. oryzae* bore into the stem of young plants and damage the growing shoot, resulting in dead hearts; the ragi pink borer-*Sesamia inferens* bores into the stem, feeds on the inner tissues and damages the central shoot; the polyphagous cut worms-*Agrotis ipsilon, A. biconica, A. flammatra,* and *A. spinifera* cut and feed on the foliage; the climbing cut worm-*Mythimna separata* cuts and damages the earheads; the leaf beetles-*Myllocerus discolor* and *M. blandus* notch and feed on the foliage; the sugarcane leaf hopper-*Pyrilla perpusilla* and other leaf hoppers-*Empoascanara maculifrons* and *Riptortus pedestris* suck the sap from the leaves; the termites-*Odontotermes obesus* and *Microtermes obesi* live underground in colonies and attack young seedlings, as well as grown-up plants and kill them; the wheat aphid-*Toxoptera (Schizophis) graminum* and other aphids-*Acyrthosiphum pisum, Myzus persicae, Rhopalosiphum maidis* and *Macrosiphum miscanthi* suck and feed on the sap from the tender leaves; the nymphs and adults of the wheat thrips-*Anaphothrips flavicinctus* lacerate the leaves and suck the sap from leaves; the pentatomid bugs-*Dolycoris indicus, Nezara hilaris* and *Menida histrio* suck the sap from the developing grains; the adults of the weevil pest-*Tanymecus indicus* cut and feed on the tips and margins of leaves, while the grubs feed on the roots; the surface grass hopper-*Chrotogonus trachypterus* feeds on the foliage and tender shoots.

PESTS OF PULSES

The pulse crops grown on a large scale include Red gram *(Cajanus cajan)*, Gram *(Cicer arietinum)*, Bean *(Phaseolus vulgaris)*, Black gram *(Phaseolus mungo)*, Green gram *(Phaseolus aureus)*, Pea *(Pisum sativum)*, Lab lab *(Dolichos lab lab)*, Cowpea *(Vigna sinensis)*, Cluster bean *(Cyamopsis tetragonoloba)* and Horse gram *(Dolichos biflorus)*.

1. **Gram pod borer** or **American boll worm**

Helicoverpa (Heliothis) armigera

Order - Lepidoptera

Family - Noctuidae

This polyphagous pest is one of the most serious and well-known pests of pulse crops. It attacks several crops such as cotton, chillies, tomato, lady's finger, tobacco, sweet potato, groundnut, soybean, cholam, maize, pearl millet, black gram, green gram, field bean, bean, Bengal gram, lab lab, pea etc.

Nature of damage. The caterpillars bore the pods and feed on the developing seeds and sometimes cause severe damage. While feeding, the caterpillar introduces its head and only a part of the body into the pod, while the rest of the body remains outside the pod. The caterpillar never enters into the pod completely. One caterpillar damages several pods. Before pod formation, the caterpillars feed on the leaves and tender shoots of the plants.

Life cycle of the pest and control measures. Refer pests of cotton (Page 159, Fig. 42).

2. Red gram plume moth or Red gram pod borer

Exelastis atomosa

Order - Lepidoptera

Family - Pterophoridae

It is also another important borer pest of red gram. It also infests soybean and field bean.

Nature of damage. The caterpillar, which is a typical borer bores the pods, puts its head inside and feeds on the developing seeds inside the pod. While feeding the major portion of the body remains outside the pod. The larva also bores the flower buds and flowers and feeds on the tissues. The affected flowers and buds drop down prematurely (Fig. 18).

Life cycle of the pest. The female moth lays minute, oval, greenish eggs singly on the flower buds and tender pods. The eggs hatch in 3-4 days. The larval stage lasts for 14-30 days. Full-grown larva is about 1.5 cm. long, green or greenish-brown in color and fringed with short hairs and spines radiating all over the body, arising from tubercles. The grown-up caterpillar pupates on the surface of the pod and rarely in the burrow of the infested pod. The pupa is also fringed with radiating spines and hairs. The adult moth emerges from the pupa in 4-8 days. The moth is small, slender, delicate, soft, light brown in color and the wings are arranged in plumes. The light brown forewings have two splits and the hindwings have three splits (Fig. 18).

Control measures

Chemical control (i) Dusting the foliage with carbaryl 10 D or endosulfan 4 D or quinalphos 1.5 D or phosalone 4 D at 10 kg./ acre, when the caterpillars are young gives good control (ii) Foliar spraying with endosulfan-600 ml. or monocrotophos-300 ml. or fenvalerate-110 ml. or cypermethrin-75 ml. in 300 litres of water per acre with a high volume sprayer controls the pest.

3. Gram cut worm

Agrotis ipsilon

Order - Lepidoptera

Family - Noctuidae

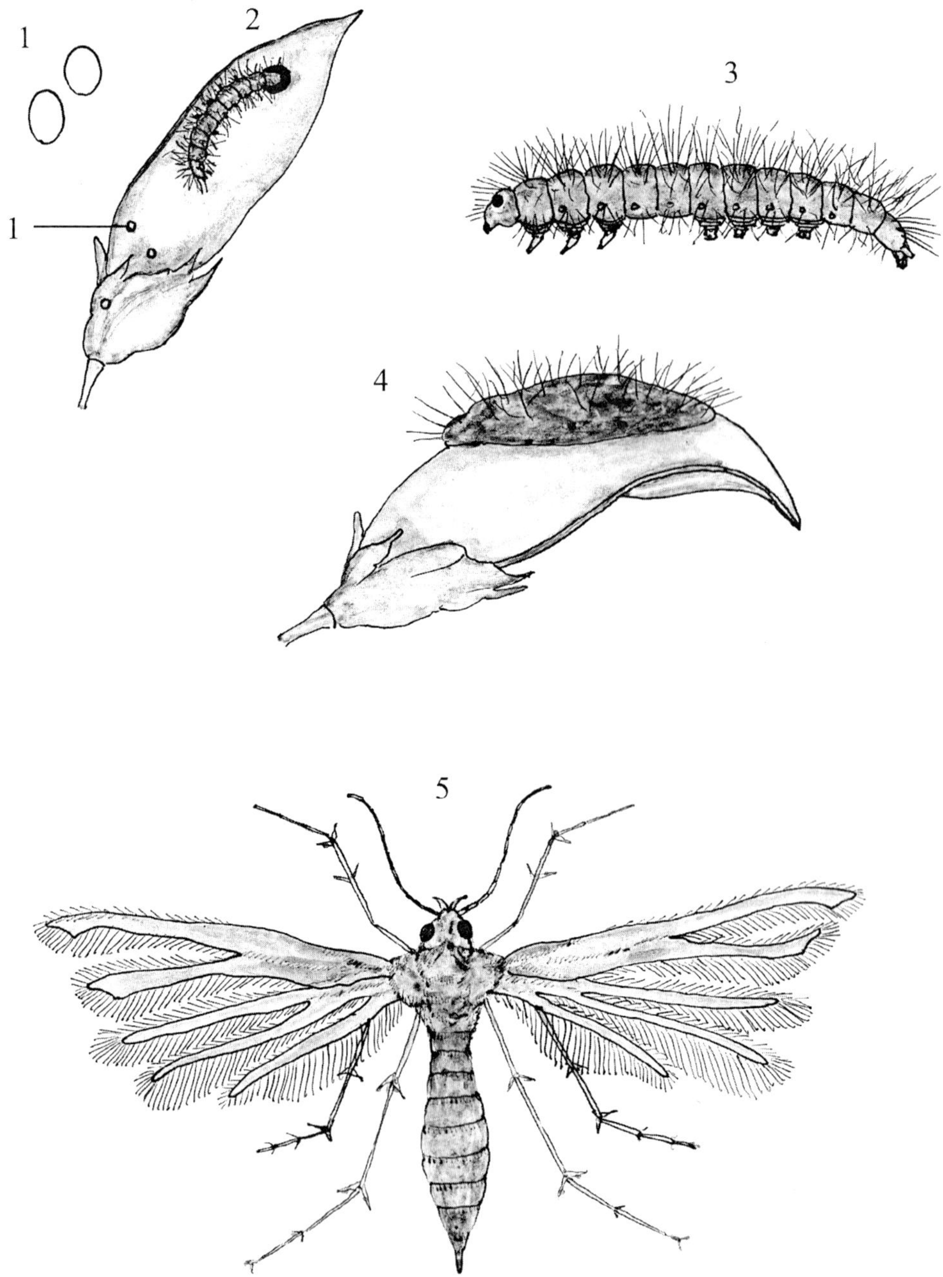

Fig. 18. Red gram plume moth-*Exelastis atomsa*

1. Eggs 2. Larva infesting a pod 3. Larva 4. Pupa inside cocoon on the surface of pod 5. Moth

This is another polyphagous, well-known pest, which is nocturnal in habit. The wide host range of this pest includes potato, tobacco, cotton, cabbage, beet root, lady's finger, cucumbers, peas, barley, oats etc.

Nature of damage. The caterpillars, which are night-feeders, hide in cracks, crevices, in-between mud clods and such hiding places in the soil at a depth of 5.0-10.0 cm. during the daytime and come out in the evening hours in swarms, feed on the leaves, cut the tender portion of the shoots and cause severe damage. The young plants are cut at the ground level or below the ground level. The tender shoots and flowers are also eaten away. The caterpillars cut and damage much more than they can eat. The larvae carry the cut portions of the plants to their hiding places and eat them during the daytime. The infested crop appears as if grazed by cattle.

Life cycle of the pest. The female moth lays yellow-colored eggs on the under surface of leaves or on the moist soil under the plant canopy in clusters of about 30 numbers. One moth lays 200-300 eggs. The eggs hatch in 4-10 days. The young caterpillars hatching out from the eggs start feeding by cutting and eating the leaves. They become full-grown in 15-30 days. The grown-up larva is about 2.0 cm. long, smooth and dark brown in color, with a red head. It makes an earthen cocoon under the soil, pupates inside it and emerges out as an adult moth in 15-30 days. The total life cycle is completed in 30-68 days. The moth is stout, grayish-brown in color, 3.0-6.0 cm. across the wings and a good flier. The forewings are brown in color, with definite markings and each wing has two, black, wavy lines near the lateral margin. The hind wings are whitish in color. There are generally 2 generations in a year (Fig. 19).

Control measures

Physical and cultural methods (i) Light traps may be set up to attract and destroy the adult moths (ii) Heaps of green grasses are kept in different places in the field during the evening hours. The caterpillars after feeding during the nighttime hide under the grass heaps during the daytime. They can be collected and destroyed (iii) Inside the field, short trenches 25 cm. deep and 10 cm. wide are dug in a few places at random. The caterpillars, which move about in the field during the night, fall into the trenches and because of their 'kenoblastic' nature, fight each other and die.

Chemical control (i) In tracts where the pest occurs regularly, chlordane 5 D or heptachlor 5 D at 20 kg./acre is applied to the soil uniformly and incorporated into the soil, prior to sowing (ii) If the pest occurrence is found during the growing stages of the crop, chlordane 5 D or heptachlor 5 D at 20 kg./ acre is incorporated into the soil.

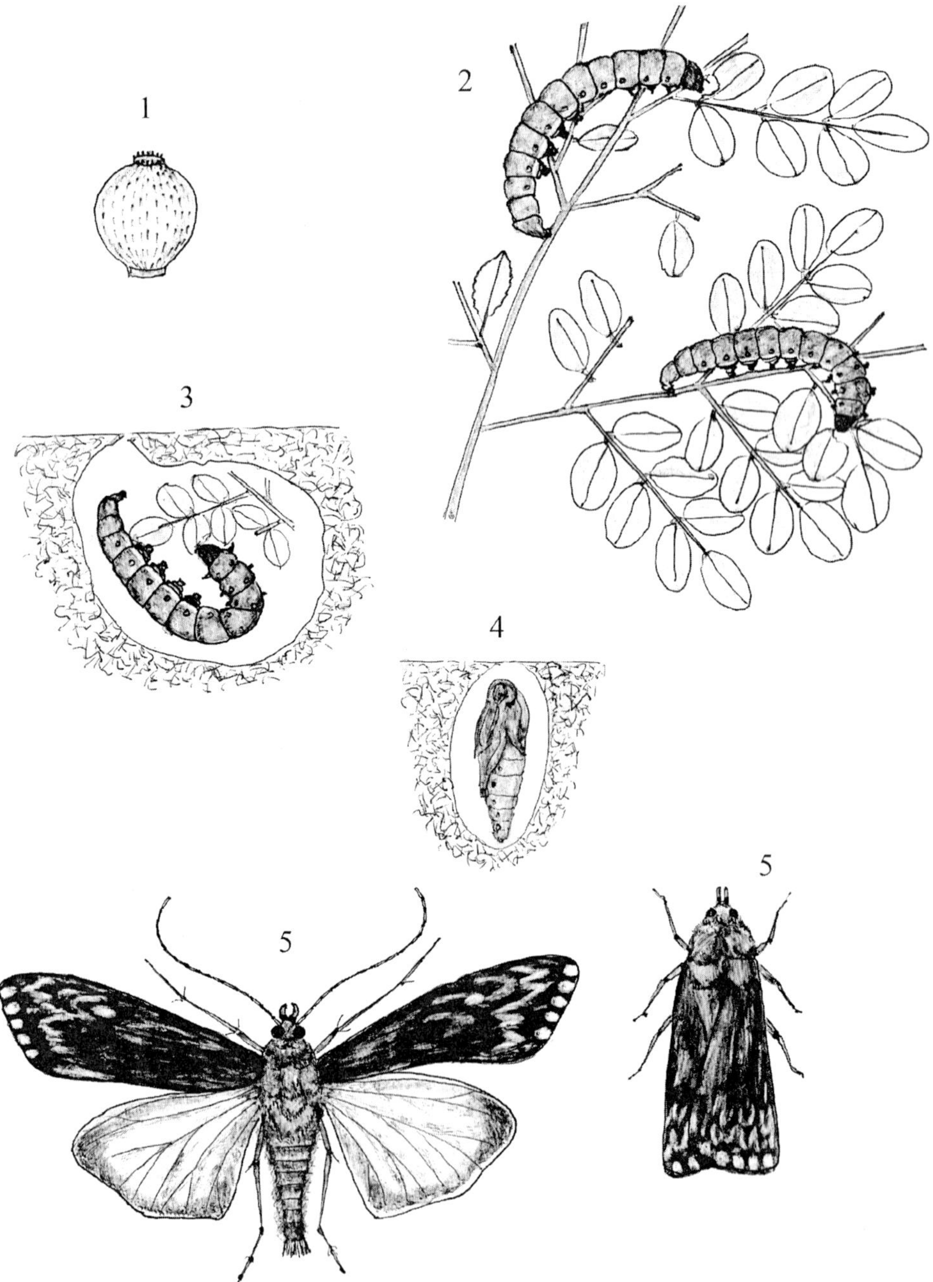

Fig. 19. Gram cut worm-*Agrotis ipsilon*

1. Egg 2. Larvae feeding on the leaves 3. Larva feeding on the leaves in its hiding place during the day time 4. Pupa 5. Moth

4. Red gram pod borer

Etiella zinckenella

Order - Lepidoptera

Family - Phycitidae

Nature of damage. The caterpillars bore the pods, enter inside and feed on the developing seeds and cause severe damage. Besides red gram, the pest also infests field bean, beans, peas and allied pulse crops.

Life cycle of the pest. The female moth lays eggs singly on the tender pods. The young caterpillars hatching out of the eggs in about 10 days bore into the pods, feed on the inner contents and become full-grown in 10-17 days. The young larva is greenish in color, with 5 black dots on the dorsal surface of the first thoracic segment. Full-grown larva becomes pink in color. It comes out of the pod and pupates in an earthen cocoon under the soil. The pupal stage lasts for 9-12 days normally. When conditions are unfavorable for carrying on the life cycle, the pupa remains inside the cocoon in a dormant stage for a considerably long time. The moth is small and grayish-brown in color. The grayish-brown forewing has a whitish, thin band near the anterior margin. The hindwings are dull grayish in color (Fig. 20).

Control measures

Cultural methods (i) The soil should be stirred well or ploughed to bring out and destroy the pupae present under the soil (ii) Application of potash fertilizer at 50 kg./acre at the time of sowing prevents pod borer attack to a large extent by inducing resistance against the pest.

Chemical control (i) Dusting the crop at the flowering stage with carbaryl 10 D or endosulfan 4 D or phosalone 4 D or quinalphos 1.5 D at 10 kg./ acre is effective. The treatment should be repeated after about 15 daya (ii) Spraying the crop with endosulfan-600 ml. or monocrotophos-300 ml. or chlorpyrifos-600 ml. or fenvalerate-115 ml. or cypermethrin-75 ml. in 300 litres of water per acre with a high volume sprayer, once at the time of flowering, followed by a second spraying 15 days later controls the pest.

5. Field bean pod borer

Adisura atkinsoni

Order - Lepidoptera

Family - Noctuidae

Nature of damage. The caterpillars bore into the pods, eat the developing seeds and destroy the pods. Besides field bean, the pest infests black gram, green gram, beans, red gram, Bengal gram and other pulses.

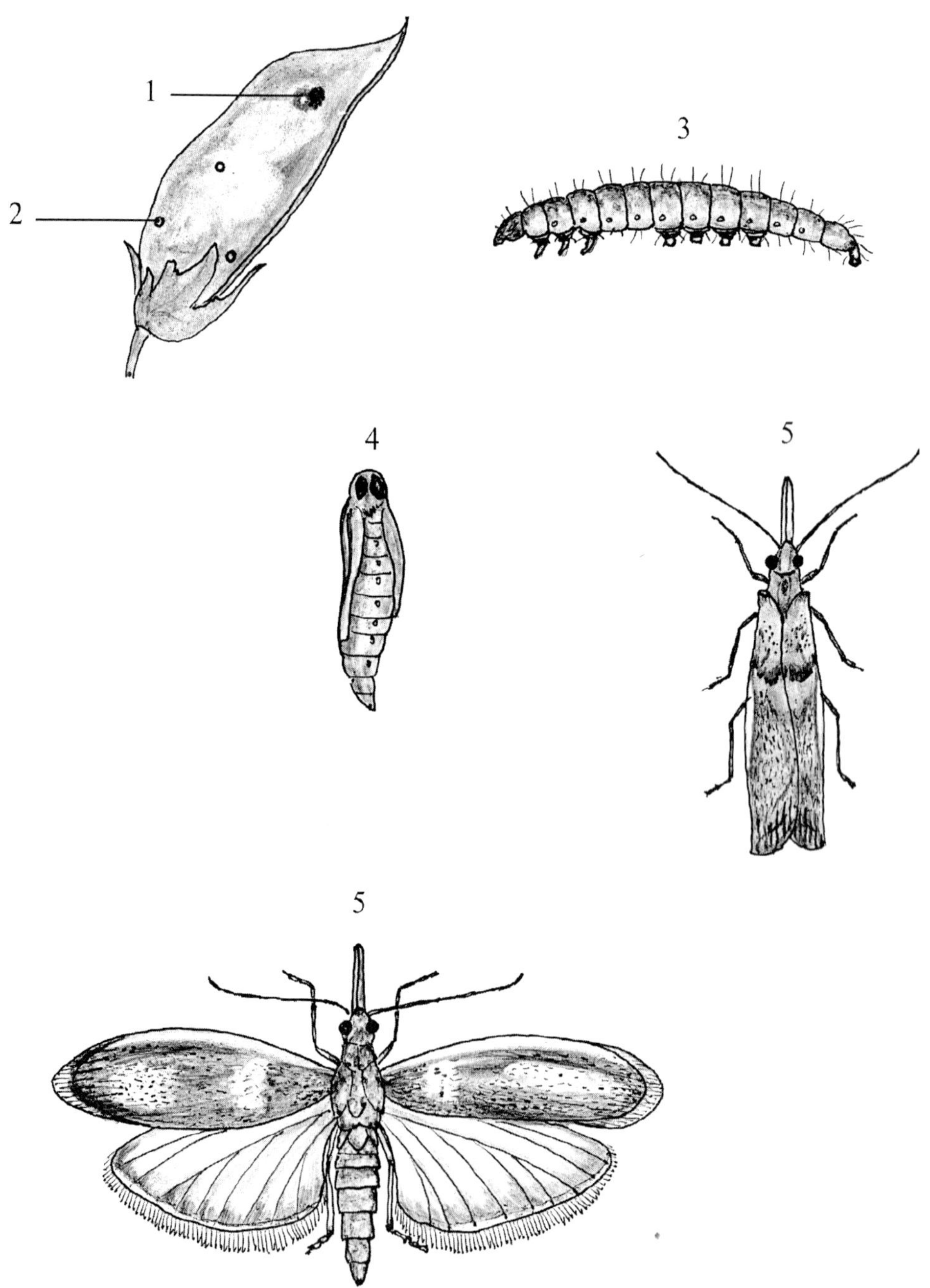

Fig. 20. Red gram pod borer-*Etiella zinckenella*

1. Bore hole in the pod 2. Eggs on the pod 3. Larva 4. Pupa 5. Moth

Life cycle of the pest. The female moth lays minute, spherical eggs, singly on the flower buds, tender pods and flowers. The eggs hatch in 2-3 days. The young larvae bore into the pods, feed on the developing seeds and become full-grown in 14-16 days. The grown-up caterpillar is about 2.5 cm. long, green-colored, with brown-colored lateral lines on the body and the last abdominal segment slightly elevated like a hunchback. The grown-up caterpillar comes out of the pod and pupates in an earthen cocoon under the soil. The adult moth emerges from the pupa in 8-16 days. The moth is yellow in color and the light brown forewings have 'V'-shaped markings. The hindwings are whitish in color, with light brown markings.

Control measures

Cultural and chemical methods. The methods suggested for the control of red gram pod borer may be followed (Page 55).

Biological control. The larval parasitoids-*Bracon hebetor* and *B. brevicornis* parasitize the larvae.

6. Lab lab aphid

Aphis craccivora

Order - Hemiptera
Suborder - Homoptera
Family - Aphididae

Nature of damage. Both the nymphs and adults of this polyphagous pest congregate in large numbers on the young leaves, tender growing shoots, flowers and tender pods, suck and feed on the sap, causing severe damage to the plants. The quality of the pods becomes very poor and harvest of the affected pods also creates problems. The pest infests several other crops such as, soybean, groundnut, red gram, field bean, peas, green gram, black gram and allied pulse crops (Fig. 21).

Life history of the pest and control measures. Refer pests of groundnut (Page 73).

7. Red gram spotted pod borer

Maruca testulalis

Order - Lepidoptera
Family - Pyraustidae

Nature of damage. The caterpillars web the flowers together with fine, silken thread, live inside and feed on the floral parts. Later, they bore the tender pods and feed on the developing seeds. The pest infests red gram, field bean, green gram, black gram and allied crops, as well as the green manure crop, daincha

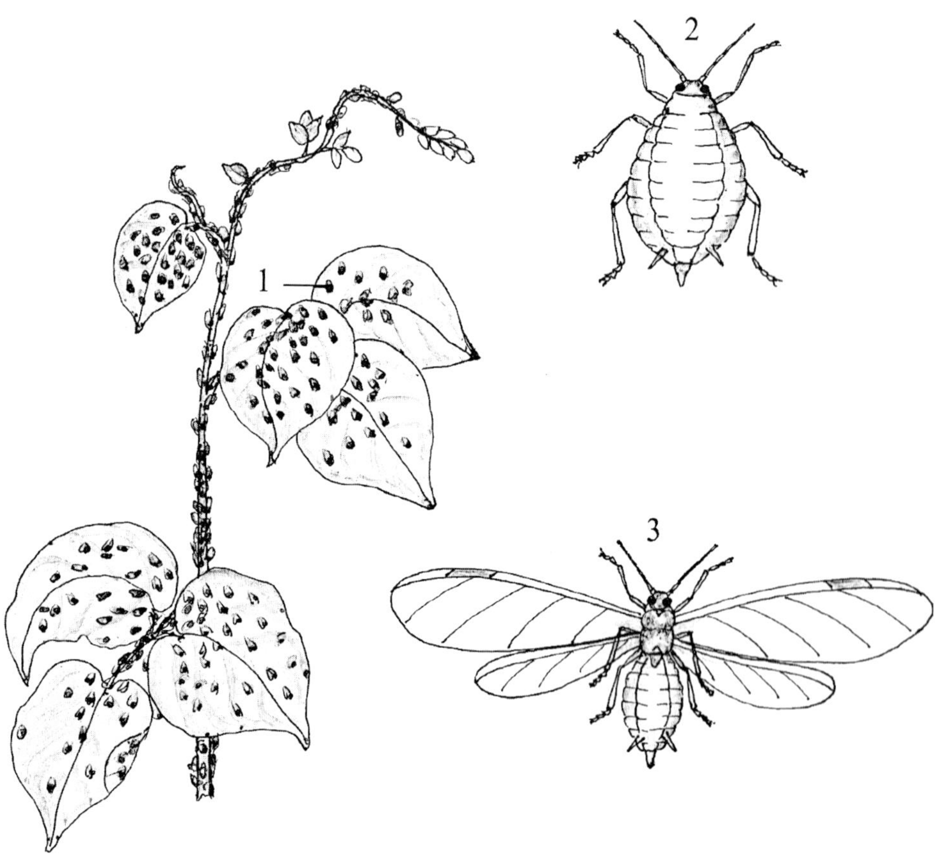

Fig. 21. Lab lab aphid-*Aphis craccivora*

1. Aphids feeding on the shoot 2. Wingless adult aphid 3. Winged adult aphid

Life cycle of the pest. The young larvae hatching out from the eggs are light green in color. The grown-up caterpillar is green in color, with a brown head and has warty, black dots on the body and short, black hairs on the warts. The full-grown caterpillar constructs a cocoon with fine, silken thread and pupates inside it. The cocoons are found on the pods or within the leaf fold. The moths are brown in color. On each of the dark brown forewing there is a white cross band. The hind wings are white in color, with a dark brown band along the margin (Fig. 23).

Control measures

Cultural and chemical methods. The methods suggested for the control of red gram pod borer may be followed.

Biological control. The larval parasitoid-*Bracon greeni* parasitizes and destroys the larvae.

8. Red gram pod fly

Melanagromyza obtusa

Order - Diptera

Family - Agromyzidae

Nature of damage. The minute maggots bore into the pod and feed on the immature seeds. The seeds are not completely eaten by the maggot, however saprophytic fungi and bacteria infect the partly eaten seeds and thus, the attacked pods are completely damaged. In case of severe infestation the loss in yield of seeds may go up to 80 per cent. The pest is serious on bean also. It also attacks lady's finger and safflower, in which the maggots mine into the stem and cause wilting of plants.

Life cycle of the pest. The female fly inserts the eggs in the skin of tender pods. About 4 eggs are thus laid on a single pod. One fly lays up to 80 eggs. The eggs hatch in about 3 days and the minute grubs bore and enter into the pods, start feeding on the developing seeds and become full-grown in 9-10 days. The grown-up maggot pupates inside the pod and in 8-9 days emerges out as an adult fly. The flies are very small, blackish in color and the wings are transparent (Fig. 22).

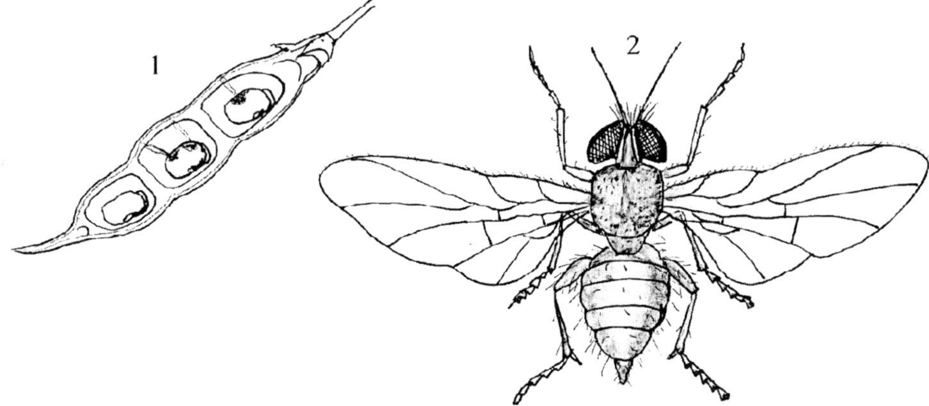

Fig. 22. Red gram pod fly-*Melanagromyza obtusa*

1. Larva inside the pod 2. Fly

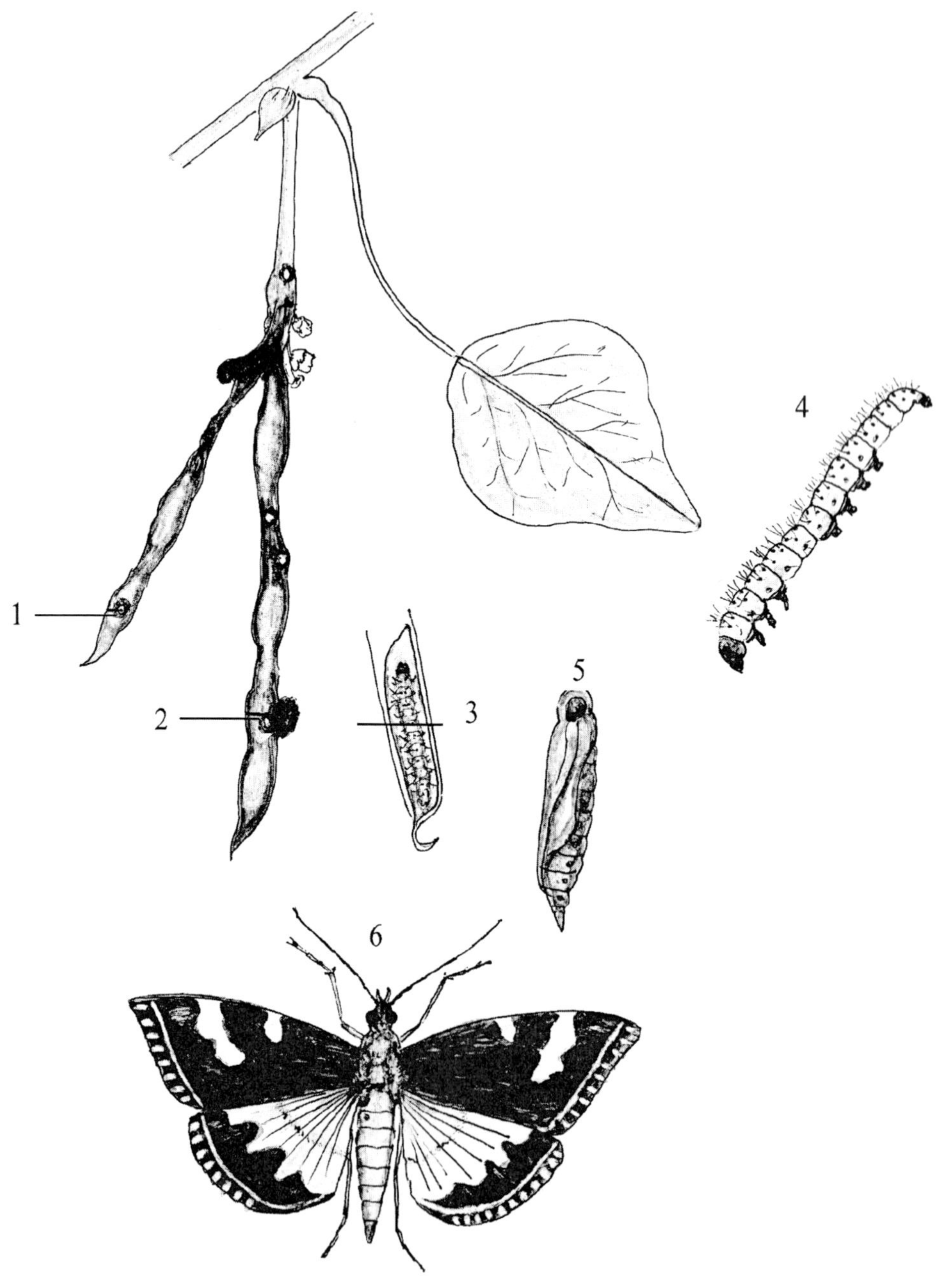

Fig. 23. Red gram spotted pod borer-*Maruca testulalis*

1. Borer infested pod 2. Excretory matter through the bore holes 3. Larva inside the pod 4. Larva 5. pupa 6. Moth

Control measures

Cultural methods. Early sowing helps to reduce the incidence of the pest by way of escape.

Chemical control. The methods suggested for the control of red gram pod borer may be followed (Page 59).

9. Pulse beetle

Bruchus chinensis

Order - Coleoptera

Family - Bruchidae

Nature of damage. Both the adult beetles and the grubs bore the matured grains in the field, as well as in the store houses, feed on the inner content and cause damage. The pest attacks peas, red gram, beans, black gram, green gram and other pulses.

Life history of the pest. Refer pests of stored grains (Page 263, Fig. 104).

Control measures (i) The grains should be properly dried, so that the moisture content is below 10-12 per cent and then stored (ii) If the grains are to be used for consumption, then the seeds are mixed with any edible oil at the rate of 1.0 kg. oil to 100 kg. of grains or 1.0 kg. of activated clay to 100 kg. of grains and then stored in airtight store rooms. In this way, the grains can be protected from pest infestation for a period of up to one year. Activated clay absorbs water from the atmosphere. If water is absorbed, then it looses its potential to provide protection from pest infestation. So, grains treated with activated clay should be stored under airtight condition (iii) If the grains are to be used for seed purposes, then they can be mixed thoroughly with malathion 5 D or endosulfan 4 D at 1.0 kg. of the insecticide per 100 kg. of seeds.

Other pests of pulses. A number of pests other than the ones described above in detail are known to attack the leaves, shoots, flower buds, flowers and pods of various pulse crops. The larvae of the webber pest-*Eublemma hemirrhoda* web the flowers of the inflorescence together with silken thread, live within and feed on the floral parts of green gram, black gram, beans and cowpea; the larvae of the blue butterfly-*Lampides boeticus* feed on the flower buds, flowers, immature pods and developing seeds of redgram, beans, soybean and other pulses; the polyphagous hairy caterpillars-*Amsacta moorei, A. albistriga* and *Diacrisia obliqua* feed gregariously and voraciously on the foliage of green gram, black gram, red gram, cowpea etc.; the pea semilooper-*Plusia nigrisigma* feed on the foliage of pea and gram; the grubs of the bud weevil-*Ceuthorrhynchus asperulus* feed on the pollen and other floral parts of several pulse crops; the galerucid beetle-*Madurasia obscurella* feed on the foliage and make small holes in the leaves of green gram, black gram, red gram cowpea etc.; the nymphs and adults of the bug pest-*Riptortus pedestris* suck and feed on the sap from leaves and tender shoots of red gram, beans, black gram, green gram and allied pulse crops; the grubs of the stem fly-*Melanagromyza phaseoli* bore the shoots and cause wilting and drying of the shoots in lab lab, cowpea, black gram and green gram; the larvae of the leaf feeding pest-*Euproctis fraterna* feed on the leaves of red gram; the larvae of the sphingid moth-*Acherontia styx* feed on the leaves of black gram and lab lab; the leaf folder-*Lamprosema indica* feeds on the leaves of lab lab, green gram, black gram and soybean; the maggots of the pea leaf miner-*Phytomyza atricornis* make zigzag mines in the leaves, feed on the green matter and cause damage; the weevil pests-*Myllocerus viridanus, M. discolor* and *M. maculosus* feed on the leaves of various pulse crops; the adults of the wasp-*Megachile anthracina* cut circular bits of leaves from red gram; the nymphs and adults of the thrips-*Thrips florum* infest the flower buds and flowers of red gram; the nymphs and adults of the white fly-*Bemisia tabaci* suck and feed the sap from the leaves and tender growing shoots of green gram, black gram, red gram and cowpea and the pest is also responsible for transmitting the virus causing 'yellow mosaic virus disease'; the nymphs and adults of the aphid pests-*Acyrthosiphum pisum, Macrosiphum pisi, Aphis craccivora* and *A. medicagenis* suck and feed on the sap from tender leaves and shoots of pea.

PESTS OF OILSEEDS

GROUNDNUT *(Arachis hypogaea)*

1. Groundnut red hairy caterpillar

Amsacta albistriga

Order - Lepidoptera

Family - Arctiidae

'Red hairy caterpillar' is a very serious and devastating pest of groundnut. The caterpillars attack groundnut crop, especially the rainfed crop and cause extensive damage. The pest occurs in an epidemic form during some seasons depending upon the weather conditions. In Tamil Nadu, the pest occurs in a serious form in Coimbatore, Pollachi, South Arcot, North Arcot and Salem districts during the months of June-July and in Madurai and Ramanathapuram districts during August-September. Severe infestation may result in heavy loss and occasionally in total loss of yield. Though groundnut is the primary host for this pest, many other crops, such as red gram, Bengal gram, sorghum, pearl millet, cotton, finger millet and castor are also attacked.

Nature of damage. The larvae hatching out of the eggs, harbor on the under surface of leaves in crowded batches, scrape and feed gregariously on the green matter of the leaves, while the upper epidermis of the leaves remains intact. As they grow up, they separate and move to other plants, bite and eat the leaves completely, leaving behind only the mid ribs, petioles and stem portions. After smothering the whole field completely within a few days, they march in thousands and swarm adjacent fields. The attacked fields appear as if grazed by cattle (Fig. 24).

Life cycle of the pest. After two or three good summer showers, the adult moths emerge from the earthen cocoons under the soil during the evening hours. Depending upon the quantum of rains received, the moths continue to emerge out in large numbers. As soon as they come out, they mate and begin to lay eggs. One female moth lays 600-700 eggs in clusters within a span of 2-6 days, on the under surface of leaves or on mud clods, rocks, dried sticks and on any such surfaces. The eggs are spherical, yellow in color, smooth and shiny. They hatch in 2-3 days. For the first few days, the young caterpillars aggregate on the under surface of leaves, scrape and feed on the green matter of leaves. They are gray-colored and without hairs. They become full-grown in 40-50 days. The young caterpillars are very active, gregarious and voracious eaters than the grown-ups and they do much of the damage to the crop. The full-grown larva is large-sized, cylindrical, about 5.0 cm. in length, reddish-brown in color, with dense, reddish-brown hairs all over the body. After a few light showers, when the soil becomes somewhat soft, the full-grown larva burrows a hole to a depth of 10-20 cm. in the soft soil and pupates in an earthen cocoon. If there are no rains during that time, most of the larvae die without being able to burrow into the soil. Most often, many of the larvae pupate near about the bunds or shady places or in moist areas, where the soil is comparatively soft. The pupal period lasts for 10-12 months. During the next year, after the summer showers, the adult moths emerge from the pupae and come out of the soil again. The moth is medium-sized, white in color, with white-colored forewings and a prominent yellow line along the anterior margin and a few brown-colored streaks. The hindwings are also white in color, with a few black, scattered spots. A yellow band is seen on the head (Fig. 24).

Control measures

Because this pest appears simultaneously in large areas and ravages the crop within a very short period, concerted efforts should be taken by all groundnut growers jointly to control the pest by adopting all integrated methods of control. When the pest appears in an epidemic form over extensive areas, the Government evokes the 'Agricultural Insects and Diseases Act, 1919'

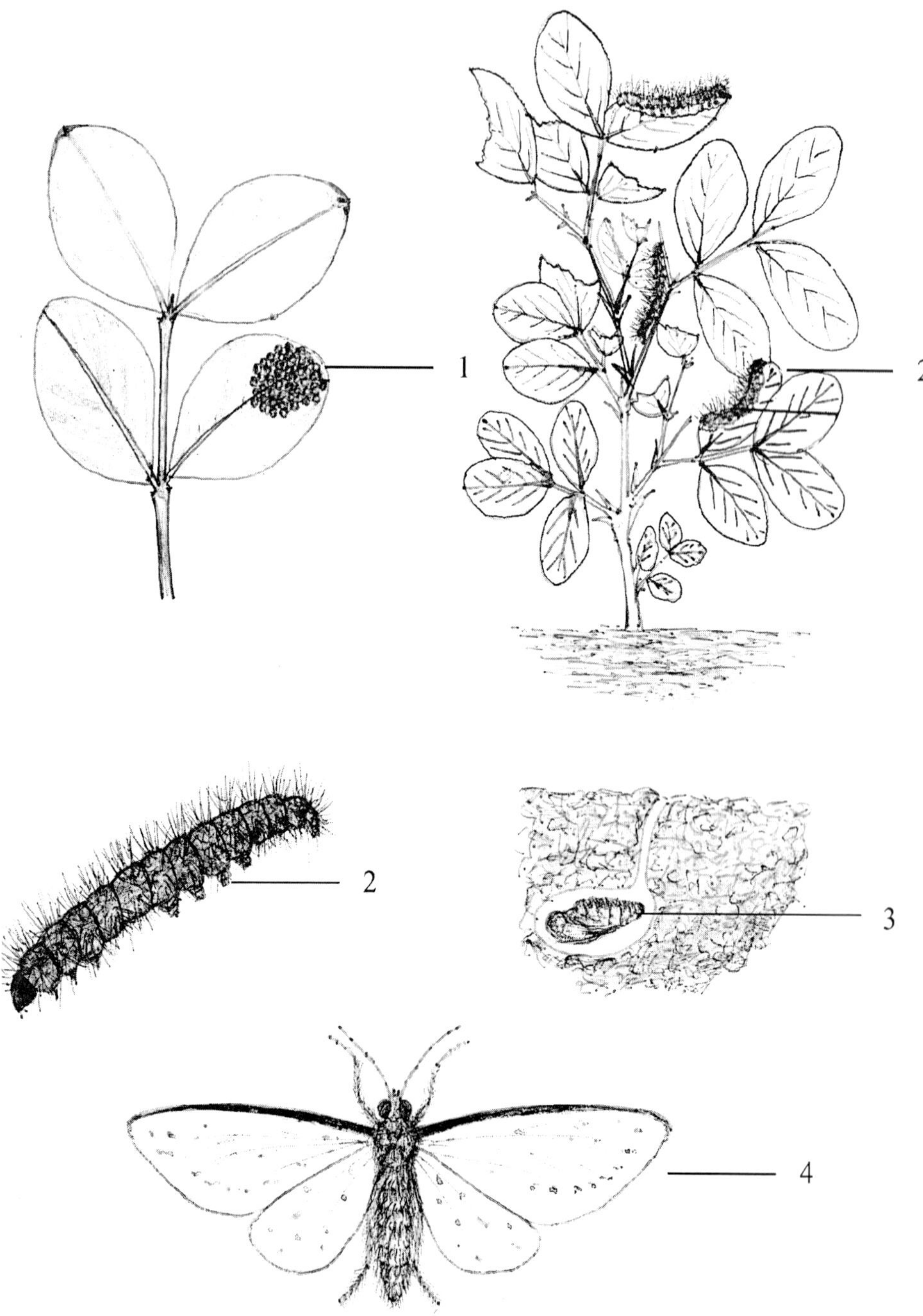

Fig. 24. Groundnut red hairy caterpillar-*Amsacta albistriga*

1. Egg mass 2. Hairy caterpillar 3. Pupa inside soil 4. Moth

to control the pest by resorting to 'Mass ground spraying' or 'Aerial spraying', so as to cover large areas within a short span of time.

Physical and cultural methods (i) Immediately after the receipt of summer showers, the fields are ploughed deeply and the pupae brought out to the soil surface to be picked up by several insectivorous birds or allowed to dry and get killed in the heat (ii) After the receipt of summer showers, light traps and bonfires are set up from 7.00-11.0 PM. to monitor the pest occurrence and also to attract and destroy the moths (iii) The egg masses, which can be easily spotted should be collected and destroyed (iv) Soybean or red gram may be grown as intercrop or border crop in groundnut fields. The eggs laid on such tall crops may be easily collected and destroyed. Further, the caterpillars, which climb and rest on these plants during the sunny hours may be easily collected and destroyed (v) The young caterpillars, which remain aggregated together on the under surface of leaves can be collected and destroyed (vi) The bigger caterpillars can be collected by hand picking and destroyed.

Chemical control (i) The young hairless caterpillars can be controlled by dusting carbaryl 10 D or quinalphos 1.5 D or phosalone 4 D or endosulfan 4 D at 10 kg./ acre , so as to cover the leaves thoroughly (ii) Slightly grown-up caterpillars can be controlled by foliar spraying with endosulfan-600 ml. or fenitrothion-600 ml. or chlorpyrifos-600 ml. or quinalphos-600 ml. or phosalone-600 ml. or dichlorvos-300 ml. in 300 liters of water per acre with a high volume sprayer (iii) It is very difficult to control fully-grown caterpillars, as they feed very little. They can be controlled to some extent by spraying with dichlorvos-600 ml./ acre, which has fumigant action.

Biological control

Egg parasitoids. The egg parasitoid-*Telenomus manolus* parasitizes and destroys the eggs.

Larval parasitoids. (i) The larval parasitoid-*Sturmia inconspicuella* parasitizes the larvae and destroys them (ii) The predatory pentatomid bug-*Eocanthecona furcellata* attacks and kills the larvae.

Fungal parasite. The fungus-*Aspergillus flavus* attacks the larvae, causes disease and kills them.

Viral parasite. Nuclear polyhedrosis virus (NPV) is capable of destroying young caterpillars. Spraying 100 larval equivalent (LE) of the virus suspension in 300 liters of water per acre is effective in controlling young caterpillars (one LE = 6×10^9 virus particles)

Preparation of nuclear polyhedrosis virus spray suspension. Primary culture containing virus particles capable of affecting red hairy caterpillars is obtained from the Tamil Nadu Agricultural University, Coimbatore or Agricultural College, Madurai or Agricultural Research Station, Paiyur. Medium-sized, young hairy caterpillars are collected and kept without any food for 12 hours in cages. The primary culture is mixed with a small quantity of water. *Calotropis procera* leaves are collected, wetted with the virus suspension, shade-dried and then fed to the starving caterpillars for two days. After two days, fresh leaves may be fed to the caterpillars. The larvae ingest the leaves along with the viruses. After about 5 days, the caterpillars develop disease symptoms such as non-feeding, color changes, dullness etc. and they climb to the top of the cages and hang head downwards and ultimately die. This is known as **'tree top disease'**. Sometimes, the body wall develops cracks and a white fluid oozes out. The dead larvae are collected and soaked in water for 2 days. Then, they are crushed and squeezed and the juice collected. The juice is filtered through a fine muslin cloth and all solid particles are removed. To this extract, some water is added and the whole thing allowed to stand for sometime. The virus particles settle down. The supermatent liquid is separated and discarded. This procedure is repeated a few more times and the fluid at the bottom, which contains the virus particles in a more or less pure form is collected, mixed with enough quantity of water and spreading agents such as 'Teepol' or 'Triton" and then sprayed on the foliage of groundnut crop during the evening hours. The virus affects the caterpillars feeding on the sprayed leaves and they die of sickness.

2. Groundnut leaf roller or Leaf miner

Aproaerema modicella (Stomopteryx subsecivella)

Order - Lepidoptera

Family - Gelechiidae

It is one of the major, endemic pests of groundnut and infests all cultivated varieties. The rainfed crops are attacked to a larger extent than the irrigated crops and bunch varieties are more prone to attack by the pest than the semi-spreading and spreading varieties. The pest is also known to attack red gram, soybean etc.

Nature of damage. The caterpillars mine the leaves and feed on the inner tissues or they web the adjacent leaflets together with silken thread to form a nest and from within scrape and eat the green matter, thus causing extensive damage. The scrapped portions appear whitish and papery in the beginning and eventually dry. The infested leaflets are also distorted. The infested plants remain stunted. Severely affected crop presents a burnt-up appearance and large number of small, adult flies can be seen flying in the field (Fig. 25).

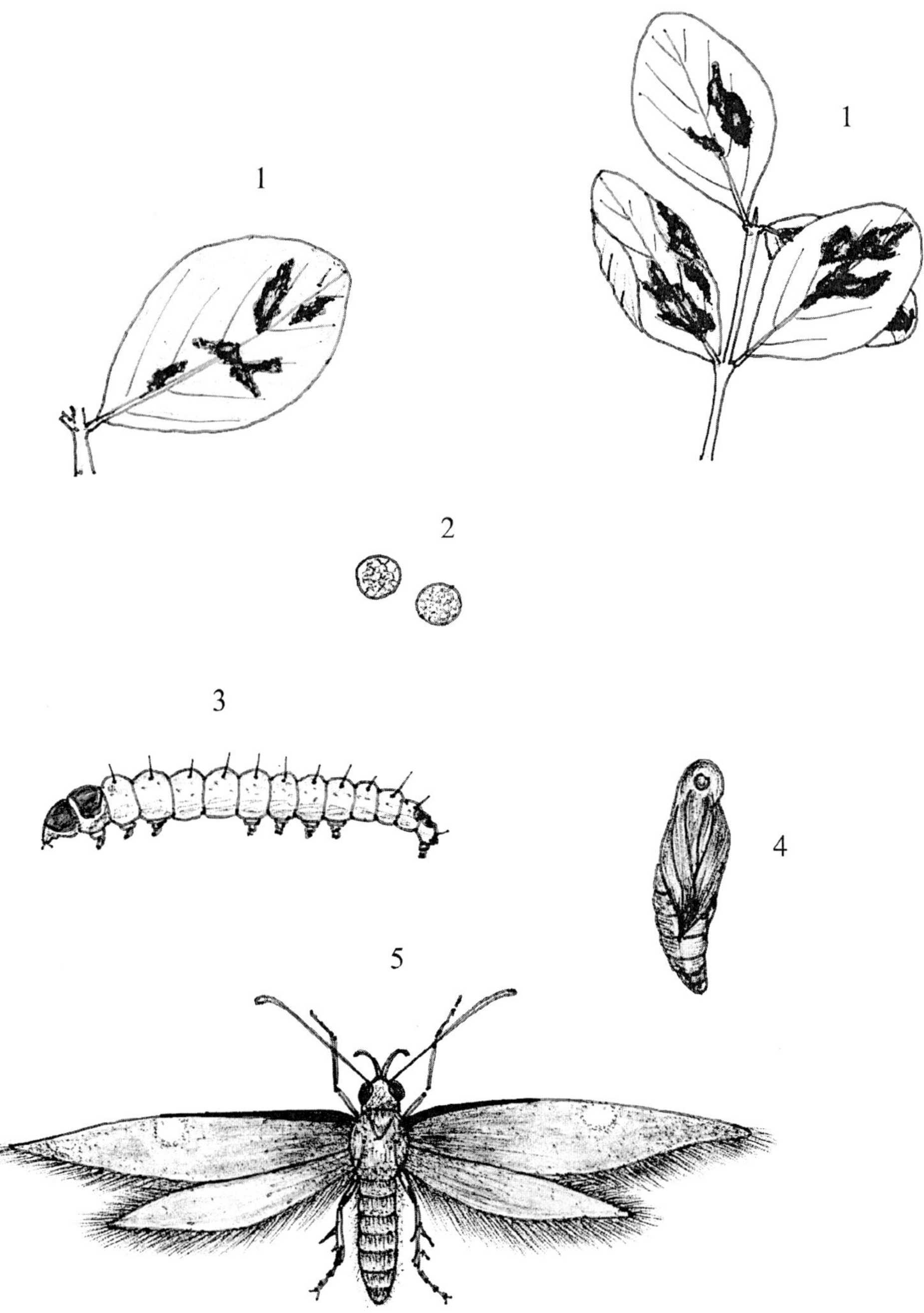

Fig. 25. Groundnut leaf miner-*Aproaerema modicella*

1. Infested leaves 2. Eggs 3. Larva 4. Pupa 5. Moth

Life cycle of the pest. The female moth lays very small, spherical, sculptured, shiny eggs singly on the tender leaves and young shoots. The young caterpillar, which hatches out from the egg in 3-4 days is grayish in color, with a black head. The caterpillar becomes full-grown in about 14 days. The grown-up caterpillar is greenish in color and smooth, with a small, black head, dark brown prothorax and measures 0.5-0.6 cm. in length. It pupates inside the webbed nest or inside the mine. The pupa is small and reddish-brown in color. The moth that emerges from the pupa in 4-5 days is small-sized, dark brown in color, with dark brown wings. Near the anterior margin of the narrow and pointed forewings, there is a whitish spot (Fig. 25).

Control measures

Physical and cultural methods (i) The crop should be sown early immediately after the receipt of monsoon rains, so that the crop may escape from pest attack (ii) Light traps should be set up to attract and destroy the moths as they are positively phototrophic.

Chemical control (i) Dusting with carbaryl 10 D or endosulfan 4 D or phosalone 4 D at 10 kg. / acre in the early stages of pest infestation, controls the pest (ii) Foliar spraying with endosulfan-300 ml. or quinalphos-600 ml. or monocrotophos-300 ml. or dichlorvos-300 ml. in 300 liters of water per acre with a high volume sprayer affords effective control of the pest.

3. Groundnut earwig or Groundnut pod borer

Euborellia stali

Order - Dermaptera

Family - Labiduridae

Nature of damage. The nymphs and adults of the pest bore the groundnut pods under the soil, eat the kernels inside the pods and cause appreciable damage and yield loss during some seasons. The shell of the infested pod shows round holes made by the pest and inside the pod, the faecal remains of the insect are found.

Life cycle of the pest. The female insect lays eggs in the soil around the groundnut plants. The young nymphs hatching out of the eggs are brown in color, flattish and without wings. Two, long, incurved, forceps-like anal cerci are found in the last abdominal segment. The forewings or tegmina of the adults are leathery and hard. The hindwings are large, thin, membranous, and foldable and are useful for flying. The insects are not capable of flying long distances (Fig. 26).

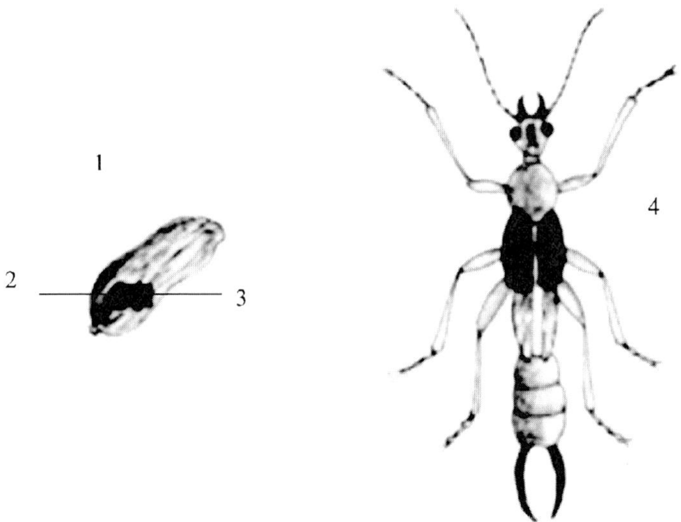

Fig. 26. Groundnut earwig-*Euborellia stali*
1. Damaged pod 2. Bore hole 3. Faecal matter 4. Adult insect

Control measures

Chemical control. Incorporating carbaryl 10 D at 20 kg./ acre into the soil at the time of second weeding or at the time of earthing up at peg formation stage controls the pest.

4. Lab lab aphid

Aphis craccivora

Order - Hemiptera
Suborder - Homoptera
Family - Aphididae

Nature of damage. The pest, which is very common on lab lab, infests soybean, black gram, green gram, red gram, Bengal gram etc., besides groundnut. Young plants of age 2-2½ months are more prone to attack by the pest. Both the nymphs and adults congregate in large numbers on the young shoots and leaves, puncture the parts, suck and feed on the sap. The infested leaves become crinkled and distorted. Because of continuous drainage of sap, the plants become stunted. Severe infestation leads to wilting of the shoots and ultimate death of the plants. The honeydew secreted by the pest attracts ants and several other kinds of insects. Further, sooty mould, a fungal disease may occur on the honey dew secretion, which hampers the photosynthetic functioning of the leaves of the plants. The insect is also the vector that transmits the virus causing **'rosette'** disease in groundnut.

Life cycle of the pest. The insects are very small and soft-bodied. The nymphs are yellowish-green in color, while the adults are greenish-black. Both winged and wingless adults are present. The female aphids are larger in size than the males and have big abdomen. They reproduce parthenogenetically and are viviparous. The adult female lays young ones one after another. The nymphs become full-grown in about 7 days and the females start laying young ones. The pest multiplies very rapidly (Fig.21).

Control measures

Chemical control. Foliar spraying with the systemic insecticide, dimethoate-300 ml. or methyl demeton-300 ml. or phosphamidon-150 ml. or monocrotophos-300 ml. in 300 liters of water per acre with a high volume sprayer controls the pest. However, the pest has developed resurgence to some of the organo phosphorus insecticides. So, repeated application of such chemicals should be avoided.

5. Groundnut stem borer

Sphenoptera perotetti

Order - Coleoptera

Family - Buprestidae

The pest is commonly known as **'ver poochi'** in Tamil and is found in the entire groundnut growing tracts of South India. The insect belongs to a group of beetles called 'jewel beetles', because of their beautiful, shiny and colorful appearance.

Nature of damage. The grubs of the beetle bore into the stem of groundnut plants close to the soil surface and feed on the inner tissues resulting in wilting and eventual death of the infested plant. Infested fields show dead and dying plants. When such a infested plant is pulled out and examined, the grub or pupa of the insect or signs of attack by the pest can be seen near the base of the stem of the plant (Fig. 27).

Life cycle of the pest. The female beetle lays eggs on the main stem of the plant. The grubs hatching out from the eggs, which are pale whitish in color burrow into the stem, feed on the inner tissues and grow. The grown-up grub is 3.75-4.00 cm. long, whitish in color, with a flattened anterior region and hence is known as the 'flat-headed grub'. The full-grown grub pupates inside the larval burrow and emerges as an adult beetle in about 10 days. The beetles are shining and dark brown in color (Fig. 27).

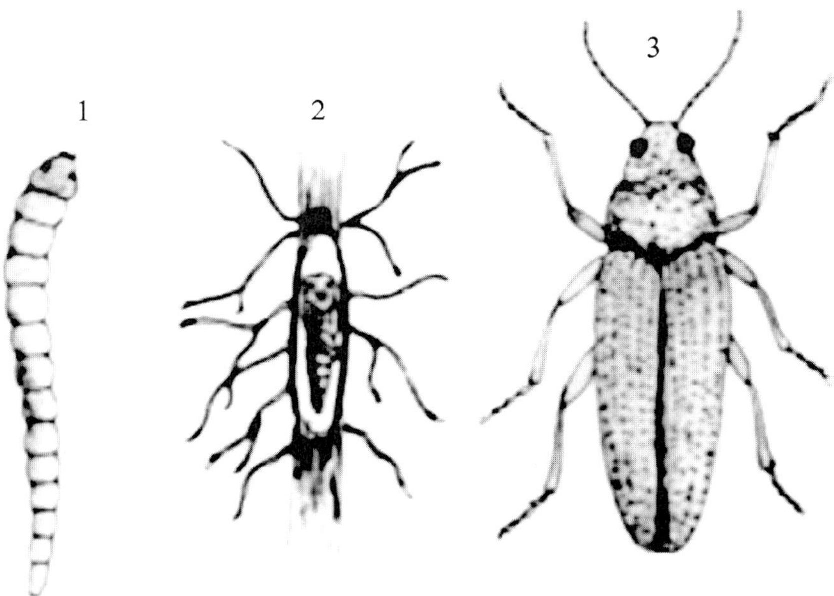

Fig. 27. Groundnut stem borer-*Sphenoptera perotetti*

1. Grub 2. Pupa inside the stem 3. Adult beetle

Control measures

Physical methods (i) The infested plants showing early symptoms of attack should be pulled out and destroyed (ii) The adult beetles may be collected by using a hand net and destroyed.

Other pests of groundnut. A number of other pests are also known to attack groundnut and some of them are considered to be of minor importance. The gram caterpillar-*Helicoverpa (Heliothis) armigera* and the cutworm-*Spodoptera litura* feed on the foliage and sometimes cause severe damage; the caterpillars of the semilooper pest-*Plusia signata* and *P. eriosoma* also feed on the foliage; the red hairy caterpillar-*Amsacta moorei* occasionally feeds on the foliage voraciously; the Bihar hairy caterpillar-*Diacrisia obliqua* also feeds gregariously on the foliage; the grass hopper-*Chrotogonus trachypterus* notches and feeds on the leaves; the adults of the weevil-*Myllocerus viridanus* cut and feed the leaflets from the edges; the blister beetle-*Mylabris pustulatus* and the chaffer beetle-*Oxycetonia versicolor* feed on the floral parts; the roots are sometimes damaged by the termite-*Odontotermes obesus* and the white grubs-*Holotrichia consanguinea* and *H. insularis;* the nymphs and adults of the chilli thrips-*Scirtothrips dorsalis* besides lacerating and sucking the sap from the tender shoots are also responsible for transmitting the virus causing **'bud necrosis'** or **'bud blight'** virus disease of groundnut plants.

GINGELLY or SESAMUM *(Sesamum indicum)*

1. Gingelly leaf and shoot webber

Antigastra catalaunalis

Order - Lepidoptera

Family - Pyraustidae

Nature of damage. It is one of the most serious pests infesting gingelly crop and is prevalent in all gingelly-growing tracts and occasionally causes extensive damage to the crop, resulting in appreciable yield loss. The young caterpillars hatching out from the eggs scrape and feed on the green matter of tender leaves. Soon, the larvae fold and web the top leaves and shoots together, bore the tender shoots and feed the inner tissues, as a result the shoots wilt. The caterpillars also bore the flower buds, tender and immature capsules and eat the inner contents, thus causing severe damage. It is a monophagous pest and infests gingelly only (Fig. 28).

Life cycle of the pest. The female moth lays 100-150 eggs, singly on the leaves or growing shoots. The eggs are flatfish and shiny. They hatch in 3-5 days and the young larvae become full-grown in 10-16 days. The grown-up larva is light brown in color, about 3.0 cm. long, with a black head and raised, black tubercles on the body. Fine, black hairs are seen on these tubercles. The larva makes a white, silken cocoon and pupates inside it within the webbed leaves. The pupa is whitish-green in color. The moth emerges from the pupa in 5-8 days. The moth is small, brown-colored and with long, yellowish-brown wings (Fig. 28).

Control measures

Cultural methods. Severely affected plant parts may be removed and destroyed

Chemical control (i) The pest can be controlled by dusting with endosufan 4 D or phosalone 4 D or quinalphos 1.5 D at 10 kg./acre (ii) Spraying the crop with endosulfan-300 ml. or phosalone-600 ml. or quinalphos-300 ml. or monocrotophos-300 ml. in 300 liters of water per acre with a high volume sprayer is also effective (iii) Spraying with neem oil-2% also gives good control of the pest.

Biological control. The ichneumonid larval parasitoid-*Cremastus flavoorbitalis* parasitizes the larvae.

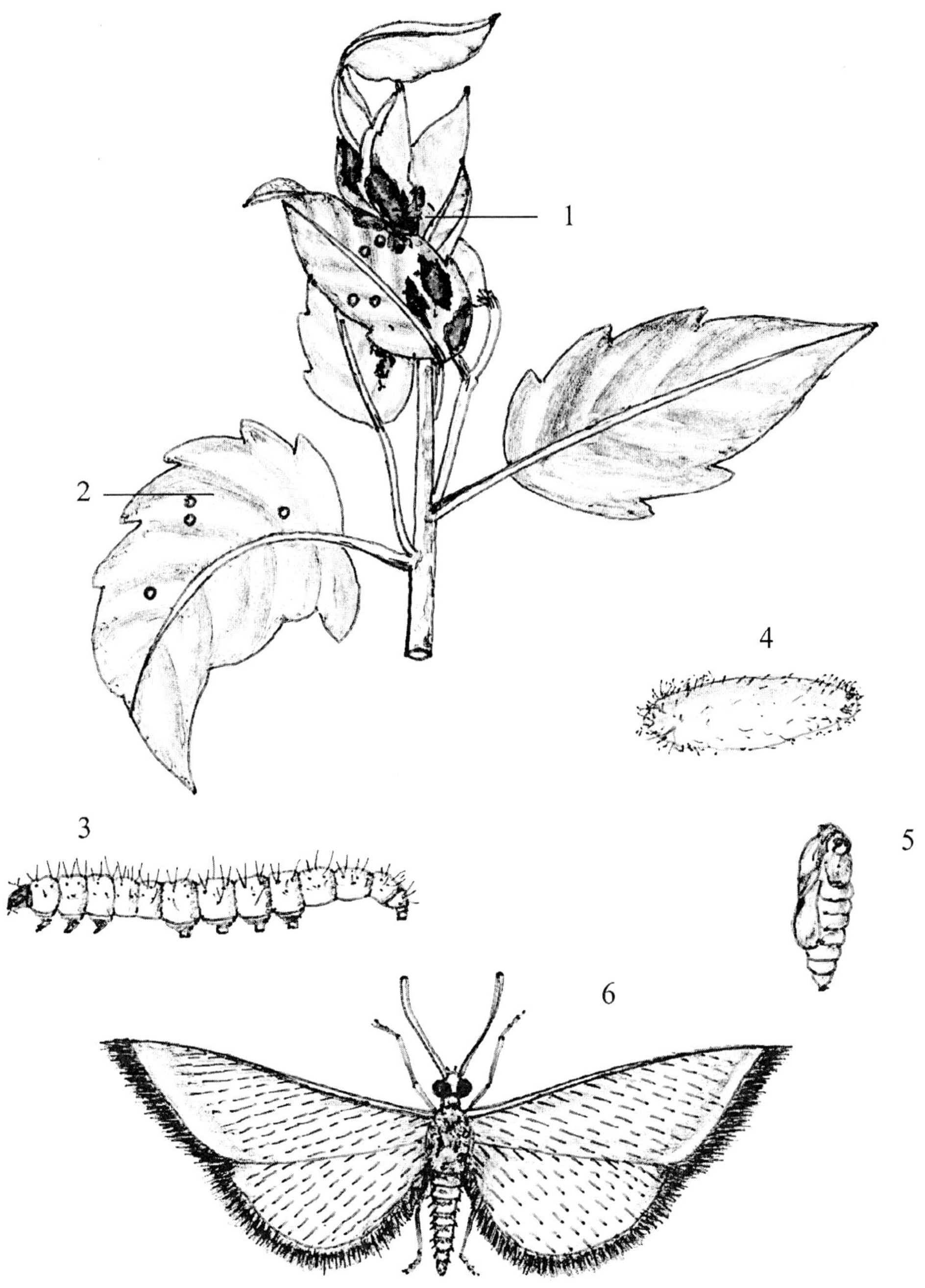

Fig. 28. Gingelly leaf and shoot webber-*Antigastra catalaunalis*

1. Infested shoot 2. Eggs 3. Larva 4. Cocoon 5. Pupa 6. Moth

2. Gingelly gall fly

Asphondylia sesami

Order - Diptera

Family - Cecidomyiidae

Nature of damage. It is also a specific pest of gingelly and occurs in all gingelly-growing tracts of South India. The larvae bore into the ovary of the flowers and feed on the inner tissues, as a result the capsules become distorted and abnormal swellings are produced in the capsules. The larvae and pupae are found in the swollen portions of the capsules. The affected capsules rarely produce any seeds (Fig. 29).

Life cycle of the pest. The female fly inserts the eggs in the ovary of the flower buds. The maggot is white in color and without legs. The grown-up larva pupates inside the swollen portion of the capsule and emerges as an adult fly in about 7 days. The fly is small and mosquito-like (Fig. 29).

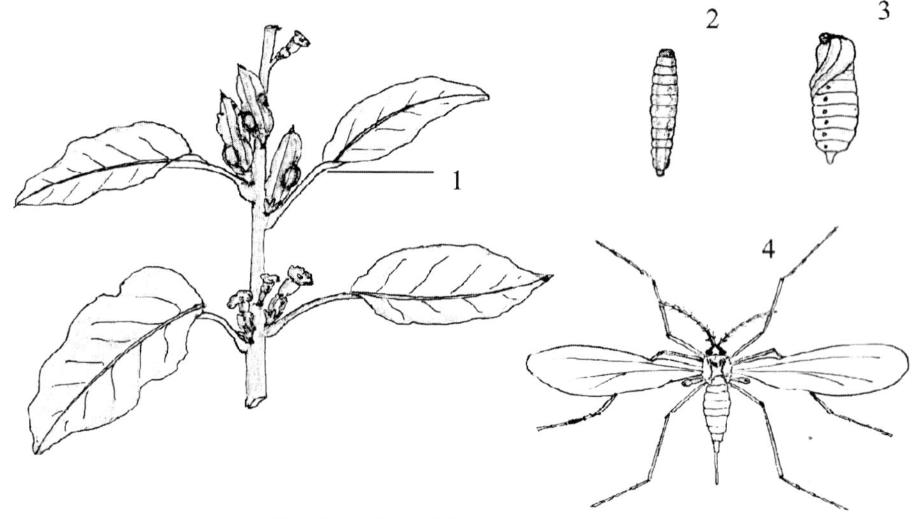

Fig. 29. Gingelly gall fly-*Asphondylia sesami*

1. Galls in the infested capsules 2. Larva 3. Pupa 4. Fly

Control measures

Cultural methods. The infested capsules should be removed and destroyed.

Chemical control (i) Dusting with carbaryl 10 D at 10 kg./acre controls the pest (ii) Spraying with endosulfan-300 ml. or quinalphos-600 ml. in 300 liters of water per acre with a high volume sprayer affords good control of the pest. Dusting or spraying should be done once at the time of flowering, followed by a second round of application 15 days later.

3. Horned caterpillar or Sphinx caterpillar

Acherontia styx

Order - Lepidoptera

Family - Sphingidae

Nature of damage. The pest is a leaf-feeding caterpillar. The caterpillars feed on the leaves and tender shoots and cause extensive damage. Besides gingelly, the pest infests soybean, lab lab, eggplant, jasmine and a few other crops (Fig. 30).

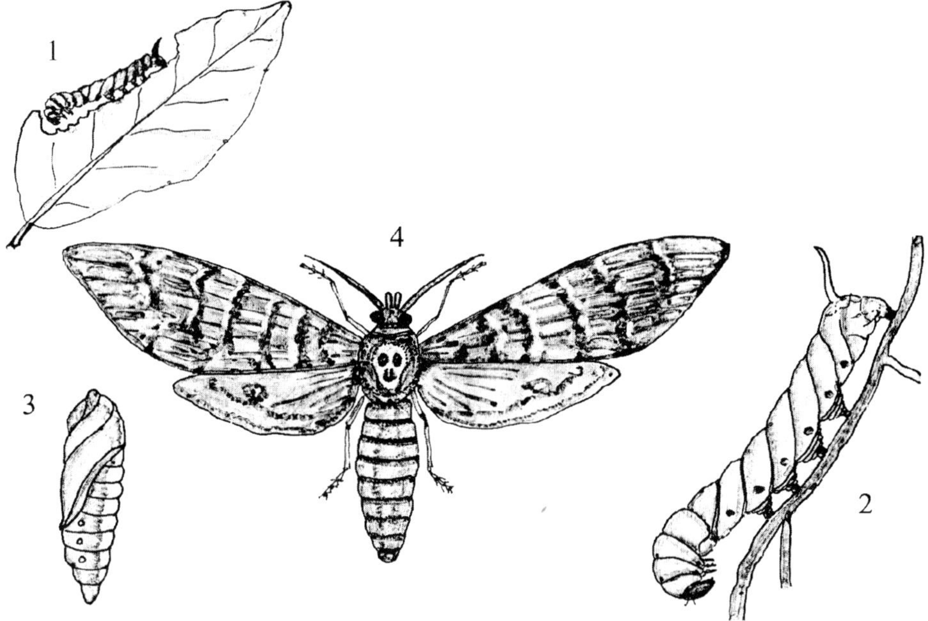

Fig. 30. The horned caterpillar-*Acherontia styx*

1. Larva feeding on the leaf 2. Caterpillar 3. Pupa 4. Moth

Life cycle of the pest. The female moth lays yellow-colored eggs, singly on the leaves. The young larvae feed on the young leaves and shoots voraciously and grow. The grown-up caterpillar is big-sized, stout, green in color with yellow-colored oblique stripes across the body. On the dorsal side of the last abdominal segment, a yellow-colored, long, horn-like anal cercus is present. The full-grown caterpillar pupates in an earthen cocoon under the soil. The moth is large in size, brown-colored, with a human skull-like marking on the dorsal side of the thorax and violet and yellow-colored cross bands on the abdomen. The wings are brown in color, the hind wings lighter in color than the forewings. The moths are known to visit beehives and suck the honey from the honeycombs.

Control measures

Physical and cultural methods (i) The caterpillars, which are quite large and easily visible can be hand-picked and destroyed (ii) The field may be ploughed deep, so as to expose the pupae to be picked up by insectivorous birds or get killed by exposure to the sun.

Chemical control. Dusting with carbaryl 10 D at 10 kg./ acre controls the pest.

4. Cotton aphid

Aphis gossypii

Order	-	Hemiptera
Suborder	-	Homoptera
Family	-	Aphididae

Nature of damage. The pest, though a serious one of cotton, infests many other crops, such as lady's finger, eggplant, chillies, guava etc., besides gingelly. The nymphs and adults appear in large numbers on the under surface of leaves and on the young, growing shoots, suck and feed on the plant sap. The infested leaves become mottled, distorted and turn yellow. The affected plants become stunted.

Life cycle of the pest *and* **control measures.** Refer pest of cotton (Page 193, Fig. 63).

Other pests of gingelly. A few other pests also infest gingelly crop, but are considered to be of lesser importance than the pests dealt with above in detail. The bug pests-*Nezara viridula* and *Dolicoris indicus* suck the sap from the leaves and tender shoots; *Orosius albicinctus* another sucking leaf hopper pest, besides causing direct damage by sucking sap from the leaves and tender shoots, is the vector of the mycoplasma causing **'phyllody disease'**; the semilooper-*Plusia signata* occasionally infest gingelly crop and feeds on the leaves and shoots; the red hairy caterpillar-*Amsacta moorei* feeds on the foliage and causes severe damage sometimes; the thrips-*Frankliniella schultzei* lacerate and suck the sap from the young leaves and tender shoots.

CASTOR *(Ricinus communis)*

1. Castor stem and capsule borer

Conogethes (Dichocrocis) punctiferalis

Order - Lepidoptera

Family - Pyraustidae

This polyphagous pest is a potential pest of castor and occurs in all castor-growing tracts of India. Besides castor, it infests ginger, turmeric, cardamom, guava, cocoa, jack, mango inflorescence, peaches, pear etc. The pest causes damage to the inflorescence and earheads of sorghum also.

Nature of damage. The pest generally appears when the castor crop is in the flowering stage. The caterpillars attack the shoots and the capsules and cause much damage. The young caterpillar bores into immature capsules through the pedicel or through the tip of the capsule and feeds on the inner contents. The grown-up caterpillar webs together many immature capsules with fine silken thread, remains inside the nest, bores the capsules and feeds on the immature beans inside the capsules. A single caterpillar attacks and damages many capsules. Young shoots and the inflorescences are also attacked similarly. Infested shoots wilt and dry. The webbed portion is covered with dark-colored faecal matter of the larva and frass (Fig. 31).

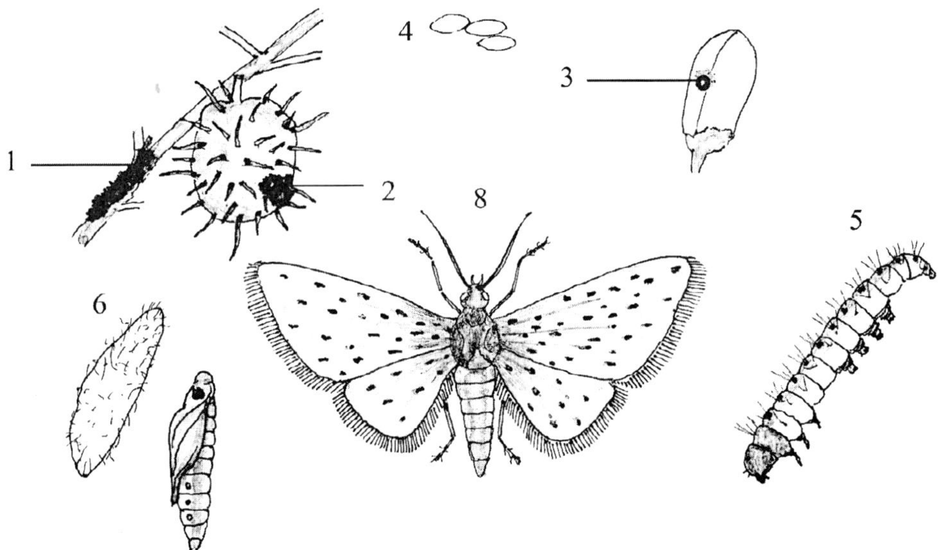

Fig. 31. Castor stem and capsule borer-*Dichocrocis punctiferalis*

1,2,3. Infested stem, capsule and seed 4. Eggs 5. Larva 6. Cocoon 7. Pupa 8. Moth

Life cycle of the pest. The female moth lays pink-colored, flattish eggs singly or in clusters of 2-3 on the under surface of flowers or on the tips of tender capsules. In a single inflorescence 20-30 eggs are laid. The eggs hatch in 6-7 days and the young larvae emerge out. The larvae become full-grown in 12-15 days. The grown-up larva is 2.0-2.4 cm. long, light green in color and the dorsal surface of the body pale pink in color. The head is brown in color. On the body surface, there are minute, raised, black warts, from which hairs appear. The dorsal surface of the first thoracic segment is covered with a brown, hard, leathery shield-like skin. The full-grown larva pupates inside a fine, silken cocoon. The cocoons are found inside the webbed area or inside the stem or inside the capsule. The adult moth emerges from the cocoon in 10-12 days. The life cycle is completed in 25-33 days. The moth is medium-sized and brown in color. The brownish-yellow wings have several black spots (Fig. 31).

Control measures

Cultural methods. Infested shoots and capsules should be removed and destroyed.

Chemical control. Spraying with malathion-600 ml. or fenitrothion-300 ml. or methyl parathion-300 ml. or carbaryl 50 WP-600 gm. or fenthion-600 ml. in 300 liters of water per acre with a high volume sprayer, so as to cover the inflorescence thoroughly controls the pest.

2. Castor semilooper

Achaea janata

Order - Lepidoptera

Family - Noctuidae

Nature of damage. The caterpillars bite and feed on the leaves and tender shoots and within a short period of time cause extensive damage to the crop. Only the leaf petioles and midribs of the leaves are left out in the infested plants. Severe outbreaks of this pest may occur during the months of August-January. The pest also attacks rose, pomegranate and a few weed plants, apart from castor.

Life cycle of the pest. The female moth lays spherical, ribbed, bluish-green eggs on the growing shoots and leaves in clusters of 1-6. One female lays about 400 eggs. The eggs hatch in 2-5 days and yellowish-green colored larvae emerge out. They feed on the leaves and tender shoots voraciously and become full-grown in 10-15 days. The grown-up larva is long, slender, about 7.5 cm. long, with a black head, a red spot on the hunched portion of the body and a bunch of long, red hairs on the dorsal side of the last abdominal segment. Generally the larva is gray-colored, with red and brown lateral lines along the entire length of the body or black-colored with whitish lateral lines. However, there are considerable color variations in the larvae. Of the 5 pairs of abdominal legs, the first two pairs are lacking and so, while they crawl, the body is raised in the middle portion of the body like a hunched back. The full-grown larva pupates under the soil or in-between dried fallen leaves. The moth emerges from the pupa in 10-15 days. The moth is stout, dull reddish-brown in color, with grayish-brown forewings and in each of the brownish hindwing, there is a central white spot and 3 white prominent spots on the outer margin (Fig. 32).

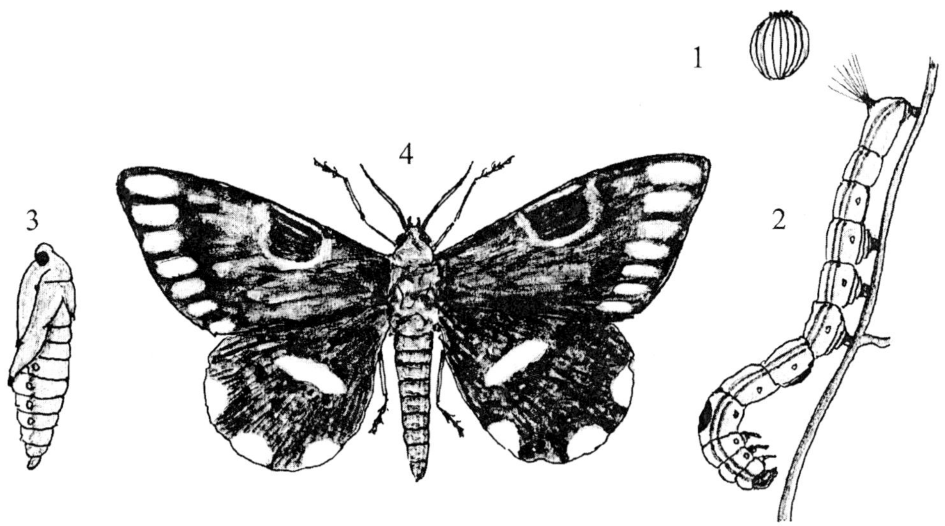

Fig. 32. Castor semilooper-*Achaea janata*

1. Egg 2. Larva 3. Pupa 4. Moth

Control measures

Physical methods. The larvae are long and big-sized and are easily visible. They can be handpicked and destroyed.

Chemical control (i) Dusting with endosulfan 4 D or carbaryl 5 D or phosalone 4 D at 10 kg./acre is effective in controlling the pest (ii) Spraying the crop with malathion-600 ml. or methyl parathion-300 ml. or fenitrothion-300 ml. or endosulfan-600 ml. in 300 liters of water per acre with a high volume sprayer, so as to cover the foliage thoroughly is also effective in controlling the pest.

Biological control (i) The naturally occurring egg parasites-*Trichogramma achaeae* and *T. australicum* parasitize and destroy the eggs (ii) The larval parasitoids-*Microplitis ophiusae* and *Tetrastichus ophiusae* destroy the larvae.

3. Tobacco caterpillar or cut worm

Spodoptera (Prodenia) litura

Order	-	Lepidoptera
Family	-	Noctuidae

Nature of damage. The caterpillars of this polyphagous pest bite and feed on the leaves and tender shoots voraciously, defoliate the plants and cause severe damage. The larvae bite and make big holes on the leaves and ultimately eat the entire leaves. In a single plant several larvae are found feeding simultaneously and within a few hours, all the leaves are eaten away and the larvae move on to the next plant. The larvae come out and feed only during the nighttime and hide in cracks and crevices and below mud clods etc. during the daytime. The infested crop appears as if grazed by cattle. The pest infests several other crops such as, tobacco, tomato, cotton, chillies, mustard, cabbage, cauliflower, knol khol, potato, Bengal gram, beans, soybean etc., besides castor.

Life cycle of the pest and **control measures.** Refer pests of cotton (Page 161, Fig. 68).

4. Castor white fly

Trialeurodes ricini

Order	-	Hemiptera
Suborder	-	Homoptera
Family	-	Aleyrodidae

Nature of damage. The nymphs and adults harbor on the under surface of leaves in large numbers, pierce, suck and feed on the sap. In case of severe infestation, the leaves turn yellow, wilt and dry. The honeydew secreted by the insects attracts several other insects. On the honey dew secretion on the surface of the leaves, sooty mould, a fungal disease appears, which interferes with the photosynthetic functioning of the leaves, thereby indirectly affects the vitality of the plants. The pest also infests cotton crop.

Life cycle of the pest. The adult female lays 100-150 eggs on the under surface of leaves. The eggs hatch in about 3 days. The young nymph, which hatches out from the egg, after crawling about for a short while on the leaf, fixes itself at one place on the under surface of the leaf in a sedentary state, sucks and feeds on the sap. The nymphs are yellow in color, scale-like, and wingless and are covered with a waxy, white, powdery coating on the body surface. The nymphs become full-grown in 6-10 days. The grown-up nymph pupates in the same place under the scale-like covering and emerges as an adult fly after 7-10 days. The adult white fly is whitish in color, with a yellow abdomen and white wings. The body surface and the forewings are covered with a waxy, white, powdery or mealy coating. The adults are capable of flying only short distances.

Control measures

Chemical control. Foliar spraying with malathion-600 ml. or methyl demeton-300 ml. or phosalone-600 ml. or acephate-600 gm. or monocrotophos-300 ml. in 300 liters of water per acre with a high volume sprayer, so as to cover the under surface of the leaves thoroughly is effective in controlling the pest.

5. Castor butterfly or Spiny caterpillar

Ergolis (Ariadne) merione

Order - Lepidoptera

Family - Nymphalidae

Nature of damage. The pest is monophagous and infests castor alone. The caterpillars feed on the leaves by remaining on the upper surface of leaves and cause severe damage to the foliage.

Life cycle of the pest. The female butterfly lays beautifully sculptured, spherical eggs on the leaves singly. The eggs hatch in 4-5 days and the young caterpillars bite and feed on the leaves and become full-grown in 20-22 days. The grown-up caterpillar is about 3.0 cm. in length, green-colored, with bunches of branched, short, spine-like hairs all over the body. It pupates in a brownish chrysalis and the butterfly emerges from the puparium within a few days. The butterfly is brown-colored, with thin wavy lines on the brown-colored wings.

Control measures

Chemical control (i) The pest can be controlled by dusting with carbaryl 10 D or endosulfan 4 D or phosalone 4 D at 10 kg./acre (ii) Foliar spraying with carbaryl 50 WP-600 gm. or methyl parathion-300 ml. or endosulfan-600 ml. or phosalone-600 ml. in 300 liters of water per acre with a high volume sprayer is also effective in controlling the pest.

6. Castor hairy caterpillar

Euproctis fraterna

Order - Lepidoptera

Family - Lymantriidae

Nature of damage. The caterpillar is a leaf-feeder and feeds voraciously on the leaves and causes damage to the foliage. The pest also infests pomegranate, rose, mango, red gram etc., besides castor.

Life cycle of the pest. The female moth lays flattish, circular, yellowish eggs in masses on the under surface of leaves and covers them with yellow hairs. The eggs hatch in 5-9 days. The larval stage lasts for 29-35 days. The full-grown larva is stout, dark reddish in color, profusely hairy, with a red head and white hairs all over the body and a tuft of long hairs on the preanal segment. It pupates in a thin, silken, yellowish hairy cocoon within the leaf folds. The adult moth emerges from the cocoon in 10-12 days. The yellowish, tussock moths with yellowish wings have pale transverse lines on the forewings. The entire life cycle is completed in 45-57 days (Fig. 97).

Control measures

Chemical control. The measures suggested for the control of groundnut red hairy caterpillar may be adopted (Page 66).

7. Castor slug caterpillar

Latoia (Parasa) lepida

Order - Lepidoptera

Family - Limacodidae (Cochlidiidae)

Nature of damage. It is a sporadic pest of castor and occurs in all castor-growing tracts of India. It appears in an epidemic form during some years and causes severe damage to castor crop. The larvae feed on the leaves, leaving only the midrib and veins. The pest attacks several other crops such as, coconut, palmyrah, pomegranate, mango, citrus, wood apple etc.

Life cycle of the pest. The female moth lays flat, scaly, shining eggs on the tender parts of the plant in batches of 20-30 and the eggs hatch in 6-7 days. The young larvae feed on the tender leaves and shoots and become full-grown in 42-45 days. The grown-up larva is flattish, short, stout, about 2.5 cm. in length and moves like a slug. It has a greenish, soft, hairy body, with pale white or bluish stripes along the body. The body is covered with rows of tubercles from which tufts of spines and hairs are given out. The hairs cause irritation and serious itching when they come into contact with the skin. The caterpillar pupates inside a thick, compact, shell-like, oval, chocolate-brown cocoon, which is convex above and flat ventrally. The cocoon is also covered with irritating spines and hairs. Sometimes a large number of cocoons are found attached in batches to the surface of the stem and main branches. The adult moth emerges from the cocoon in about 20-22 days. The moth is stout, and short, with green wings and a prominent dark patch at the base of each forewing.

Control measures

Chemical control. The measures suggested for the control of groundnut red hairy caterpillar may be followed (Page 66).

COCONUT *(Cocos nucifera)*

1. Coconut rhinoceros beetle

Oryctes rhinoceros

Order - coleoptera

Family - Scarabaeidae

'The rhinoceros beetle' is one of the serious pests of coconut and trees of all ages are subjected to attack by the pest. However, young plants between 3-10 years of age are more vulnerable to attack by the pest. The pest infestation is more severe, especially during and after the monsoon rains. Unlike several other pests, in this case only the adult beetles are capable of attacking the trees, while the grubs live on decaying organic matter in the soil. The pest has a wide distribution and is found all over Peninsular India, especially along the coastal regions, Southeast Asian countries, Sri Lanka, Philippines and Southern China. Besides coconut, the pest attacks palmyrah, oil palm, date palm, sago palm, pineapple and sugarcane.

Nature of damage. The adult beetle burrows through the unopened, tender fronds, biting the soft fibrous portions, chewing the juice from the fibrous tissues and throws out the remnants as a dry, fibrous mass through the burrow. This dry fibrous mass and the faecal matter of the beetle are commonly seen at the mouth of the burrow. The injury thus made is seen clearly as a series of holes on the fronds when they open up. When the growing frond is attacked, a 'V'- shaped cut is seen when the leaflets open and the terminal portion above the point of attack appears as a fan. Sometimes the tip of the frond breaks at the burrowed point and hangs down. The spathes are also burrowed, as a result the buds dry and are shed prematurely. The growth of the infested trees is also adversely affected. Severely affected young trees may even die (Fig. 33).

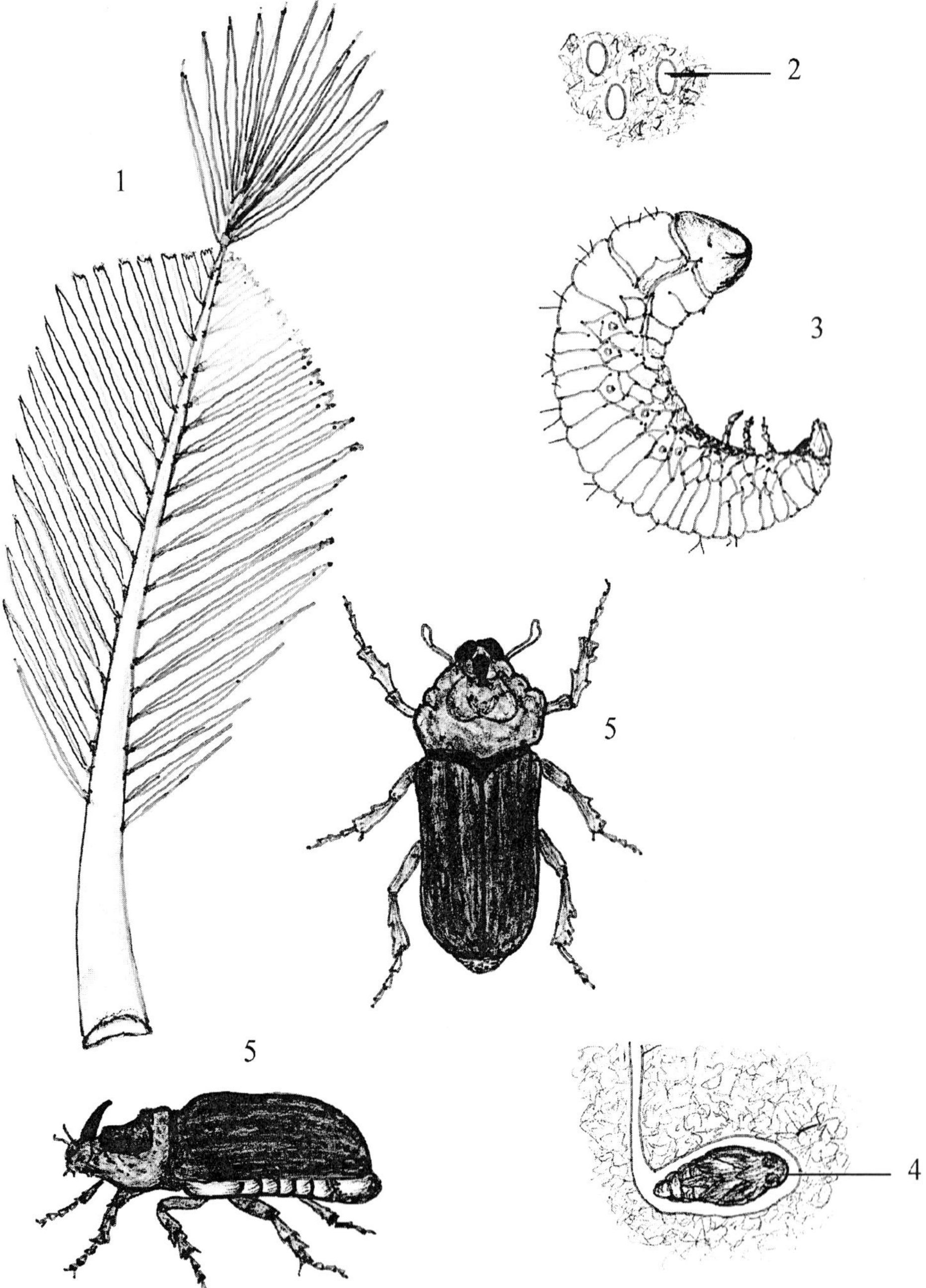

Fig. 33. Coconut rhinoceros beetle-*Oryctes rhinoceros*

1. Symptoms on leaf 2. Eggs 3. Grub 4. Pupa 5. Beetle

Life cycle of the pest. The female beetle lays yellowish-white, oval eggs in manure pits or heaps of organic debris at a depth of 5.0-15.0 cm. One beetle lays 100-150 eggs. The eggs hatch in 10-15 days and the young grubs emerging out eat the rotting organic matter and grow. During the growth, the grub moults 3 times and become full-grown in 100-180 days. The full-grown grub is dirty white in color, smooth, shiny, stout and fleshy with wrinkled body segments and measures about 10.0 cm. in length. It has a light-brown, small head and a darker tail end. In its natural position, it is always curved up ventrally in the form of 'C' and is found at a depth of 5.0-25 cm. The full-grown grub burrows to a depth of about 30.0-100.0 cm., construct a pupal chamber in the manure pit and pupates inside it. The pupal period varies from 10-60 days, at the end of which period the adult beetle emerges from the pupa. On emerging from the pupa, the beetle flies away to the nearest palm tree and starts damaging the tree. The beetle is black or reddish-brown in color, stoutly built with elongated, cylindrical body and measures 3.5-5.0 cm. in length. The dorsal side is covered by a smooth, shining, black, thick and hard elytra. The ventral side is saw dust-colored and hairy. It has well-developed wings and can fly long distances. On the dorsal side, near the tip of the head, the beetle has a pointed horn, bent slightly backwards, which is longer in the male and much shorter or rudimentary in the female and hence the insect is referred to as the **'rhinoceros beetle'**. There is a depression on the dorsal side of the first thoracic segment. In a single tree, there may be 5-8 beetles at a time. The adult beetle can live for a period of about 300 days. This perennial pest is found to be more severe during and just after the monsoon seasons. The beetles fly by night and are attracted by light. There is only one generation in a year (Fig. 33).

Control measures

Physical *and* **cultural methods** (i) Manure pits should not be located nearabout coconut gardens (ii) The manure pits should be stirred once in a month and the grubs present brought out and killed by exposure to sun or devoured by carnivorous birds (iii) The grubs can live in rotting coconut leaves dumped in coconut gardens under the trees. These should be removed immediately (iv) A long iron needle with fish hook-like pointed tip is used to poke through the burrows made by the beetle and the beetle hooked out or speared to death (v) The beetles being positive phototrophic, can be attracted, trapped and killed by setting up light traps in the coconut gardens after the receipt of monsoon rains (vi) The beetles are known to be attracted by the smell of cowdung. In 5.0 liters of water cowdung is mixed to make a slurry. To this slurry 1.0 kg. of castor cake or 1.0 kg. of crushed castor seeds is added and allowed to ferment. the fermented wash is poured in mud pots and placed in the coconut garden to attract and kill the beetles. the fermented wash is replaced every month with fresh fermented wash. (vii) Of late pheromone traps have been introduced to attract and destroy the beetles.

Chemical control (i) Carbaryl 50 WP solution is prepared at a concentration of 0.05% (1.0 gm. of carbaryl / liter of water) and sprinkled in the manure pit to wet the content up to a depth of about 30 cm. to destroy the grubs that may be present in the manure heap. This treatment is repeated every 2 months (ii) Sand and carbaryl 10 D or chlordane 5 D or aldrine 5 D are mixed in 1:1 proportion and this mixture is placed in-between the axils of the topmost fronds at 150 gm. per tree. This treatment is repeated every 3 months (iii) Sand and carbofuran 3 G or phorate 10 G are mixed in the proportion 150 : 25 and this mixture is placed in-between the axils of the topmost fronds at 150 gm. per tree. This treatment is repeated every 6 months (iv) Neem seed kernel powder and sand are mixed in the proportion 1:2 and this mixture is placed in-between the topmost fronds at 150 gm. per tree. This treatment is repeated once in 3 months (v) For small plants 3 numbers of naphthalene balls are placed in-between the axils of the topmost leaves of the crown once in 45 days.

Biological control (i) The fungal parasites *Metarrhizium anisopliae* and *Beauveria bassiana* are capable of infecting and destroying the grubs. These microorganisms are multiplied in suitable media and incorporated in the manure pits (ii) The parasitoid-*Platymeris laevicollis* can attack and destroy the beetles (iii) A few soil nematodes are also known to parasitize and destroy the grubs in the manure pits (iv) Mole crickets, frogs, rats, squirrels and some insectivorous birds are predatory on the grubs.

2. Coconut red palm weevil

Rhynchophorus ferrugeneus

Order - Coleoptera

Family - Curculionidae

'The red palm weevil' is another important pest of coconut that attacks the stem region of trees of all ages, which often results in the death of affected trees. Unlike the rhinoceros beetle, the adult red palm weevil does not cause damage to the trees, while the grubs alone cause damage. The pest is prevalent in most of the coconut growing countries such as, Papua, New Guinea, Indonesia, Thailand, Philippines and Burma besides India. In India, it is one of the most destructive pests of coconut in Tamil Nadu, Kerala, Maharashtra, Assam, Mysore and Orissa. Besides coconut, the pest attacks oil palm, palmyrah, sago palm, date palm and wild dates. The pest causes extensive damage in young coconut plantations.

Nature of damage. Infested trees show small holes in the top portion of the stem. Through these holes, fibrous waste materials and faecal matter are thrown out and during the early stages of infestation a brownish-colored, gummy liquid also oozes out, which dries out and remains like small beads at the mouth of the holes. This is quite characteristic of the pest infestation. Severely affected trees show wilt symptoms. The young grubs burrow the top, tender portion of the stem or bore through wounds in the stem and enter into the stem portion of the trees. After entering into the stem, the grubs tunnel the inner portion up and down, eat the inner tissues and cause damage. The inner tissues in the affected portion rot and emanate a foul odor, at the same time a brownish liquid also oozes out through the holes. The stem of infested trees becomes hollow inside, the trees become weak, stunted, wilt and ultimately die. When young plants are attacked, the top portion characteristically bends slightly above the point of attack, wilt and die soon.

Life cycle of the pest. The female weevil scoops out a small hole with its snout at the top portion of the stem region covered by the leaf base, which is comparatively soft and lays eggs singly in each hole. The eggs may be laid in wounds found in the tree trunks also. In each day the weevil may lay about 5 eggs. In all one female weevil lays 250-275 eggs during its life span of 3-4 months. The eggs are whitish in color and oval-shaped. The eggs hatch in 3-5 days. The newly emerged grub is soft, with pale-yellowish body and reddish head. The grub has no legs and moves by muscular contractions of the body. It feeds on the soft, inner tissues of the trunk and becomes full-grown in 60-90 days. The grown-up grub is 3.5-5.0 cm. long, stout, fleshy, and pale-yellowish in color and with reddish head. It constructs an elongate, oval, cylindrical cocoon with fibrous materials, pupates inside it and becomes an adult weevil in 12-21 days. The adult weevil remains quiescent inside the pupal case for 11-18 days and after that period comes out of the cocoon. The full-grown weevil is reddish-brown in color with thick elytra. It has 6 black spots on the dorsal side and a long snout slightly bent downward. The male weevil has a tuft of hairs on the dorsal surface of the snout. The whole life cycle of the pest is completed inside the tree trunk. The full-grown weevils come out of the tree trunk through a hole made with the snout. Soon after the weevils come out of the tree, they mate and from the very next day the females start laying eggs. The weevils are good fliers and are diurnal in habit. The weevils have a long life span of 108-112 days and the males live longer than the females (Fig.34).

Control measures

Cultural methods (i) The dead and infested trees in coconut gardens may contain grubs and pupae at all stages of development. So, such dead and severely affected trees should be cut and destroyed immediately (ii) The adult female weevils generally remain in the leaf axils and lay eggs. So, old and ripe leaves should be cut and the top stem portion kept clean to avoid egg laying in the top portion of the stem (iii) Care should be taken, not to cause any kind of injuries or wounds on the stem portion of trees. Any wounds found on the trunk should be sealed with cement or coal tar mixed with sawdust (iv) Pheromone traps are available to trap and kill adult weevils. One pheromone trap is sufficient to attract the weevils within an area of 5-10 acres. The sex hormone in the pheromone traps remains effective for a period of about 3 months (v) A bait consisting of a mixture of acetic acid-5.0 ml. + yeast-5.0 gm. + sugar factory waste-2.5 lit.is kept in mud pots in coconut gardens. The weevils, which are attracted by the smell of the bait, fall into the mixture and are killed.

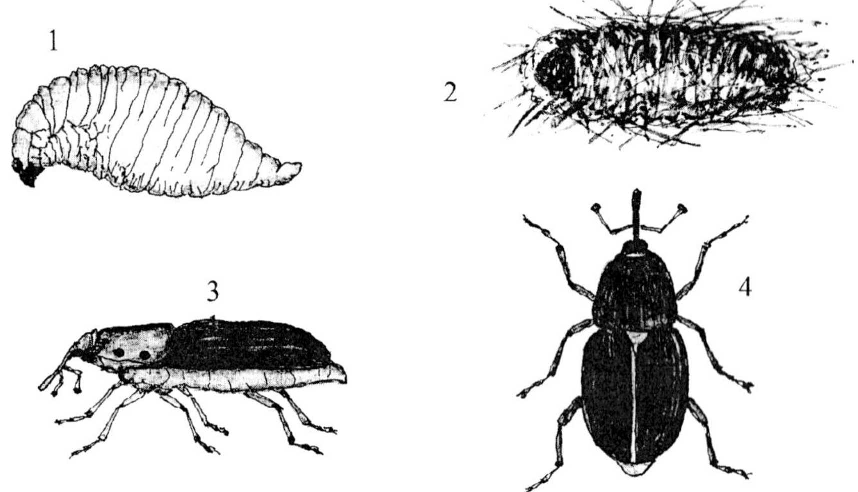

Fig. 34. Coconut red palm weevil-*Rhynchophorus ferrugeneus*

1. Grub 2. Cocoon made of fibre 3,4. weevil

Chemical control (i) The burrows made by the grubs are first sealed with clay. Then a fresh hole is drilled just above the affected portion. Through this hole pyrocone-1.0% or carbaryl-1.0% solution is poured inside the trunk at 1000-1500 ml. per tree and then the hole sealed with clay mixed with copper oxychloride fungicide (ii) A hole is drilled just above the affected portion and through this hole ethylene dichloride-carbon tetrachloride (EDCT) at 10 ml. or monocrotophos-10 ml. or a mixture of monocrotophos-5.0 ml. + dichlorvos-5.0 ml. per tree is injected and the hole sealed with a mixture of clay and copper oxychloride fungicide (iii) Through a hole drilled above the affected portion, Aluminium phosphide 3 G tablet is put at the rate of one tablet per tree and the hole sealed (iv) Fine sand and lindane-0.65 D are mixed in the ratio 1 : 1 and 150 gm. of this mixture is placed in the leaf axils of each tree, which will repel the weevils (v) Monocrotophos and water are mixed in equal proportion and 60 ml. of this mixture is fed through two active feeder roots at 30 ml. per root. This systemic insecticide effectively kills vigorously feeding grubs. However, this treatment is not effective in killing the pupae or weevils in the quiescent state.

3. Coconut black-headed caterpillar

Opisina arenosella (Nephantis serinopa)

Order - Lepidoptera

Family - Cryptophagidae

'The coconut black headed caterpillar' is one of the most important leaf pests of coconut and is prevalent all over Peninsular India, especially along the coastal regions. It is a serious pest in most of the states of South India. In Tamil Nadu, besides the coastal belts, the pest is of common occurrence in the interior regions of many districts, such as Coimbatore, Salem, Tiruchirapalli, South Arcot, North Arcot, Chinglepet etc. The peak activity of the pest is mostly during the dry months of March-June and it may extend if the monsoon is delayed. It is a well-known pest in Sri Lanka also. The pest attacks palmyrah trees also.

Nature of damage. The pest infests the leaves and causes severe damage. The caterpillars live on the under surface of leaflets, hidden inside galleries made of silken thread and frass and feed by scraping the green matter of the leaflets. When the pest occurs in large numbers, the leaflets dry up and when many fronds of the tree are similarly attacked, the tree is affected badly and the nut production is drastically reduced. Severely affected trees present a burnt-up appearance from a distance (Fig. 36).

Life cycle of the pest. The female moth lays yellowish-white, scale-like eggs on the under surface of leaflets in small batches, especially near the old larval galleries. One female lays 125-250 eggs. The eggs hatch in 5-7 days. On hatching, the young caterpillar starts feeding by scraping the green matter of the leaflets, at the same time constructs a gallery with fine, silken thread and frass, inside which it generally remains and feeds on the green matter. As the caterpillar grows, the gallery is also extended gradually. The larval period extends for 30-40 days. The full-grown larva is slender, elongated, about 2.5 cm. long, greenish-brown in color with black head and fine, thin hairs on the body. The first thoracic segment is dark-brown in color and the second thoracic segment is reddish in color. The full-grown larva changes into a brownish pupa inside the same gallery. The moth emerges from the pupa in 10-12 days.

The moth is small and whitish-gray in color. There are several overlapping generations in a year (Fig. 36).

Control measures

Cultural methods. The older leaves, which are severely affected should be cut and burnt to destroy the egg masses, larvae and pupae.

Chemical control (i) Young trees can be sprayed with carbaryl-0.1% or toxaphene-0.05% or malathion-0.05% or neem oil-3% (neem oil-6.0 lit. + liquid soap-300 ml.), so as to give a thorough coverage of the under surface of the leaflets (ii) It is very difficult to take up foliar spraying in grown-up and tall trees. So, root feeding with insecticides may be done. For this, monocrotophos-10 ml. is mixed with 100 ml. of water and the solution fed through the root. Before adopting this method, all the ripe coconuts are harvested and for the next 40 days harvesting should be avoided.

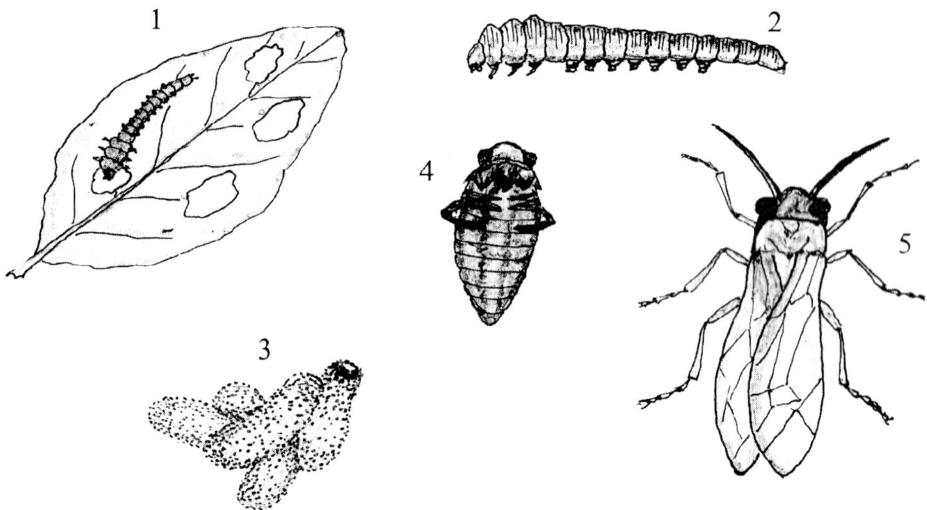

Fig. 35. Mustard saw fly-*Athalia lugens proxima*

1. Larva feeding on the leaf 2. Larva 3. Earthen cocoon 4. Pupa 5. Saw fly

Biological control (i) Repeated release of parasitoids a few times at the appropriate time affords effective control of the pest. The Bethylid-*Perisierola nephantidis,* the Braconid-*Bracon brevicornis,* the Eulophid-*Trichospilus pupivora* and the Ichneumonid-*Eriborus trochanteratus* are capable of controlling the pest effectively. The parasitoids belonging to the Braconid, Bethylid and Eulophid are released in the ratio, 10:10:20 per tree, per release. They should

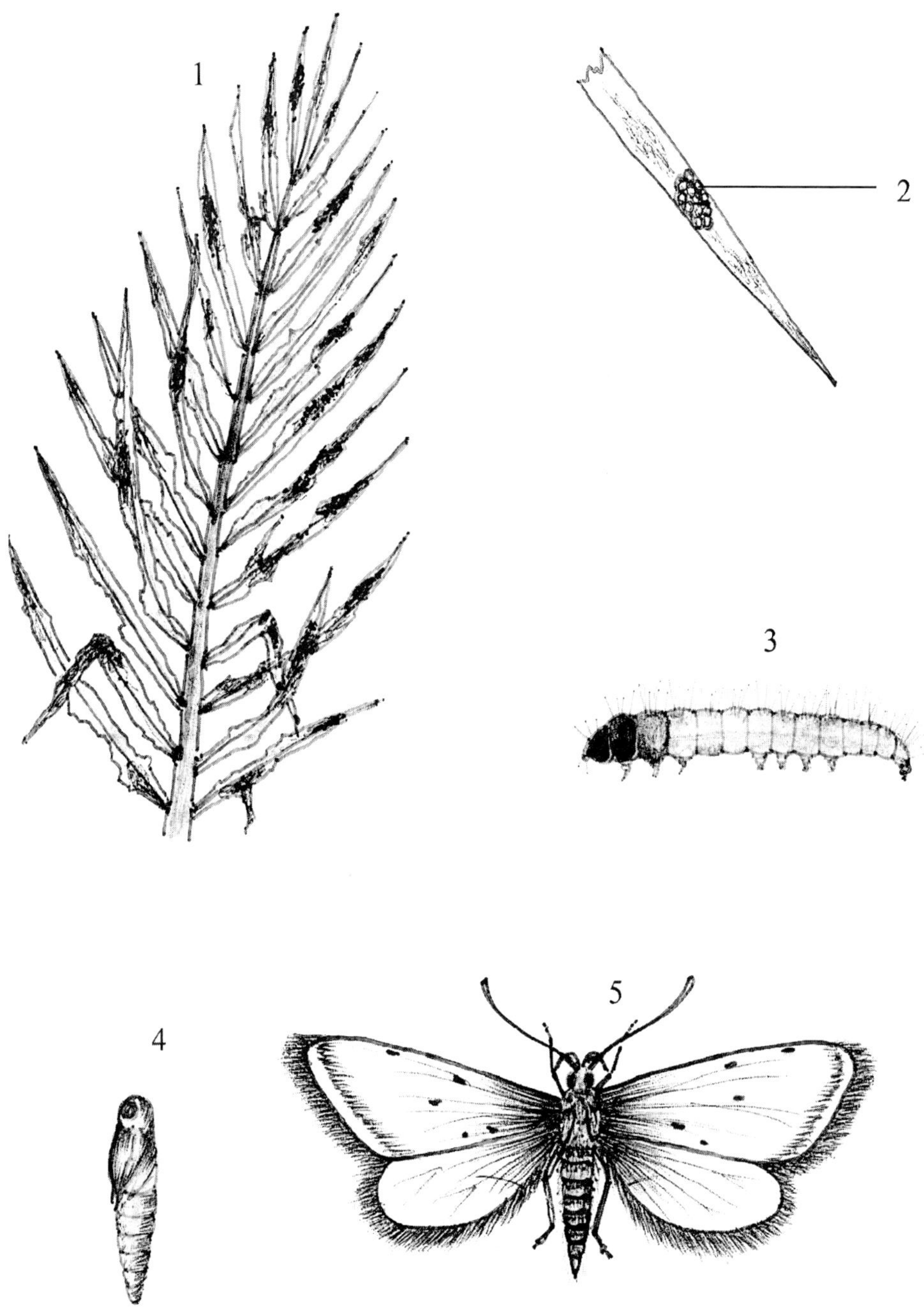

Fig. 36. Coconut black-headed caterpillar-*Opisina arenosella* (or) *Nephantis serinopa*
1. Infested leaf 2. Egg mass 3. Larva 4. Pupa 5. Moth

be released at the second or third instar larval stage and at the early stages of the pupae of the pest to obtain good results. The parasitoids should be released near the canopy of the trees and not at the ground level. For this, a parasitoid release trap may be used. After the release of parasitoids, insecticides should not be sprayed on the tress for the next 3 weeks. If insecticides have to be necessarily used, 5-10% of the trees in the garden should be left out unsprayed to safeguard the parasitoids present in the garden (ii) The fungal parasite-*Aspergillus flavus* infects the larvae and destroys them (iii) The bacterial parasite-*Serretia marcescens* is also capable of infecting and destroying the larvae.

4. **Coconut white grub** or **Root grub**

Leucopholis coneophora

Order	-	Coleoptera
Family	-	Melolonthidae

'The coconut white grub' is a sporadic pest of coconut and is found in parts of Kerala and Tamil Nadu. Sometimes the infestation is serious during the months of March-May. The pest also attacks tapioca, sugarcane, sweet potato, arecanut, banana, yam etc.

Nature of damage. The grubs, which remain under the soil cut, bite and feed on the roots and cause damage. When young plants are attacked, their growth is adversely affected. When grown-up palms are attacked, they show retarded growth and the nut production is also considerably reduced

Life cycle of the pest. During the months of June-July, after the onset of the monsoon rains, the beetles emerge out and soon after mating, the female beetles lay eggs in the soil near the root zone at a depth of 7.0-15.0 cm. The eggs hatch in about 20 days. The grubs emerging from the eggs start biting and feeding on the roots. The first, second and third instar grubs are found in the soil at depths of 15-45 cm., while the older grubs live at depths ranging from 30-100 cm. The grubs are white in color and fleshy. The larval period lasts for 10-11 months. The full-grown grub, after a pre-pupal, quiescent period of 9-12 days pupates under the soil at a depth of 45-50 cm. and in about 25 days the beetle emerges out. The emergence of the beetles almost coincides with the next monsoon season, as monsoon showers are necessary for the emergence of the beetles from the soil. The beetles are medium sized, oval in shape, grayish-black in color with ash-colored abdomen and they live for 20-45 days. There is only one generation in a year.

Control measures

Cultural methods (i) Coconut gardens should be ploughed deep to bring out the grubs and pupae from under the soil to the soil surface so that they may be picked up and destroyed by insectivorous birds (ii) The beetles are positive phototrophic in nature. So, light traps should be set up in coconut gardens to attract and destroy the beetles (iii) By using a bait, the beetles can be attracted and destroyed. In 5.0 lit. of water, cowdung is added to make a slurry. To this slurry, 1.0 kg. of castor cake or 1.0 kg. of crushed castor seeds is added and allowed to ferment. This fermented wash is poured in wide-mouthed mud pots and kept in the coconut garden. The beetles attracted by the smell of this bait fall into the slurry and are killed.

Chemical control. In case of severe incidence of the pest, chemical control measures may be adopted. For this carbaryl 10 D or chlordane 5 D or heptachor 3 D or endosulfan 4 D is incorporated into the soil at the rate of 20 kg. per acre by ploughing, twice in a year, once in April and again in August.

5. Coconut slug caterpillar

Contheyla rotunda

Order - Lepidoptera

Family - Limacodidae (Cochlidiidae)

'The slug caterpillar', a sporadic leaf pest of coconut is found in parts of Kerala and Tamil Nadu and during some years assumes epidemic proportions and causes considerable damage to the leaves. The aftereffect of severe infestation may last for a period of about 3 years till new leaves are put up and the tree recovers completely. As a result of pest infestation, production of inflorescence and development of nuts are also drastically affected.

Nature of damage. The young larvae scrape and feed on the chlorophyll portion of the leaflets. As they grow, they start eating the entire lamina of the leaflets, leaving behind only the hard mid vein, as a result the normal functioning of the leaves are arrested. This leads to weakening and loss of vigor of the trees and consequent loss in yield of nuts (Fig.37).

Life cycle of the pest. The adult female moth lays small, slightly elongated eggs on the under surface of leaflets. One female lays 62-324 eggs during its lifetime, which lasts for a few days. The eggs hatch in about 5 days. The larvae are thin, elongated and appear like slugs and are commonly known as slug caterpillars. The head, legs and sucker legs are retractable, the ventral surface forming a peculiar, smooth, sucker-like attachment along which the larva slides like a slug. The larva bears fine, spiny tubercles. They become full-grown in 25-42 days. The grown-up larva pupates inside a white-colored cocoon. The cocoons are found sticking on to the surface of the leaflets. The moth emerges from the cocoon after 10-12 days. The moth is very small, ashy-gray in color with somewhat rounded forewings and shorter, rounded hind wings and measures about 1.0 cm. across the wing span. When at rest, the moth keeps its wings folded over its back like a roof. The whole life cycle is completed in about 60 days. There are many overlapping generations in a year (Fig. 37).

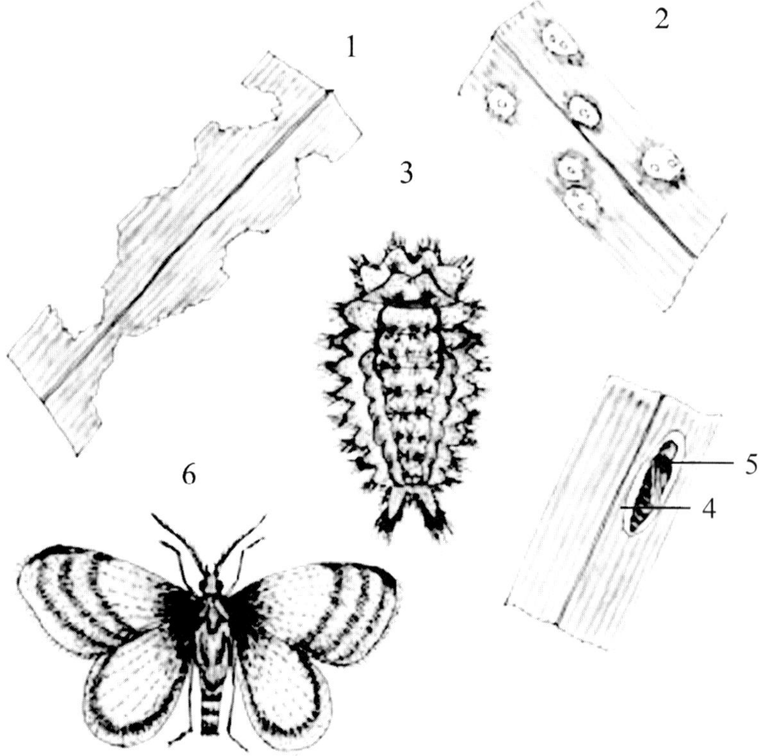

Fig. 37. Coconut slug caterpillar-*Contheyla rotunda*

1. Damaged leaflet 2. Eggs 3. Caterpillar 4. Whitish pupal case attached to the leaf 5. Pupa 6. Moth

Control measures

Cultural methods. (i) Egg masses on the leaflets of young plants can be collected and destroyed (ii) Light traps can be set up in coconut gardens to attract and kill the moths.

Chemical control. (i) If pest infestation is noticed in young plants, spraying with monocrotophos-0.04% or phosphamidon-0.05% or phosalone-0.1% can be taken up (ii) In the case of grown-up and tall palms, where spraying is difficult or not possible, root feeding with systemic insecticides such as, monocrotophos can be done.

6. Palm scale or Palm hard scale

Aspidiotus destructor

Order - Hemiptera

Suborder - Homoptera

Family - Diaspididae

This sucking pest is found in coconut gardens in some parts of Tamil Nadu and Kerala, especially in badly maintained gardens and in case of severe infestation may cause severe damage to the leaves and yield loss. The infestation is more during the months of February and March

Nature of damage. The nymphs and adults aggregate in large numbers and colonies of the scales are found covering the infested leaflets, inflorescence, developing nuts and leaf base at the point of attachment to the stem. With their piercing and sucking type of mouthparts, they pierce the outer skin and suck the plant sap from the tissues of infested parts. The affected leaflets turn sickly, yellow and ultimately dry. The inflorescence remains stunted and in most of the cases the fertilized female flowers and young developing nuts are shed one after another. While feeding, the insects inject some toxins along with the saliva into the plant tissues, which destroy the chlorophyll, as a result the affected regions turn yellow.

Life cycle of the pest. The female adult lays 60-200, scale-like eggs on the aerial plant parts. The young nymphs emerging from the eggs start sucking the sap from the very early stages of growth. They remain within a hard, ash-colored sheath. The females have a pale-brown, circular body. They are sedentary and do not develop wings. The grown-up males develop a pair of transparent wings (Fig. 38).

Control measures

Cultural methods. Severely infested leaves should be cut and destroyed.

Chemical control (i) The pest can be controlled by spraying with monocrotophos-0.04% or dimethoate-0.06% (ii) Root feeding with monocrotophos at the rate of 10 ml. of the insecticide in 100 ml. of water through two active feeder roots at 50 ml. per root is also effective in controlling the pest.

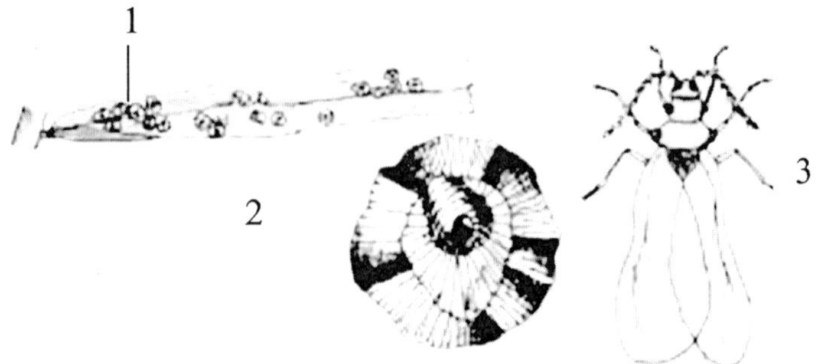

Fig. 38. Palm hard scale-*Aspidiotus destructor*

1. Scales on leaflet 2. Adult female scale 3. Winged adult male

7. Coconut mealy bugs

Pseudococcus coccotis and *P. longispinus*

Order	-	Hemiptera
Suborder	-	Homoptera
Family	-	Pseudococcidae

Nature of damage. Among the sucking pests of coconut, mealy bugs are also responsible for causing considerable damage. These pests colonize in large numbers near the tender central shoots of trees, especially young trees, pierce and suck the sap from the tissues, thereby adversely affecting the normal growth. Both the nymphs and adults are responsible for causing damage.

Life cycle of the pest. Only the early stage nymphs emerging from the eggs, called crawlers have functional legs and are capable of movement. The later stage nymphs and adult females are stationary and remain in a particular place being attached to the host with their mouthparts. The nymphs and adult bugs have a soft, slightly reddish body. The body of the mealy bug is covered with a waxy, mealy coating. Ants are attracted to these bugs for the honeydew secreted by them. The ants usually transport the bugs from one place to another and help in their multiplication. The males are alate and the forewings are larger and transparent, while the hindwings are reduced to haltere-like structures.

Control measures

Chemical control. Spray application of systemic insecticides such as, monocrotophos-0.04% or phosphamidon-0.05% or methyl demeton-0.05% or dimethoate-0.06% controls the pest.

8. Termites or White ants

Odontotermes obesus

Order - Isoptera

Family - Termitidae

Nature of damage. The white ants scrape and feed on the roots and stem portions of young coconut trees and seedlings and cause damage. The infestation may lead to death of up to 20 per cent of seedlings, especially under dry conditions. In older trees, the insects construct many, long, earthen galleries, live inside them in large numbers as a colony, scrape and feed on the bark and roots and cause appreciable damage. It is a serious pest of sugarcane also (Fig. 102).

Control measures

Chemical control (i) Incorporating chlordane 5 D or heptachlor 3 D or lindane 0.65 D into the soil in the nursery area at 20 kg. per acre prevents occurrence of white ants (ii) Mixing coal tar and kerosene oil in the ratio 1 : 3 and painting on the basal portion of the stem of trees up to a height of 3 feet prevents attack of white ants.

9. Coconut rat or Common Indian rat

Rattus rattus wroughtoni

Class - Mammalia

Order - Rodentia

Family - Muridae

The coconut rats are vertebrate pests that do direct harm by damaging tender coconuts. They burrow tender coconuts in the trees and drink the sweet coconut water. The burrowed coconuts drop down and are seen scattered below the trees. The rats live in nests built at the top of the trees between the fronds with fibrous materials from the coconut trees. In closely planted coconut gardens, the rats jump from one tree to another easily and cause damage in a number of trees. The rats are reddish or yellowish-brown in color and the belly is whitish in color.

Control of coconut rats. Coconut rats can be killed by the use of poison baits containing zinc phosphide as the poison and fried rice or pounded rice as bait material. Zinc phosphide is mixed with the bait material in the proportion of 1 : 20. The bait material is ground into a coarse powder and then mixed with the poison. A small quantity of groundnut oil or coconut oil is added to the mixture, so as to give an attractive smell to the poison bait.

Further, the oil makes the poison to stick on to the bait. Then the poison bait is packed in small packets. For this, small, thin polythene sheets of size 3" x 3" are cut and rolled into tubes. One end of the tube is folded and sealed by stapling. Through the open end, a teaspoon full of the poison bait is put inside the tube. Then this end is also folded and sealed by stapling. To prevent the poison bait from spilling out, the tubes are stapled in the middle portion also at 2 places. These tubes containing the poison bait are placed in the crown region of the trees, which are frequented by the rats. The rats eat the poison bait and are killed very quickly.

Instead of this poison bait, a readymade poison bait containing bromodiolone as the poison and other palatable food materials as bait is available in the form of cakes under the trade names 'Roban', 'Rat kill' etc. These cakes are placed in the crown region of coconut trees, which are frequented by the rats. The rats, which are attracted by the smell of the cakes, eat them. The poison causes internal bleeding and the rats are killed within 3 or 4 days.

Squirrels also cause damage to tender coconuts. They knaw and burrow the tender coconuts leading to button shedding. They also knaw the flowers and damage them. They can be controlled by using poison bait consisting of carbofuran 3 G as the poison and food materials, such as coarsely ground fried rice or pounded rice as bait. The squirrels eat the poison bait containing carbofuran and are killed quickly.

Pests of minor importance. Besides the pests described above in detail, many other pests, some of which may assume serious proportions under conditions favorable for the pests, infest coconut crop. The bag worm-*Monatha albipes,* the caterpillars of which build conical bag-like nests, hang on to the leaflets, make holes in the leaflets, feed on the leaf tissues and cause damage to the leaves; the nut borer caterpillar pest-*Cyclodes omma* burrows tender coconuts and buttons and causes button shedding; the nymphs and adults of the lace wing bug-*Stephanitis typicus* causes damage to the leaflets by sucking and feeding on the sap from the under surface of leaflets causing white spots on the upper surface; the castor slug caterpillar-*Parasa lepida* feeds on the leaf tissue and causes damage; the aphid pest-*Hysteroneura setariae* sucks sap from the green tissues and causes damage; the thrips-*Haplothrips ceylonicus* sucks and feeds on the sap from the floral parts; the coconut mite-*Raoiella indica* belonging to Family-Phytoptipalpidae also attacks coconut. This reddish mite lives on the under surface of the leaves in colonies, sucks the sap from the leaf tissues causing yellow or whitish blotches; another mite-*Tetranychus hindustanicus* also infests coconut; the shot hole borer-*Xyleborus parvulus* bores the stem and causes death of young trees; the red tree ant-*Oecophylla smaragdina* causes annoyance to the tree climbers by its stinging bite.

MUSTARD *(Brassica juncea)*
RAPE *(Brassica campestris)*

1. Mustard saw fly

Athalia lugens proxima

Order - Hymenoptera

Family - Tenthredinidae

Nature of damage. This is one of the very few Hymenopterous pests found in India. The grub of the pest is a leaf folder. It remains inside the leaf fold, bites holes in the leaves, feeds on the leaf tissues and sometimes causes severe damage. The pest occurs only during the colder months. Besides mustard, the pest attacks rape, radish and allied cruciferous crops (Fig. 35).

Life cycle of the pest. The female saw fly inserts the eggs near the margin of the leaves. One insect lays about 60 eggs. The eggs hatch in 4-5 days and the young grub bites, makes holes, feeds on the leaf tissues and becomes full-grown in 13-18 days. The grown-up grub is tubular, about 2.0 cm. in length, dark gray to black in color, with 10 pairs of legs, 3 pairs on the thoracic region and 7 pairs on the second to eighth abdominal segments. It constructs an earthen cocoon and pupates inside it and emerges as an adult saw fly in 10-15 days. The adult saw fly is a soft, dark chocolate-red wasp-like fly, with yellow colored thoracic region and slightly bigger than a honeybee. The females have long, saw-like ovipositor. The pest is capable of multiplication parthenogenetically also (Fig. 35).

Control measures

Chemical control. Foliar spraying with methyl parathion-300 ml. or carbaryl-1000 gm. or endosulfan-300 ml. in 300 liters of water with a high volume sprayer controls the pest.

Minor pests. Many other insects attack mustard and rape crops, however they are considered to be of minor importance. The mustard aphid-*Lipaphis erysime* and the polyphagous aphid-*Myzus persicae,* the white fly-*Bemisia tabaci* and the painted bug-*Bagrada cruciferarum* suck the sap from the leaves and tender shoot region; the leaf caterpillar pests viz., the cabbage borer-*Hellula undalis, Agrotis ipsilon,* cabbage diamond back moth-*Plutella xylostella* and the cabbage butterfly-*Pieris brassicae* feed on the foliage; the larvae of the cabbage leaf webber-*Crocidolomia binotalis* scrape and feed on the leaves; the pea leaf miner-*Phytomyza atricornis* mines the leaf and feeds on the inner tissues; the flea beetle-*Phyllotreta cruciferae* sucks and feeds on the sap from the plants.

SUNFLOWER *(Helianthus annuus)*

A few pests are found to attack sunflower in India and cause considerable damage to the crop during some seasons. The cut worm-*Spodoptera (Prodenia) litura* and the gram caterpillar-*Helicoverpa (Heliothis) armigera* bite and feed on the foliage; other caterpillar pests viz., *Spilosoma obliqua, Estigmene lactinea, Trichoplusia ni* and *Herse convolvuli* also feed on the leaves; the weevils-*Myllocerus maculosus, Myllocerus blandus* and *Tanymecus indicus* cut and feed on the leaves from the edges; the sucking insect pests-*Nezara viridula, Amrasca biguttula biguttula* and the safflower aphid-*Uroleucon carthami* suck the sap from the foliage; the thrips-*Microcephalothrips abdominalis* and the plant bug-*Dolycoris indicus* infest the floral heads, suck and feed on the sap.

SOYBEAN *(glycine max)*

Many pests that infest other oilseed crops and pulses are found to attack soybean crop also. The groundnut leaf miner-*Aproaerema modicella (Stomopteryx subsecivella)* mines into the leaves and feeds on the inner tissues; the tobacco caterpillar-*Spodoptera (Prodenia) litura* and the gram caterpillar-*Helicoverpa (Heliothis) armigera* bite and feed on the leaves and young shoots; the hairy caterpillar-*Diacrisia obliqua* and the lucerne caterpillar-*Laphygma exigua* feed on the foliage; the amaranthus leaf caterpillar-*Hymenia recurvalis* folds the leaves and top shoots, scrapes and feeds on the green matter; the gray weevils-*Myllocerus* spp. nibble the leaf margins and feed on the leaf tissues; the white fly-*Bemisia tabaci* and the aphids-*Aphis craccivora* and *Lipaphis erysime* suck and feed on the sap from the leaves and tender shoots; the larva of the stem fly-*Melanagromyza obtusa* bores into the stem, feeds on the inner tissues and causes wilting of the shoots; the pod borer or plume moth-*Exelastis atomosa* burrows the pods, feeds on the seeds and causes damage; the thrips-*Scirtothrips dorsalis* lacerates the leaves, sucks the sap and feeds on it.

SAFFLOWER *(Carthamus tinctorius)*

1. Safflower caterpillar

Perigoea capensis

Order - Lepidoptera
Family - Noctuidae

Nature of damage. The caterpillars feed on the leaves and damage the tender capsules. The pest attacks jute and a few other crops. The pest is active during the months November to February.

Life cycle of the pest. The female moth lays eggs singly on the under surface of leaves during the nighttime. The eggs hatch in 4 days and the young caterpillars start feeding on the leaves and become full-grown in about 14 days. The full-grown larva is smooth, green in color with purple markings on the body. The head has brown markings and the anal segment is slightly humped. It pupates in the soil and emerges as an adult moth in about 8 days. The moth is dark brown with dark brown forewings and light brown hindwings.

Control measures

Physical method. The larvae may be handpicked and destroyed.

Cultural methods. The field may be ploughed after harvest of the crop to expose and kill the pupae present in the soil.

Chemical control. Dusting with cabaryl 10 D at 10 kg./acre or spraying with carbaryl 50 WP-0.2% or chlorpyrifos-0.04% or endosulfan-0.035% controls the pest.

Other pests. Safflower is subjected to attack by a few other pests also. The larvae of *Helicoverpa (Heliothis) armigera* feed on the leaves and later on bore into the flower heads and feed on the developing seeds; the leaf caterpillar pests-*Heliothis pertigera* and *Spodoptera exigua* feed voraciously on the foliage; the safflower aphids-*Uroleucon carthami* suck the sap from leaves and young twigs; the nymphs and adults of the leaf thrips-*Frankliniella sulphurea* lacerate, suck and feed on the sap from the leaves; the maggots of the gram pod fly-*Melanagromyza obtusa* and *Melanagromyza phaseoli* bore into the stems, resulting in small swellings at the point of attack and the attacked stems wilt and die; the maggots of the safflower bud fly-*Acanthiophilus helianthi* damage the flower buds and flowers; the safflower semilooper caterpillars-*Eublemma rivula* bore into the stem and flower buds and feed on the contents; the lucerne caterpillar-*Laphygma exigua* bites and feeds on the foliage; the wingless grass hopper-*Neorthacris simulans* notches and feeds on the leaves; the maggots of the linseed gall midge-*Dasyneura lini* feed on the flower buds; the nymphs and adults of the plant bugs-*Dolycoris indicus* and *Nezara graminea* suck and feed on the sap from the leaves; other sucking pests viz., the cotton aphid-*Aphis gossypii,* the mealy bug-*Ferrisia virgata* and the flower thrips-*Thrips florum* also attack safflower.

LINSEED *(Linum usitatissimum)*

Linseed crop is also attacked by many pests, which usually occur on other oilseed crops. The cut worm-*Agrotis ipsilon,* the gram caterpillar-*Helicoverpa (Heliothis) armigera,* the tobacco caterpillar-*Spodoptera (Prodenia) litura,* the hairy caterpillar-*Euproctis scintillans,* the lucerne caterpillar-*Laphygma exigua* and the semilooper-*Plusia orichalcea* feed on the foliage and capsules; the maggots of the linseed gall midge-*Dasyneura lini* feed on the flower buds; the nymphs and adults of the aphid-*Myzus persicae* suck and feed on the sap from the leaves and tender shoots.

PESTS OF VEGETABLES
EGGPLANT OR BRINJAL *(Solanum melongena)*

1. Brinjal shoot and fruit borer

Leucinodes orbonalis

Order - Lepidoptera

Family - Pyraustidae

Nature of damage. This is a very important and serious pest of brinjal and causes extensive damage to the crop. The larvae bore into the shoots of young plants and into the fruits when the plants have grown up. The caterpillars feed on the inner tissues of the stem portion of tender shoots, as a result the shoots present a faded, drooping appearance and eventually these shoots dry. The larvae bore into the flower buds and tender fruits and they drop off. In the later stages, the larvae bore into the maturing fruits, feed on the inner contents and cause severe damage. The bored holes are plugged with the excreta of the larvae. When the infestation is serious, about 70 per cent of the fruits are affected. The affected fruits are of poor quality, have irregular and distorted shapes, and have very poor market value. The pest also infests potato, *Datura stramonium, Solanum nigrum, S. xanthocarpum, S. indicum* etc. (Fig. 39).

Life cycle of the pest. The female moth lays eggs in clusters on tender shoots and on tender fruits. The eggs are whitish and flattened like scales. One insect lays about 280 eggs. The eggs hatch in 3-4 days and the young caterpillars emerging out, start feeding by boring into the tender shoots or tender fruits and grow. The larval period lasts for 15-20 days. The grown-up caterpillar is short and stout, about 2.0 cm. long, pinkish in color with brown head and wart-like, raised minute black dots on the body and a few hairs on these dots. The full-grown larva constructs a dirty brown-colored, boat-shaped cocoon with fine silken thread and pupates inside the stem or fruit or on the plant parts or in-between dry leaves and flowers, which had fallen on the ground. The moth comes out of the cocoon in about 10 days. It is grayish-brown in color with whitish wings and with pinkish-brown markings on them (Fig. 39).

Control measures

Cultural method. The affected shoots and fruits should be removed and destroyed.

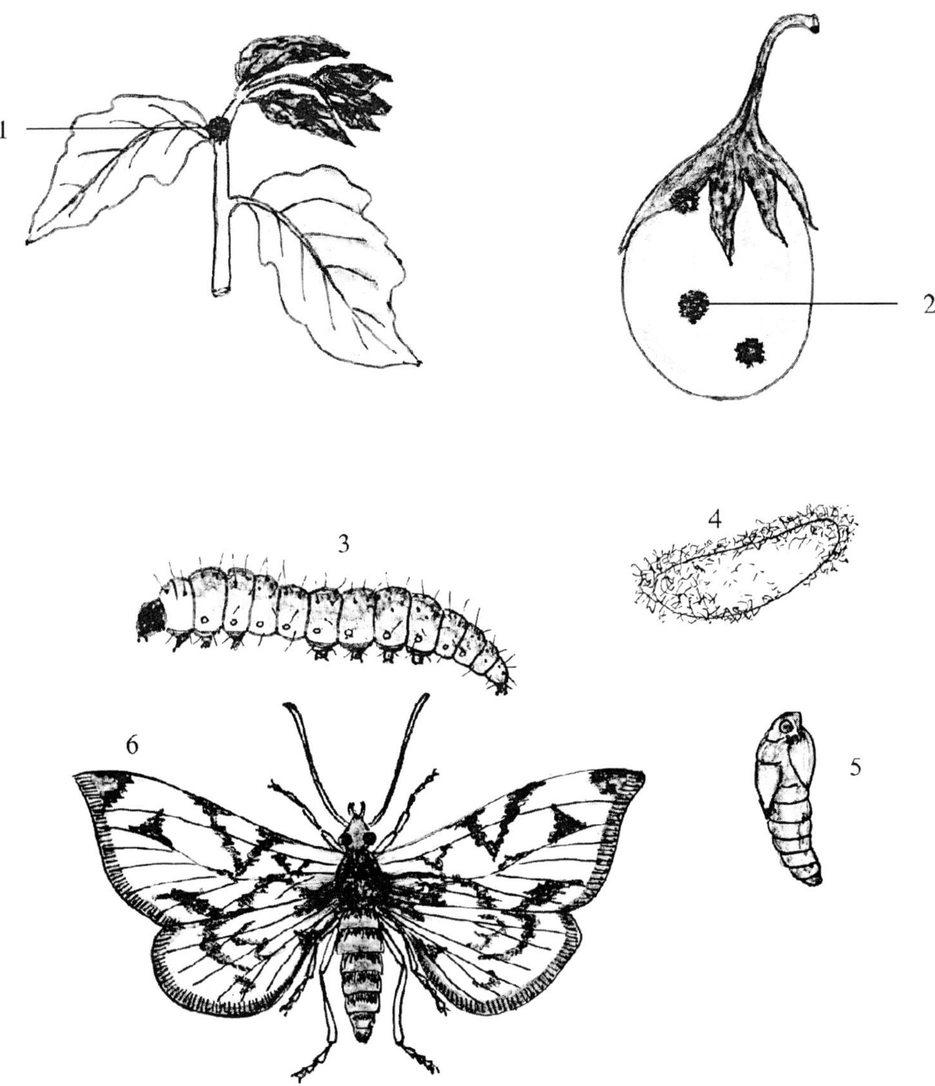

Fig. 39. Brinjal shoot and fruit borer-*Leucinodes orbonalis*

1. Infested shoot 2. Infested fruit 3. Larva 4. Cocoon 5. Pupa 6. Moth

Chemical control. Insecticide sprayings should be given at 15 days interval, commencing from 30 days after planting. Spraying with carbaryl 50 WP-600 gm. or endosulfan-600 ml. or quinalphos-600 ml. or dimethoate-600 ml. or decamethrin-150 ml. in 300 liters of water with a high volume sprayer is effective in controlling the pest. Use of different insecticides every time affords better control of the pest.

Biological control. The larval parasitoid-*Bracon greeni* is capable of parasitizing and destroying the caterpillars.

2. Brinjal stem and shoot borer

Euzophera perticella

Order - Lepidoptera

Family - Phycitidae

Nature of damage. This is also a serious pest of brinjal, which affects the stem and shoot portions alone. The larvae generally enter the stem by boring in-between the places where the shoots branch off or at the junction of the leaf petiole and stem. After entering into the stem, the larva feeds on the inner tissues of the shoot, as a result the shoot wilts, droops down and ultimately die. Severely affected plants are stunted and finally die. The bored holes are plugged with the excreta of the larvae. Generally the larvae attack the plants, which are 3 months or more than 3 months old (Fig. 40).

Life cycle of the pest. The female moth lays eggs on tender leaves, petioles and young shoots. The young caterpillars emerging from the eggs bore into the stem, feed on the inner tissues and grow. The grown-up larva is slightly bigger than the fruit borer, pale yellowish in color with brown head. Pupation takes place in a silken cocoon inside the infested stem. The moth, which emerges from the pupa, is of medium size, slightly bigger than the fruit borer moth, grayish-brown in color and the grayish-brown forewings have a few lines across them in the middle portion. The hindwings are whitish in color. The life cycle is completed in 25-40 days (Fig.40)

Control measures. Refer brinjal shoot and fruit borer.

3. Brinjal spotted beetle or Lady bird beetle

Epilachna (Henoseptilachna) vigintioctopunctata

Order - Coleoptera

Family - Coccinellidae

Nature of damage. This is the most serious leaf-feeding pest of brinjal and is found in all the countries where brinjal is cultivated. All the stages of the pest viz., eggs, grubs, pupae and adults are found in brinjal plants. The grubs, as well as the beetles scrape and feed on the green matter from both the surfaces of leaves. The affected portions of the leaves appear like a mesh with only the veins and veinlets left out. Severely affected leaves turn brown and dry leading to stunting of the plants and reduction in yield. Besides brinjal, the pest attacks potato, tomato, *Datura stramonium, Solanum nigrum, S. xanthocarpum, S. torvum* etc. (Fig. 41).

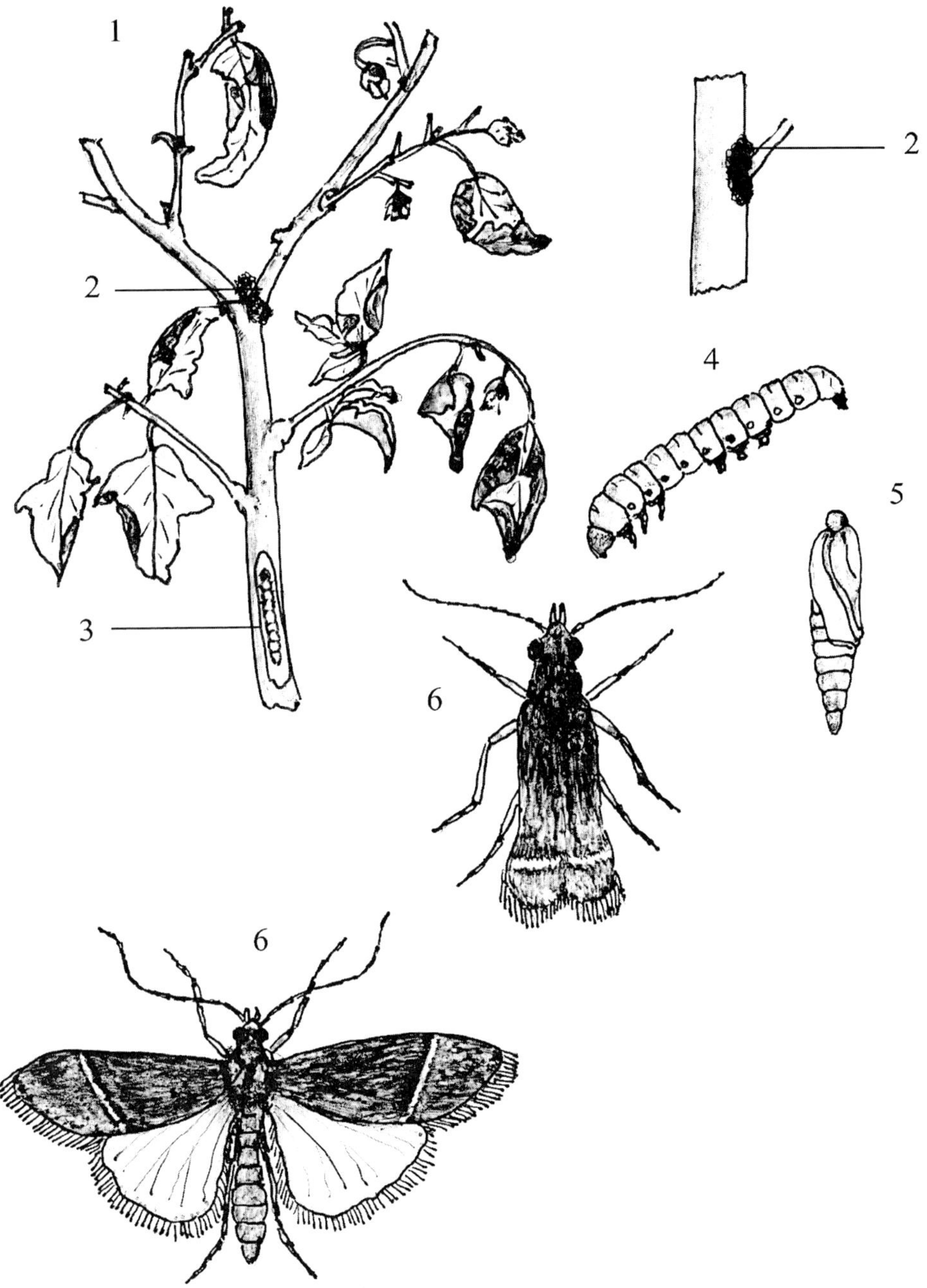

Fig. 40. Brinjal shoot borer-*Euzophera perticella*

1. Infested plant 2. Excretory matter through the bore hole 3. Larva inside the stem 4. Larva 5. Pupa 6. Moth

Life cycle of the pest. The female beetle lays bright yellow- colored, cigar-shaped eggs in clusters of 10-20 on the under surface of leaves. One female lays 120-180 eggs. The eggs hatch in 2-4 days. The young grubs are yellowish in color, flattish with black, spiny hairs on the body. The grub crawls about on the leaves, scrapes and feeds on the green matter and becomes full-grown in 10-35 days. The grown-up grub is fleshy, short, somewhat flat and spiny. It pupates on the leaf or on the stem portion. The pupa is yellowish in color, hemispherical in shape with spines on the rear portion. The pupal stage lasts for about 7 days. The adult beetle emerging out of he pupa is hemispherical in shape, orange-brown in color with 12-28 black spots on the elytra (Fig.41)

Control measures

Physical methods. The grubs, pupae and the beetles can be handpicked and destroyed.

Chemical control. Spraying the crop with malathion-600 ml. or fenitrothion-300 ml. or carbaryl 50 WP-600 gm. or endosulfan-300 ml. in 300 liters of water with a high volume sprayer is effective in controlling the pest.

Biological control. The egg parasitoid-*Tetrastichus ovulorum* parasitizes and destroys the eggs.

Minor pests. Besides the pests described above, the bud worm-*Phthorimaea blapsigona* bores into the tender flower buds and causes flower shedding; the yellow hairy caterpillar-*Selepta celtis* feeds on the leaves; the nymphs and adults of the lace wing bug-*Urentius hystricellus* suck the sap from the leaves and cause yellowing of leaves; the ash weevil-*Myllocerus subfasciatus* makes holes in the leaves and feeds on the tissues, while the grubs of this pest cut and feed on the roots resulting in the death of plants; the aphid-*Aphis gossypii* and the leaf hopper-*Amrasca biguttula biguttula* suck and feed on the sap from the leaves and tender shoots; the brown leaf hopper-*Cestius phycitis*, not only sucks sap from the leaves thereby causing direct damage is also the vector of the mycoplasma, which causes **'little leaf disease'**; the mealy bug-*Centrococcus insolitus*, which generally appears in the later stages of crop growth, sucks and feeds on the sap from the leaves, petioles and stem portions and causes severe damage; the horned caterpillar-*Acherontia styx* bites and feeds on the leaf tissues; the caterpillars of the leaf webber pest-*Eublemma olivacea* roll the leaves, remain inside, scrape and feed on the leaf tissues; the mites-*Tetranychus cinnabarinus* and *Paratetranychus indicus* make fine webbings on the ventral surface of leaves, remain inside, suck and feed on the sap, resulting in the development of whitish patches and eventual drying of leaves.

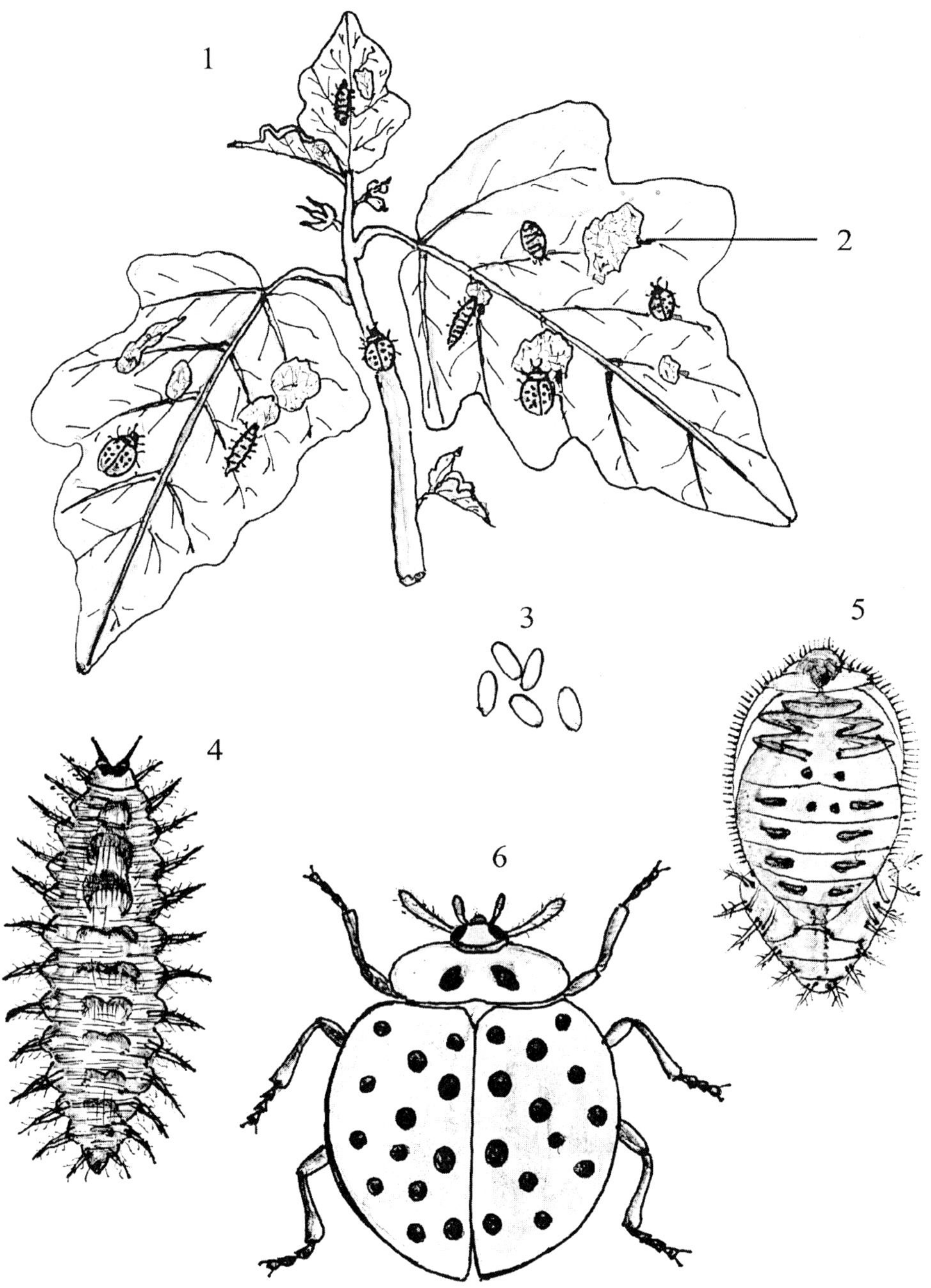

Fig. 41. Brinjal spotted beetle-*Epilachna vigintioctopunctata*

1. Larvae and adults feeding on the plant 2. Netting caused by scraping of leaves 3. Eggs 4. Larva 5. Pupa 6. Beetle

TOMATO *(Lycopersicum esculentum)*

1. Gram caterpillar or Tomato fruit borer

Helicoverpa (Heliothis) armigera

Order - Lepidoptera

Family - Noctuidae

Nature of damage. The caterpillar bores a big hole in the fruit, puts its head inside and feeds on the fleshy contents with its body remaining outside. It moves from fruit to fruit and damages many fruits. The quality of the attacked fruits becomes very poor and unfit for consumption and such fruits have very low market value. Before formation of fruits, the caterpillars feed on the leaves and young shoots (Fig. 42).

Life cycle of the pest and control measures. Refer pests of cotton (Page 159).

2. Tobacco caterpillar

Spodoptera (Prodenia) litura

Order - Lepidoptera

Family - Noctuidae

Nature of damage. The caterpillars feed on the leaves and young shoots and cause severe damage. When fruits are formed, they bore into the fruits, feed on the inner fleshy portion and cause damage (Fig. 68).

Life cycle of the pest and control measures. Refer pests of cotton (Page 161).

Pests of minor importance. A few other pests, which are of minor importance also attack tomato, but they may become serious pests under favorable environmental conditions and cause extensive damage to the crop. The serpentine leaf miner-*Liriomyza trifolii* mines the leaves and causes severe damage. It is becoming a serious pest of cotton, brinjal, cucumber, potato and several other crops besides tomato; the spotted beetle or lady bird beetle-*Henosepilachna (Epilachna) vigintioctopunctata* and its grubs scrape and feed on the green matter of leaves; the mealy bug-*Ferrisia virgata,* the cotton white fly-*Bemisia tabaci,* the jassid-*Amrasca biguttula biguttula* and the scale insect-*Pseudococcus virgatis* suck and feed on the sap from the leaves and tender shoots; the mites-*Tetranychus cucurbitae* and *Tetranychus telurius* also suck and feed on the sap from the leaves.

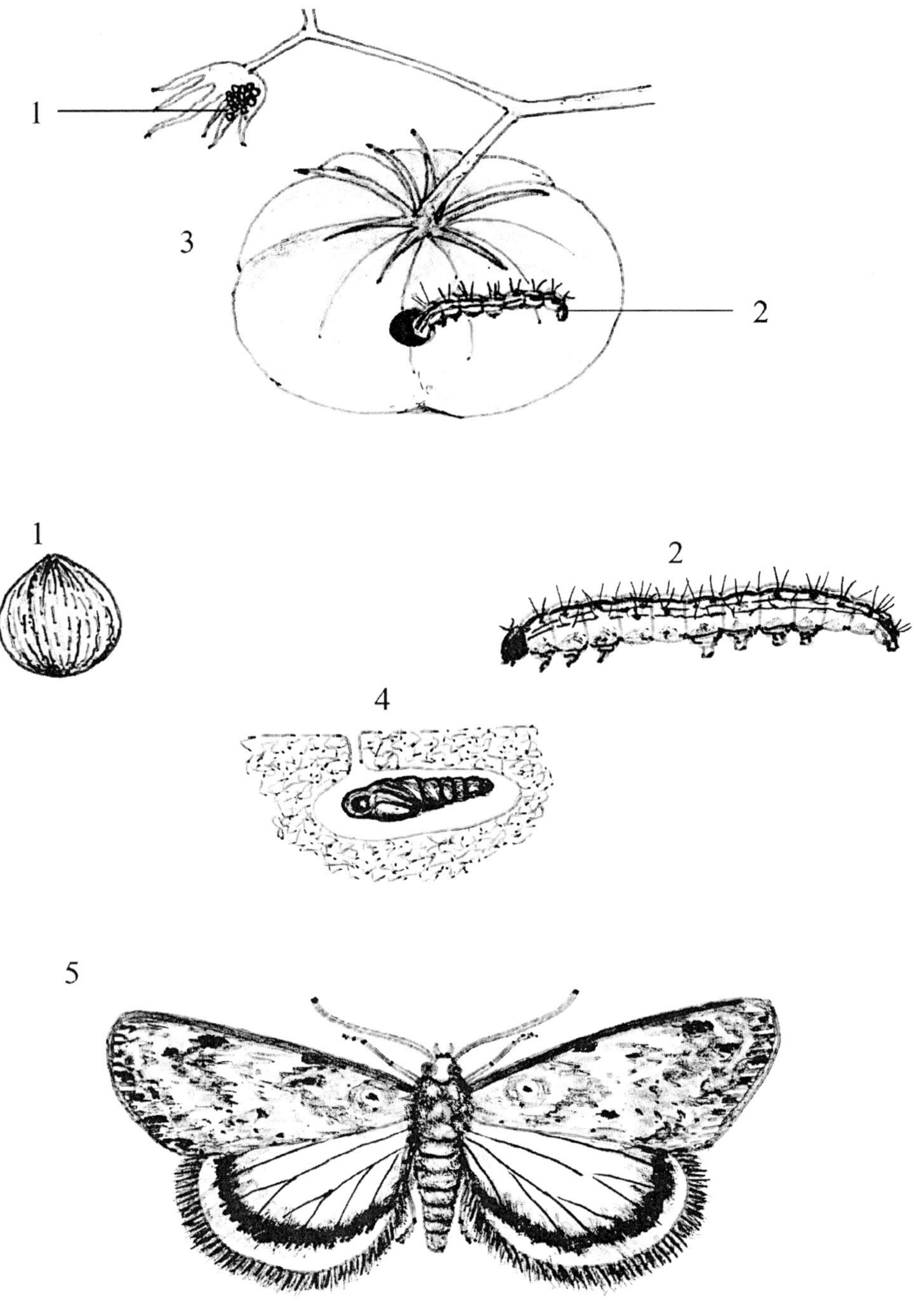

Fig. 42. Tomato fruit borer-*Heliothis (Helicoverpa) armigera*

1. Eggs 2. Larva 3. Infested fruit 4. Pupa 5. Moth

LADY'S FINGER OR BHENDI *(Abelmoschus esculentus)*

Most of the pests, which infest cotton crop, are found to infest bhendi crop also. The cotton spotted boll worms-*Earias insulana, E. vitella* and the cotton pink boll worm-*Pectinophora (Platyedra) gossypiella* bore into the tender shoots and cause wilting, drooping and drying of the shoots. Later these pests burrow the flower buds, flowers and fruits, feed on the inner contents and cause extensive loss in yield; the cotton white fly-*Bemisia tabaci* sucks and feeds on the sap from the leaves. This pest is also the vector for the virus, which causes **'vein clearing disease'** that is very serious in bhendi; the leaf hopper-*Amrasca biguttula biguttula* sucks and feeds on the sap from the leaves and young shoots, as a result the leaves become mottled and distorted, the plants become stunted and may even die; the nymphs and adults of the cotton aphid-*Aphis gossypii* and the red cotton bug-*Dysdercus cingulatus* suck and feed on the sap from the leaves and immature fruits; the American boll worm-*Helicoverpa (Heliothis) armigera,* besides biting and feeding on the leaves and young shoots, bores into the fruits and feeds on the fleshy portion and seeds; the leaf roller pest-*Sylepta derogata* rolls the leaves, lives inside, bites and feeds on the leaf tissues; the leaf weevils-*Myllocerus maculosus, M. viridanus* and *M. discolor* bite holes in the leaves or bite the leaves from the edges and feed on the tissues; the maggots of the fly-*Melanagromyza obtusa* bore through the stem of young plants and branches of grown-up plants and cause wilting of the entire plant or branches; the grubs of the weevil-*Alcidodes affaber* bore into the tender shoots or leaf bases and cause withering of shoots; the mites-*Tetranychus telarium* make fine webbings on the under surface of leaves, remain inside, suck and feed on the sap.

CUCURBITS

The cucurbits cultivated on a large scale include Snake gourd *(Trichosanthes anguina),* Bitter gourd *(Momordica charantia),* Ridge gourd *(Luffa acutangula),* Cucumber *(Cucumis sativus),* Pumpkin *(Cucurbita moschata),* Bottle gourd *(Laginaria siceraria),* Water melon *(Citrullus vulgaris),* Summer squash *(Cucurbita pepo),* Sponge gourd *(Luffa cylindrica),* Ash gourd *(Benincasa hispida)* and Musk melon *(Cucumis melo).* Among the cucurbitaceous plants, bitter gourd and snake gourd are infested by many pests, while a few pests attack pumpkin and other crops.

1. Cucumber fruit flies

Bactrocera (Dacus) cucurbitae and *Bactrocera (Dacus) ciliatus*

Order - Diptera

Family - Tephritidae

Nature of damage. The maggots bore into the young fruits, as well as the matured fruits, feed on the inner fleshy pulp and cause extensive damage.

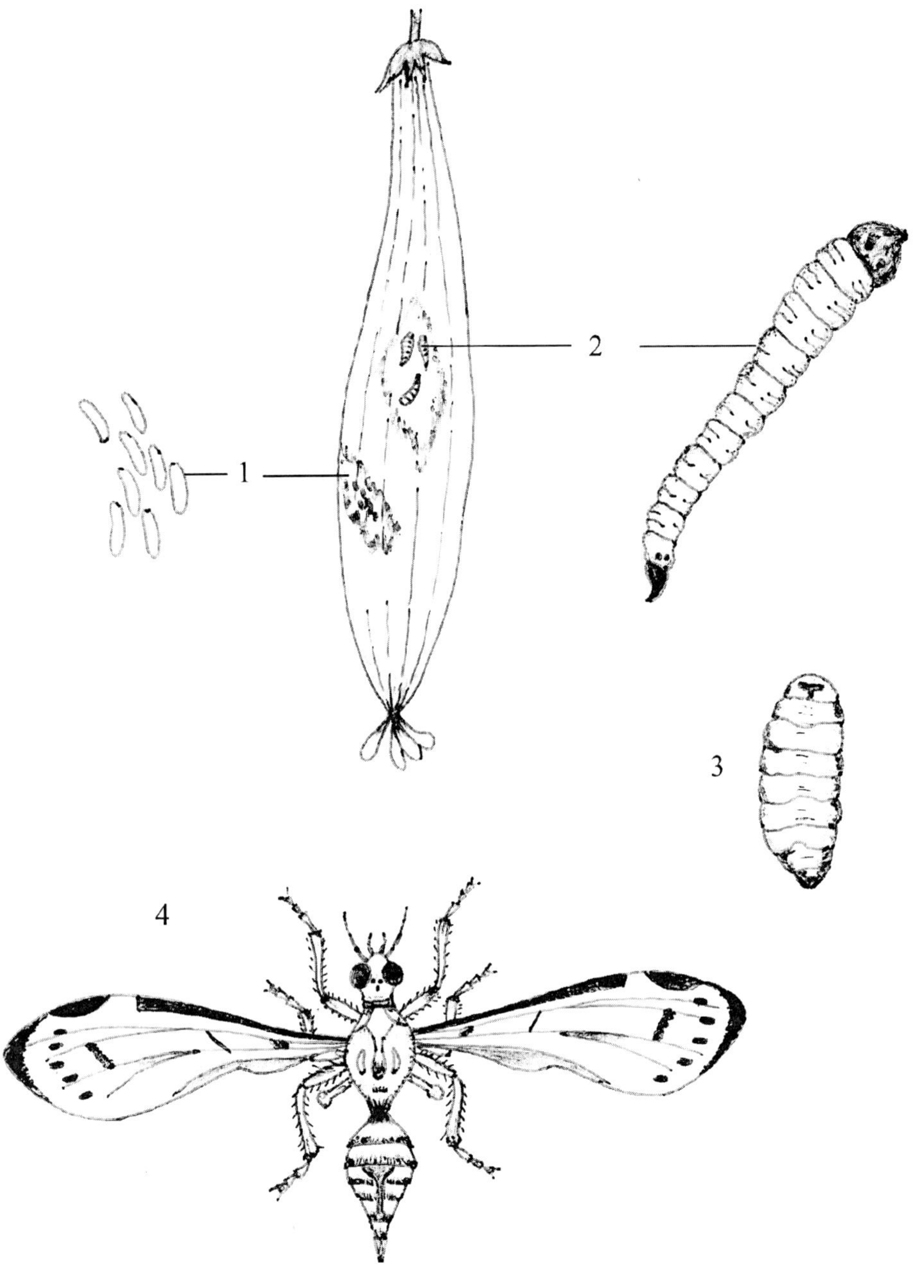

Fig. 43. Cucurbit fruit fly-*Dacus cucurbitae*

1. Eggs 2. Larva 3. Pupa 4. Fly

When the young fruits are attacked, they drop off prematurely. The older fruits, when attacked, rot and become unfit for consumption. The pest infests snake gourd, bitter gourd, pumpkin, cucumber and allied cucurbits (Fig. 43).

Life cycle of the pest. The female fly lays shiny white, cigar-shaped eggs singly or in batches of 4-10 and inserts them into the floral parts or young fruits. One insect lays up to 200 eggs. The eggs hatch in 2-3 days. The young, footless maggots emerging out of the eggs, bore into the fruits, start feeding on the fleshy pulp and seeds, and become full-grown in 5-9 days during the hot season and in 20-22 days.in the cold season. The grown-up maggot is footless and the anterior portion slightly broader than the posterior portion. Before pupation it comes out of the fruit, drops to the ground and pupates under the soil at a depth of 1.5-15.0 cm. The pupa is slightly elongated and seed-like. The fly emerges from the pupa in 7-8 days. During the cold season, the pupal period lasts for 25-30 days. The flies are very active during the hot season and hibernate in the colder months. The fly is reddish in color with black and white dots on the dorsal side. The transparent, triangular wings have brown bands and the margins of the tips have black dots. *Bactrocera (Dacus) cucurbitae* is slightly bigger, while *Bactrocera (Dacus) ciliatus* is comparatively smaller and slender (Fig.43).

Control measures

Cultural methods (i). The affected fruits should be removed and destroyed (ii) The soil may be stirred or ploughed to bring out the pupae to the soil surface.

Chemical control. The pest can be controlled by spraying with malathion-0.1 % or carbaryl 50 WP-0.1 % or fenthion-0.1 % or methyl demeton-0.025 % or profenofos-0.05 % or cypermethrin-0.025 %, initially at the flowering time, followed by 3-4 sprayings at fortnightly intervals. Before each spraying, the matured fruits should be harvested.

Poison bait. Poison baiting with fermented palm juice or protein hydrolysate mixed with any one of the insecticides is also effective in attracting and destroying the flies.

2. Red pumpkin beetle

Raphidopalpa foveicollis

Order - Coleoptera

Family - Galerucidae

Nature of damage. The beetles bite roundish holes in the leaves, flower buds and flowers, and feed on the tissues. Young plants are more prone to attack by the pest and when the leaves are severely damaged, the plant may

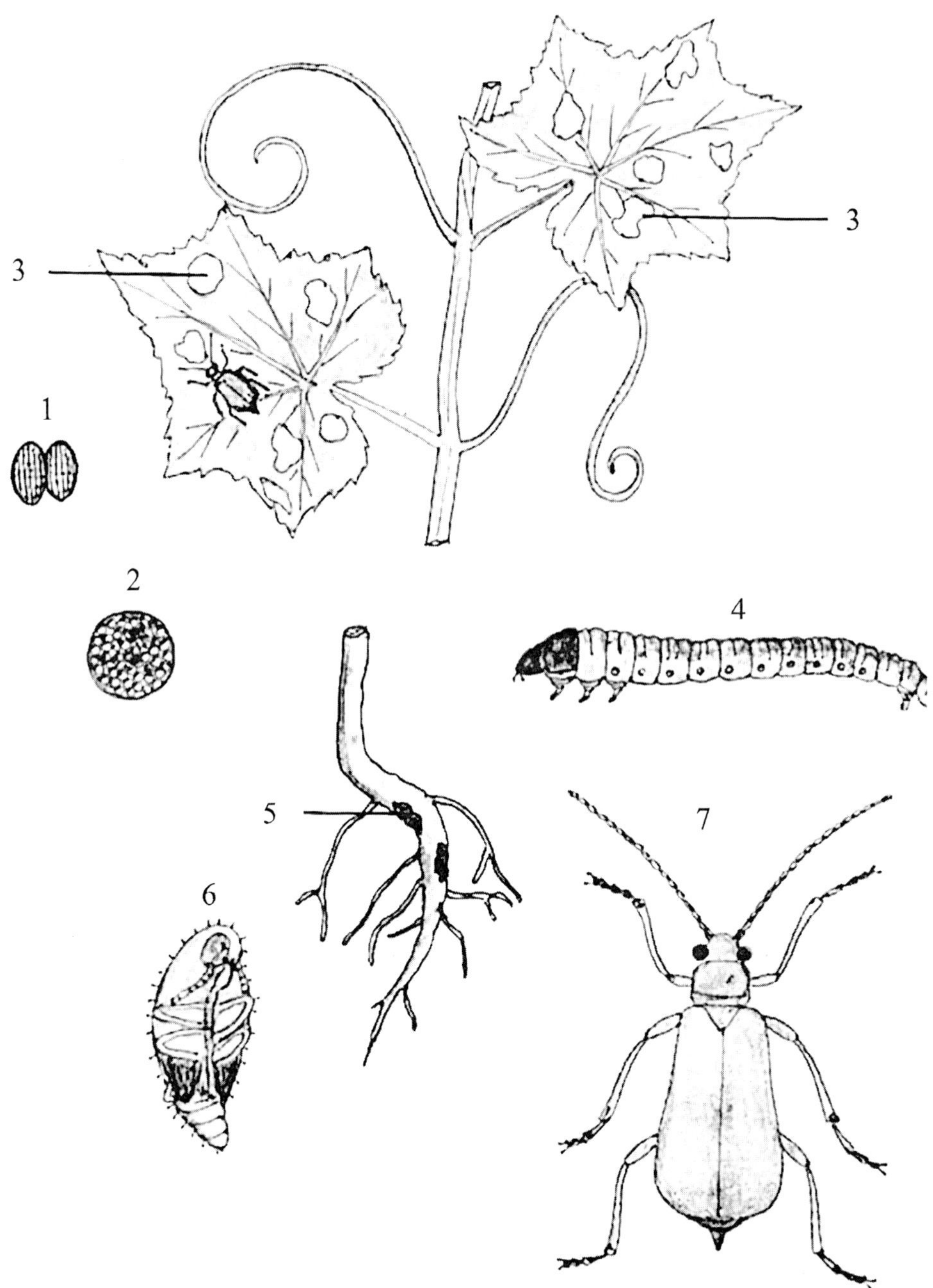

Fig. 44. Red pumpkin Beetle-*Raphidopalpa foveicollis*

1. Eggs 2. Egg 3. Infested leaves 4. Larva 5. Roots infested by the larva 6. Pupa 7. Beetle

die. In the older plants, even though numerous holes are made by the pest, the plant may not die, but the growth of the plant as well as the yield from the plant are adversely affected. The grub of the pest, which lives in the soil, bites and feeds on the roots and causes damage. The fruits, which come in contact with the soil, are also burrowed by the grubs (Fig.44).

Life cycle of the pest. The female beetle lays orange-colored, long, cylindrical eggs singly or in small clusters under the canopy of the host plant on the moist soil. One beetle lays 150-300 eggs. The eggs hatch in 3-5 days. The grubs become full-grown in 13-25 days. The grown-up grubs are pale yellow in color, with the head and first thoracic segment black in color. The full-grown grub pupates under the soil and emerges as an adult beetle in 7-15 days. There may be 5-8 generations in a year. The beetles, which are very active, are orange-red in color, with the ventral side blackish in color and fringed with fine, white hairs. The forewings are orange-red in color (Fig.44)

Control measures

Physical and cultural methods (i) The adult beetles can be collected by using a hand net and destroyed (ii) The soil may be stirred or the field may be ploughed and the pupae brought to the soil surface.

Chemical control. Foliar spraying with malathion-0.1% or dimethoate-0.06% or fenthion-0.1% or methyl demeton-0.025% or methyl parathion-0.05% affords adequate control of the pest.

3. Snake gourd semilooper

Anadevidia (Plusia) peponis

Plusia signata

Order - Lepidoptera

Family - Noctuidae

Nature of damage. It is a specific pest of snake gourd. The caterpillars roll and bind the leaves, live within it, scrape, cut holes and feed on the leaf tissues resulting in defoliation. When the pest occurs in large numbers, severe damage is caused (Fig. 45).

Life cycle of the pest. The female moth lays whitish, spherical eggs on the under surface of leaves. The eggs hatch in 3-4 days. The young larvae become full-grown in 10-15 days. The grown-up larva is slender, long, 3.4-4.0 cm. in length, whitish-green in color with lateral white lines and fringed with black, spiny hairs on the body. The hind end of the larva is slightly elevated and the first two pairs of abdominal legs are lacking. The caterpillar constructs a cocoon with fine silken threads and pupates inside it within the

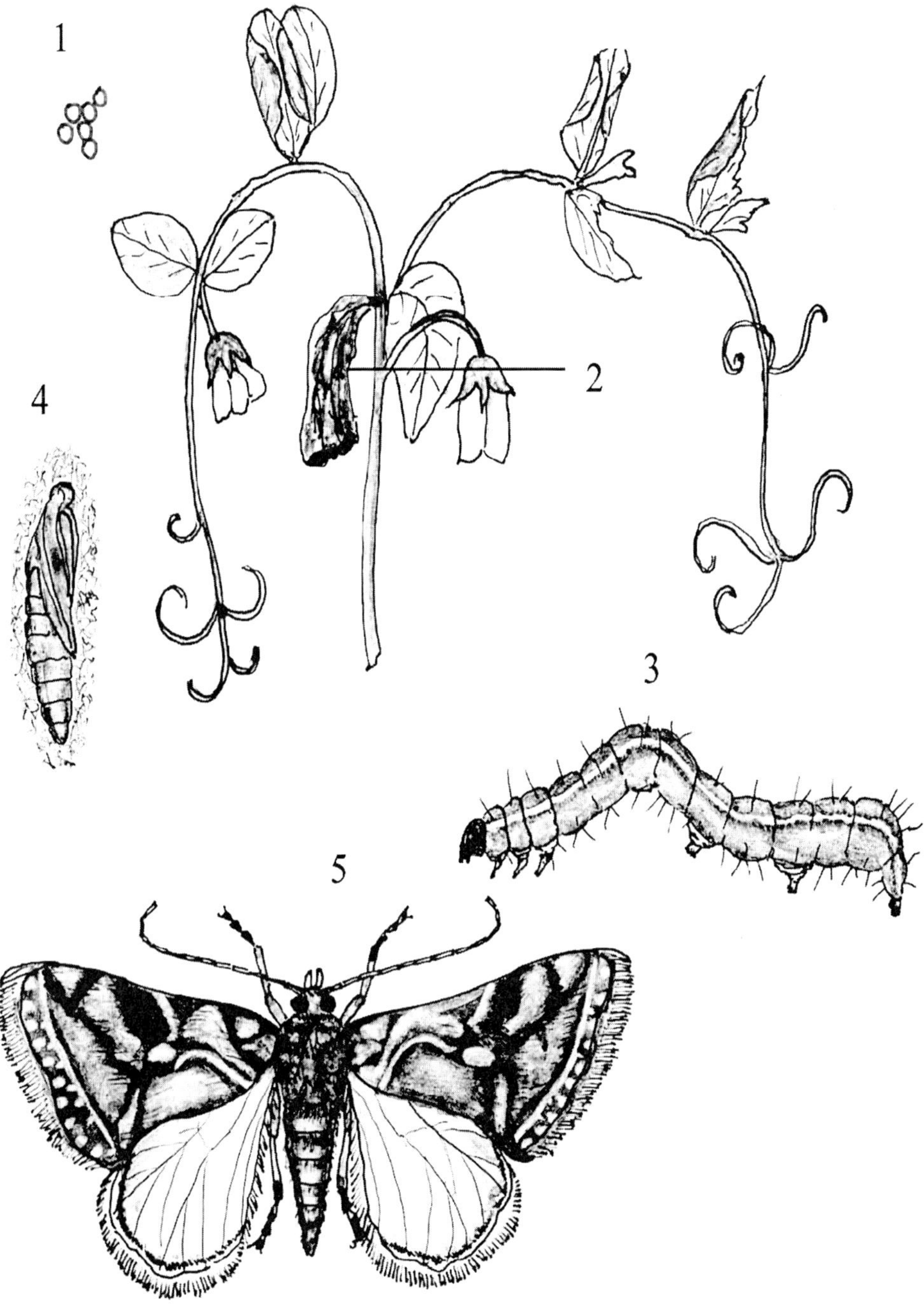

Fig. 45. Snake gourd semilooper-*Plusia signata*

1. Egg 2. Larva inside leaf roll 3. Larva 4. Pupa inside cocoon 5. Moth

leaf roll. The moth emerges from the pupa in 8-10 days. The moth is large sized, brown in color and the brownish, shiny forewings have golden markings. The hindwings are whitish in color (Fig. 45).

Control measures

Physical method. The leaf rolls with the caterpillars or pupae may be removed and destroyed.

Chemical control. Spray application of malathion-0.1 % or methyl parathion-0.05 % is effective in controlling the pest.

4. Cucumber plume moth

Sphenarches caffer

Order - Lepidoptera

Family - Pterophoridae

Nature of damage. The caterpillars bite and feed on the leaves, flower buds and flowers, and cause appreciable damage. Besides cucumber, the pest attacks other allied cucurbits also (Fig. 46).

Life cycle of the pest. The female moth lays very small, light green, spherical eggs on the leaves and flower buds. The young caterpillars emerging from the eggs feed on the leaves and flowers and grow. The green caterpillars have spiny hairs on the body. The grown-up caterpillar pupates on the leaf. The pupa is also fringed with spines. The moth, which emerges from the pupa is brownish in color, small and slender, with the wings arranged in plumes. The light brown forewings have two lobes and the hindwings have three lobes.

Control measures

Chemical control. (i) Dusting the foliage with endosulfan 4 D or malathion 5 D at 10 kg./acre controls young caterpillars (ii) Foliar spraying with malathion-0.1% or methyl parathion-0.05% or endosulfan-0.035% or fenthion-0.1% effectively controls the pest.

Other pests. A few other pests are also known to attack cucurbits. The grubs and adults of the spotted beetle or lady bird beetle-*Epilachna* spp. *(Epilachna septima, E. chrysomelina* and *E. implicata)* scrape and feed on the green matter of the leaves; the adults of the gray beetle pest-*Aulacophora cincta* and the blue beetle pest-*Aulacophora lewisii (intermedia)* bite holes on the leaves and also feed on the flowers, while the grubs cut and feed on the roots, stem and fruits that come in contact with the soil; the aphid-*Aphis malvae* is a common and destructive pest of bitter gourd. The nymphs and adults of the

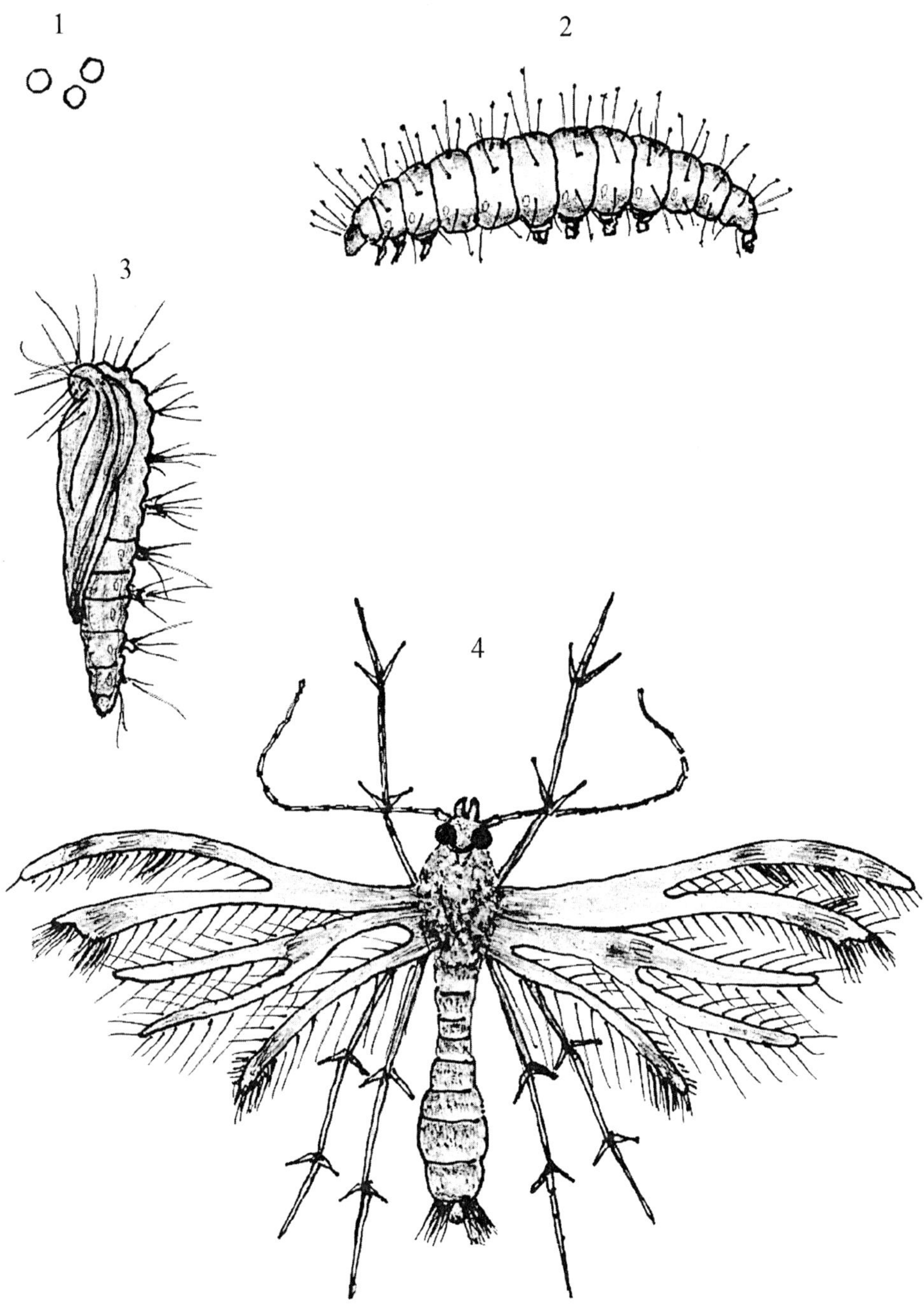

Fig. 46. Bottle gourd plume moth-*Sphenarches caffer*

1. Egg 2. Larva 3. Pupa 4. Moth

pest suck the sap from the leaves and tender stem region; the cotton aphid-*Aphis gossypii* is also a sucking pest of many cucurbits; the greenish caterpillars of *Eudioptes indicus* fold the leaves, stay inside the fold and feed on the leaf tissues of bitter gourd, pumpkin etc.; the adults of the snake gourd stem weevil-*Baris trichosanthis* feed on the green matter of leaves, while the grubs bore into the stem or petiole resulting in the withering of leaves; the flea beetle-*Phyllotreta cruciferae* feed on the leaves; the grubs of the gall fly-*Lasioptera falcata* bore into the shoots and produce elongated gall-like thickenings on the shoots of bitter gourd leading to withering of shoots; the banded blister beetle-*Zonabris pustulata* feed on the pollen and flowers and affects fruit formation; the mites-*Tetranychus cucurbitae* suck and feed on the sap from the leaves.

CRUCIFERS

The important cruciferous vegetable crops cultivated include Cabbage *(Brassica oleracea* var. *capitata),* Cauliflower *(Brassica oleracea* var. *botrytis)* and Knolkhol *(Brassica oleracea* var. *gongylodes)*

1. Cabbage diamond-back moth

Plutella xylostella (maculipennis)

Order - Lepidoptera

Family - Plutellidae

Nature of damage. This moth pest is considered to be one of the most important pests of cabbage and is prevalent in all the countries where cabbage is cultivated. The caterpillars bite holes in the leaves and feed on the tissues causing severe damage. The pest attacks cauliflower, radish, knolkhol and other allied crops (Fig. 47).

Life cycle of the pest. The female moth lays very small, yellowish-white, spherical eggs singly on the under surface of leaves near the midrib. One moth lays about 57 eggs. The eggs hatch in 3-6 days and the young caterpillars feed on the leaves and become full-grown in 14-21 days. The grown-up larva is about 1.0 cm. in length, light green in color, somewhat stout and fringed with short, fine hairs. It pupates inside a loosely spun cocoon made with silken threads. The pupal period lasts for 7-11 days, after which the moth emerges from the pupa. The moth is small-sized, about 1.5 cm. in size, grayish-brown in color with brown-colored long and narrow forewings and whitish hindwings. On the lower margin of the forewings, dull white markings are present. When the wings fold over the back of the moth when it rests, the white markings form diamond-shaped spots. The life cycle of the pest is completed in 24-35 days. (Fig. 47).

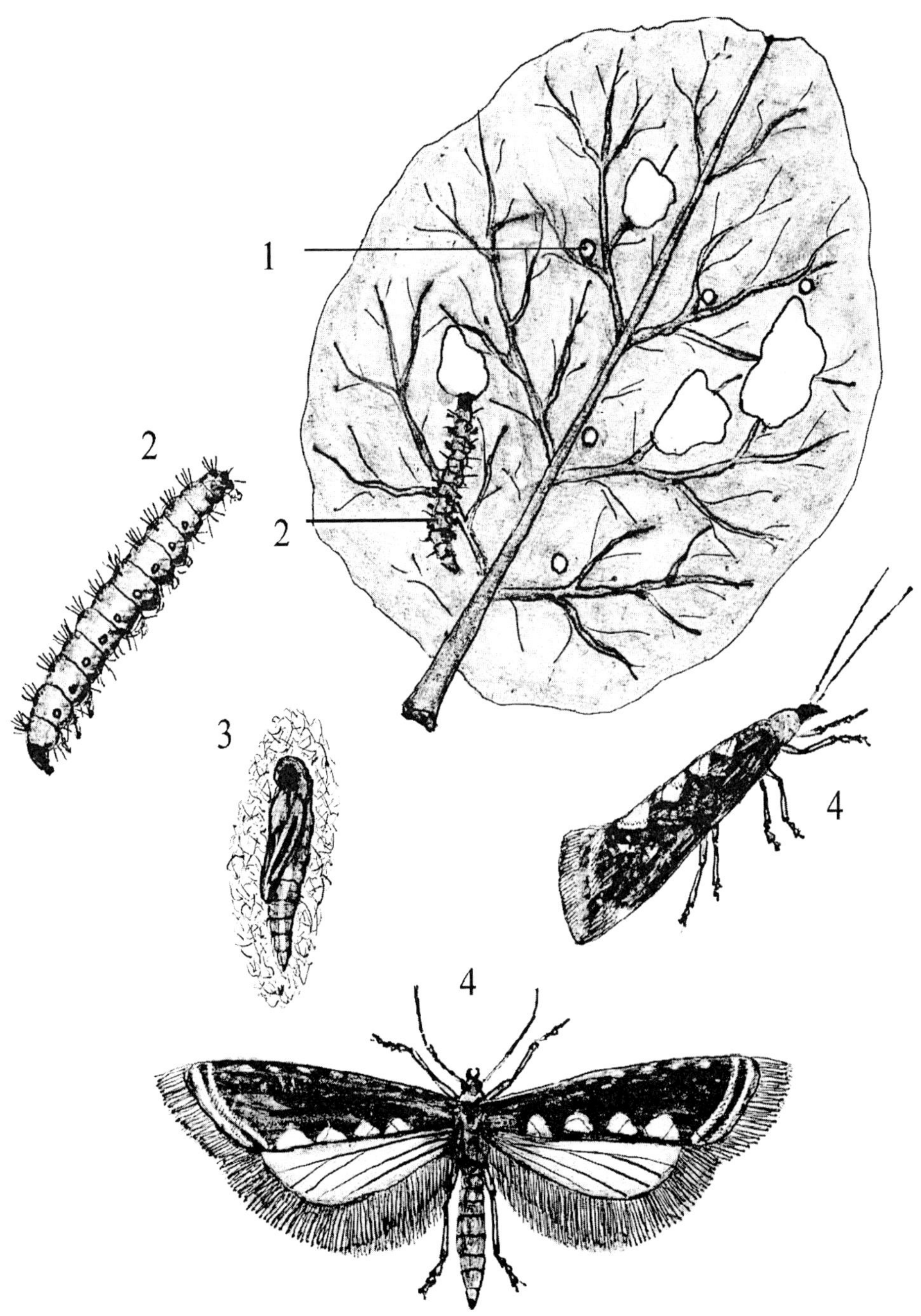

Fig. 47. Cabbage diamond back moth-*Plusia signata*

1. Egg 2. Larva 3. Pupa inside cocoon 4. Moth

Control measures

Chemical control. Spraying with malathion-400 ml. or endosulfan-400 ml. in 200 liters of water per acre with a high volume sprayer controls the pest effectively. Of late the pest has developed resistance to quinalphos, fenitrothion, cypermethrin, fenvalerate and deltamethrin.

Biological control. Spray application of a spore suspension of *Bacillus thuringiensis* var. *kurstaki* is effective in controlling the pest.

2. Cabbage aphid

Lipaphis Erysimi

Order - Hemiptera
Sub order - Homoptera
Family - Aphididae

Nature of damage. Both the nymphs and adults congregate in large numbers on the under surface of young leaves and on the tender shoot portions, suck and feed on the sap, as a result the leaves turn yellow in color and later dry and die. The affected plants become stunted. The honeydew secreted by the pest invites many other insects and ants, which come to feed on the sugary secretion. On the surface of the leaves, where the honeydew spreads, sooty mould, a fungal disease appears, which further affects the plants adversely. The pest attacks mustard, cauliflower, radish and other allied cruciferous crops.

Life cycle of the pest. The female reproduces parthenogenetically and lays young ones one by one (viviparous). The young nymphs become full-grown adults within a few days and start reproduction and thus the pest multiplies very rapidly. Both winged and wingless adult forms are present. Once the multiplication of the pest reaches a certain limit, then mostly winged females begin to appear, which fly off to neighbouring plants or areas and infest other plants. The pest occurs in large numbers during the months of December-February, when the weather is cool and during this period, both winged and wingless forms are produced. In the cooler months, many overlapping generations usually appear. The aphids are very small, less than 0.25 cm. long, soft-bodied, green in color, with two short tubes, known as the honey tubes projecting from the dorsal side of the hind end of the abdomen. The wings when present are long and white. The antennae are very short (Fig. 48).

Control measures

Physical methods. The plant parts where the pest is found in large numbers may be removed and destroyed.

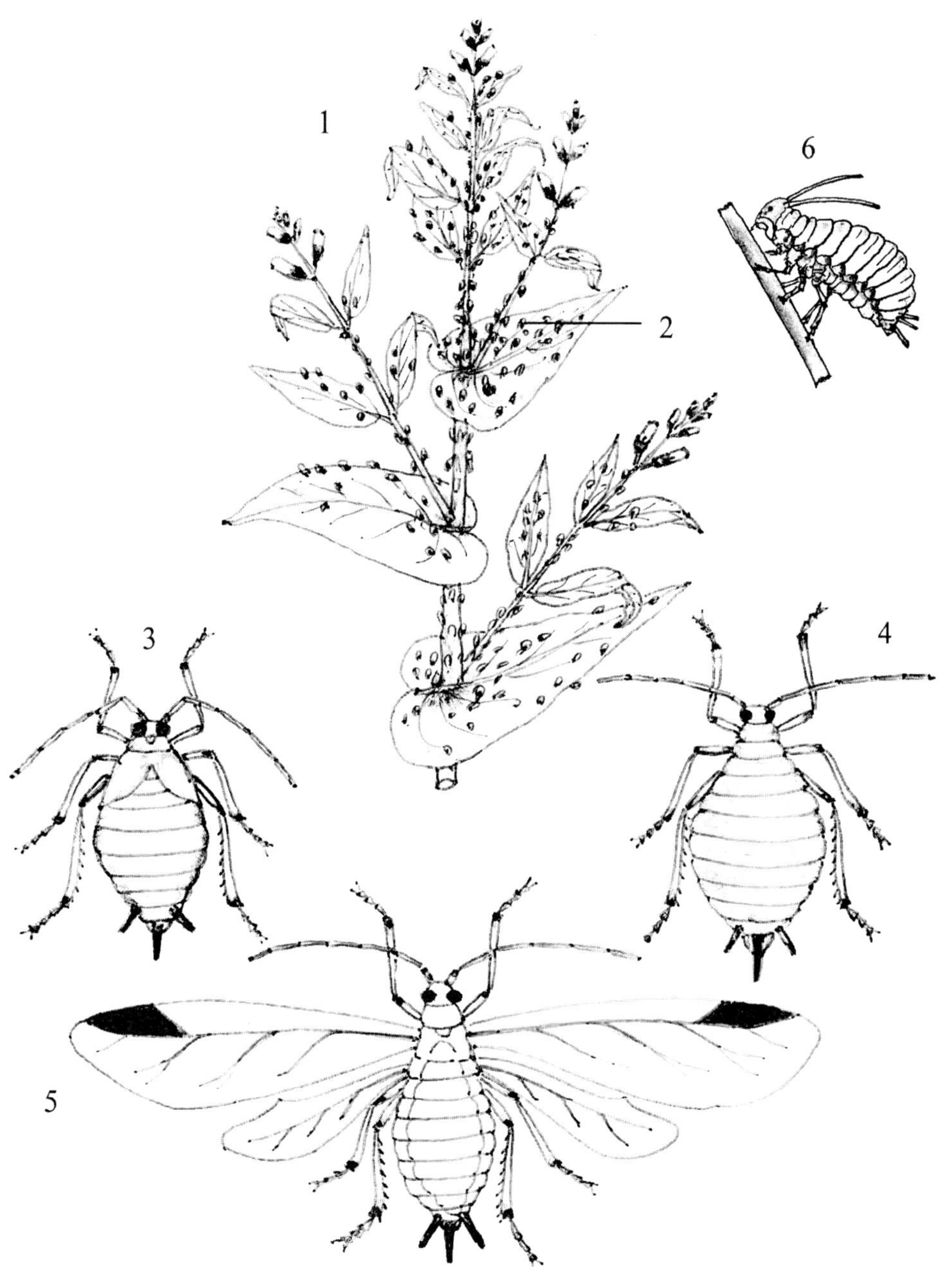

Fig. 48. Cabbage aphid-*Lipaphis erysimi*

1. Infested plant 2. Nymphs and adults feeding on a shoot 3. Nymph 4. Wingless adult 5. Winged adult 6. Aphid feeding on a tender branch

Chemical control. (i) Spraying the crop with methyl demeton-300 ml. or dimethoate-600 ml. or malathion-600 ml. or profenofos-600 ml. in 300 liters of water per acre with a high volume sprayer gives adequate control of the pest (ii) Spraying tobacco decoction is very effective in controlling the pest. To prepare tobacco decoction, 2.0 kg. of dry, cured tobacco is boiled in 20 liters of water. Then the whole thing is cooled, crushed, squeezed and the extract is filtered. To this concentrated extract 10-15 times of water is added. To this diluted filtrate 1.0 kg. of liquid soap is added and mixed thoroughly and then used for spraying Tobacco decoction has no ill-effect on human beings and animals (iii) Spraying Nicotine sulfate 40 S-750 ml. + liquid soap-1.2 kg. in 300 liters of water per acre with a hand operated sprayer has also been found to be effective.

The pest is known to have developed resistance to Organo phosphorus insecticides.

Biological control (i) The spotted beetle-*Coccinella (Epilachna) septum punctata* is predatory on the nymphs and adults of the pest (ii) The fungal parasites-*Entomophthora coronata* and *Cephalosporium aphidicola* infect and destroy the pest.

3. Cabbage borer

Hellula undalis

Order - Lepidoptera

Family - Pyraustidae

Nature of damage. The pest attacks the crop in the field, as well as the cabbage head after harvest. The young caterpillars mine the leaves, enter inside and feed on the inner tissues. Later they bore the midrib of leaves or the stem portion or the cabbage head and feed on the tissues. Even after harvest, the caterpillars remain inside the cabbage head, continue to feed on the inner tissues and cause severe damage. The pest also infests cauliflower, radish, knolkhol, beet root and other cruciferous crops (Fig. 49).

Life cycle of the pest. The female moth lays yellow-colored, shiny, spherical eggs on the surface of leaves or on the stem portions. The young caterpillars, which hatch out of the eggs in 3-4 days, become full-grown in 10-12 days. Grown-up caterpillar is about 1.0 cm. long, pale brown in color, with a black head and four longitudinal black lines on the lateral sides of the body. The full-grown caterpillar pupates in the burrow or sometimes in the soil and the moth emerges from the pupa in about 7 days. The moth is small-sized, yellowish-brown in color, with brown-colored forewings. The forewings have gray-colored, wavy lines and a elliptical marking in the central portion. The hindwings are whitish in color (Fig. 49).

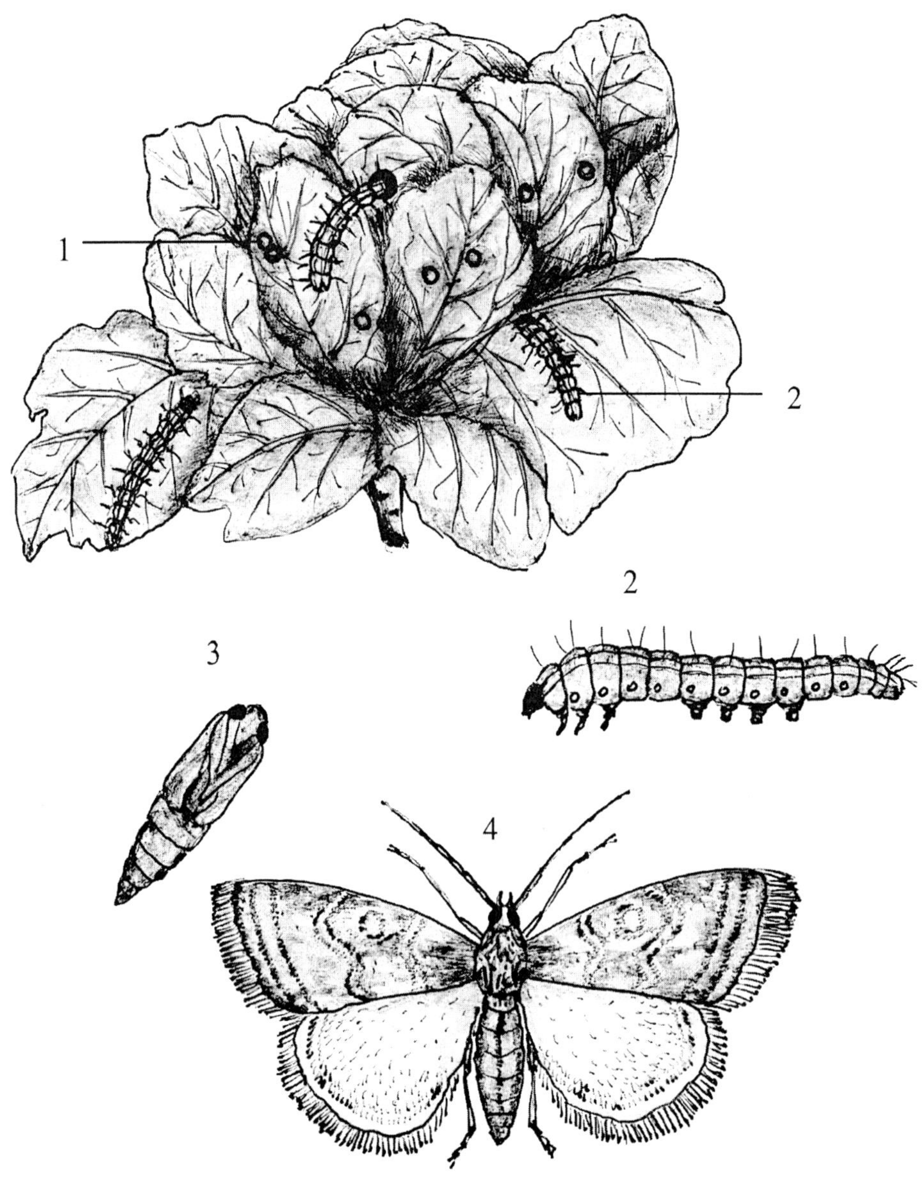

Fig. 49. Cabbage borer-*Hellula undalis*

1. Egg 2. Larva 3. Pupa 4. Moth

Control measures. Because the caterpillars are found deep inside the cabbage heads, it is very difficult to control the pest.

4. Cabbage painted bug

Bagrada cruciferarum

Order - Hemiptera

Sub order - Heteroptera

Family - Pentatomidae

Nature of damage. Both the nymphs and adults remain on the leaves and other plant parts in large numbers, pierce, suck and feed on the sap. Severely affected plants become weak and stunted, and may even die ultimately. The pest is found all over India and besides cabbage infests mustard, cauliflower, knolkhol. radish and other cruciferous crops.

Life cycle of the pest. The female bug lays barrel-shaped, light yellowish eggs singly on the leaves, stem portions etc. One insect lays 90-218 eggs. The eggs hatch in about 7 days. The young nymphs hatching from the eggs, moult 4 times during their growth and become adult bugs in about 21 days. There may be 6-7 overlapping generations in a year. The adult bugs are about 0.6 cm. long, flattish, black in color, with reddish-yellow markings on the body (Fig. 50).

Control measures

Physical methods. The nymphs and adults may be caught by using hand nets and destroyed

Chemical control (i) Dusting on the foliage with carbaryl 10 D or malathion 5 D or endosulfan 4 D at 10 kg./acre controls the pest (ii) Spraying with malathion-600 ml. in 300 liters of water per acre with a high volume sprayer is also effective.

5. Cabbage leaf webber

Crocidolomia binotalis

Order - Lepidoptera

Family - Pyraustidae

Nature of damage. The caterpillars of the pest web nests with fine silken threads on the under surface of leaves, remain inside the webs, bite and feed on the leaf tissues. They also attack the flowers and pods and cause severe damage. Besides cabbage, the pest attacks cauliflower, radish, mustard and other crucifers.

Life cycle of the pest. The female moth lays 40-100 eggs in masses. The eggs are flattish and are laid overlapping one another. The eggs hatch in

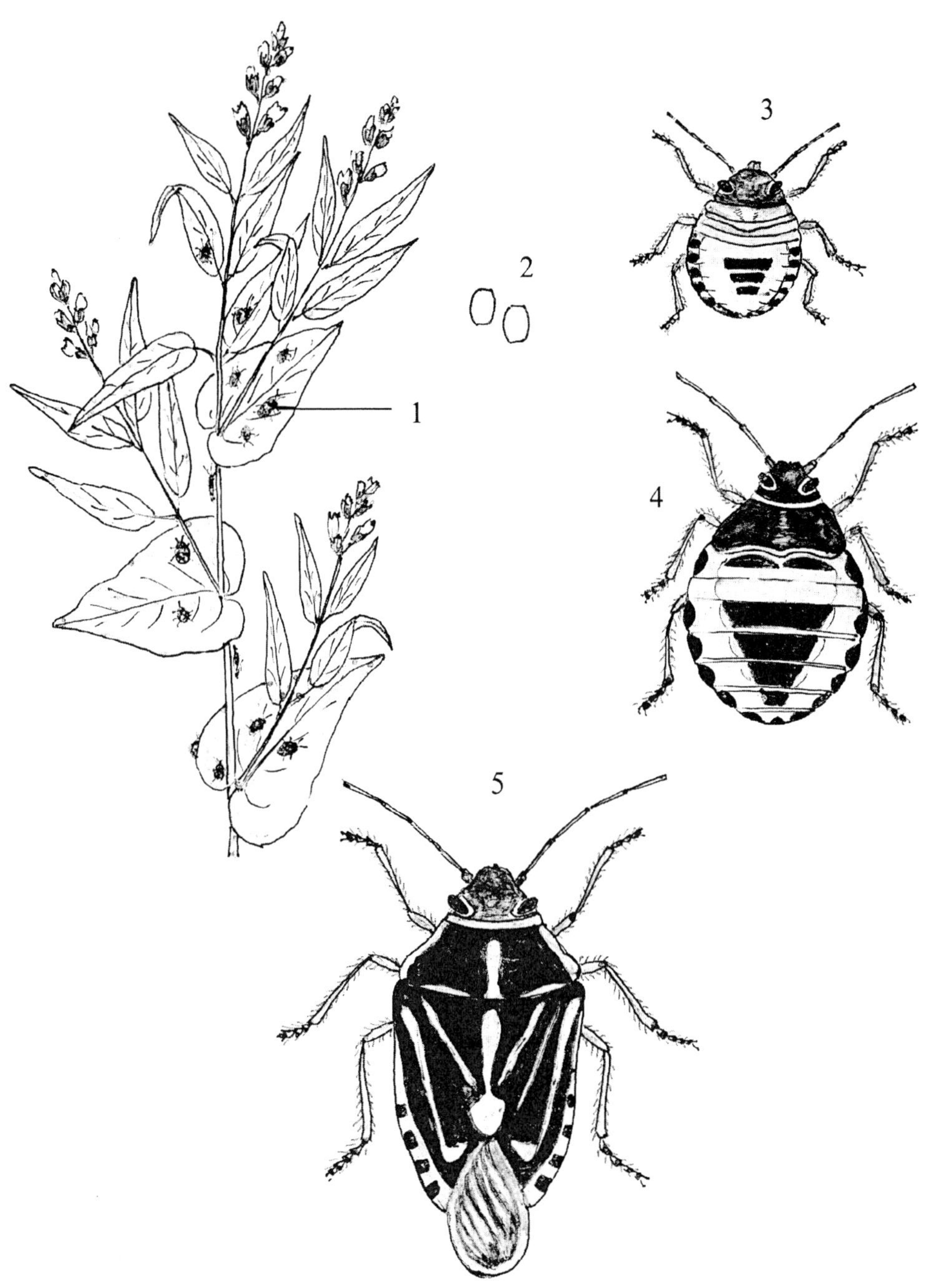

Fig. 50. Cabbage painted bug-*Bagrada cruciferarum*

1. Nymphs and adults feeding on a shoot 2. Eggs 3,4. Nymphs 5. Full grown adult bug

5-15 days. The larval stage lasts for 24-27 days in the summer season and about 50 days in the winter season. The larva is pale violet in color, with a red head. and has brown longitudinal stripes on the body, and rows of tubercles with short hairs. Full-grown larva pupates in the soil in an earthen cocoon. The moth emerges from the cocoon in 14-20 days. The moth is small sized, has light brownish forewings and whitish hindwings.

Control measures

Physical and cultural methods (i) The infested leaves with the larva or pupa may be removed and destroyed. (ii) Raising mustard as a trap crop 12-14 days prior to planting cabbage helps to control the pest. Insecticidal spraying can easily destroy the caterpillars, which climb up the mustard plants.

Chemical control. Spraying with malathion-600 ml. or endosulfan-600 ml. or carbaryl 50 WP-1000 gm. in 300 liters of water per acre with a high volume sprayer affords control.

Minor pests. Besides the pests described above, the cabbage green semilooper-*Trichoplusia ni* bites holes, feeds on the green matter and skeletonizes the leaves; the polyphagous pests viz., the tobacco caterpillar-*Spodoptera litura,* the gram cut worm-*Agrotis ipsilon* and *A. segetum* bite and feed on the leaves of cabbage, cauliflower, radish, beet root and other crucifers; the mustard saw fly-*Athalia lugens proxima* bites holes and feeds on the leaves of cabbage, cauliflower, turnip, radish, mustard and allied crops; the thrips-*Thrips tabaci* and *Caliothrips indicus* lacerate the young leaves, suck and feed on the sap from the leaves of cabbage, cauliflower etc.; the plant lice-*Myzus persicae* suck the sap from the under surface of leaves of cabbage, cauliflower, radish etc.; the caterpillars of the cabbage butterfly-*Pieris brassicae* feed on the foliage of cabbage, cauliflower, mustard etc.; the pea leaf miner-*Phytomyza atricornis* mines into the leaves and feeds on the inner tissues of cabbage, cauliflower, knolkhol. radish, turnip and carrot; the mites-*Tetranychus cucurbitae* suck and feed on the sap from the tender parts of crucifers and cucurbits; the Bihar hairy caterpillar pest-*Diacrisia obliqua* feeds voraciously on the foliage of crucifers; the radish flea beetle-*Phyllotreta downsei* bites and feeds on the leaves; the nymphs and adults of the aphid-*Myzus persicae* also suck and feed on the sap from the foliage.

CHILLIES *(Capsicum annuum)*

1. Chilli thrips

Scirtothrips dorsalis

Order - Thysanoptera

Family - Thripidae

Nature of damage. It is one of the most serious pests of chillies. Both the nymphs and adults remain on the under surface of young leaves in large numbers, lacerate the leaves, suck and feed on the sap from the leaves. The infested leaves are mottled, twisted and distorted, turn brownish in color and the leaves dry from the tips. The affected plants become stunted. When the flower buds and flowers are attacked, they drop off resulting in severe yield loss. The pest also infests several other crops such as, cotton, castor, grapevine, tea, soybean etc.(Fig.94)

Life cycle of the pest. The female insects lay eggs either parthenogenetically or after mating with the males. The eggs are inserted into the midrib of leaves. One female lays 40-48 eggs. The eggs hatch in 3-4 days. The young nymphs coming out of the eggs become full-grown in 10-15 days and start reproducing, thus the pest multiplies very rapidly. The adult is minute, fragile and yellow-colored. The wings are fringed with hairs like a comb

Control measures

Chemical control. Spraying the foliage with fenthion-300 ml. or endosulfan-300 ml. or phosalone-600 ml. or phosphamidon-150 ml. or carbaryl 50 WP-1000 gm. in 300 liters of water per acre with a high volume sprayer, so as to cover the under surface of leaves thoroughly controls the pest effectively.

Biological control. The thrips-*Franklinothrips vespiformis* and *Erythrothrips asiaticus* are predaceous on the pest.

2. Tobacco caterpillar

Spodoptera (Prodenia) litura

Order - Lepidoptera

Family - Noctuidae

Nature of damage. The larvae cut and feed on the leaves and young shoots of the plants. Later they bore into the fruits and feed on the seeds and cause considerable damage. This polyphagous pest attacks many other crops besides chillies.

Life cycle of the pest and control measures. Refer pests of cotton (Fig. 68, Page 161).

3. American boll worm

Helicoverpa (Heliothis) armigera

Order - Lepidoptera

Family - Noctuidae

Nature of damage. This polyphagous caterpillar pest feeds on the leaves and young shoots. After fruits are formed, the caterpillars burrow the fruits and feed on the inner content. Besides chillies, the pest attacks several other cultivated crops.

Life cycle of the pest and control measures. Refer pests of cotton (Fig. 42, Page 159).

Other pests. A few other pests also attack chillies. The leaf eating caterpillar pest-*Laphygma exigua* occurs sporadically and causes defoliation; the nymphs and adults of the aphids-*Aphis gossypii* and *Myzus persicae* suck and feed on the sap from the leaves and young shoots. *Aphis gossypii* also transmits **'mosaic virus disease'**; the mite-*Hemitarsonemus latus* sucks and feeds on the sap from leaves.

MORINGA or DRUMSTICK *(Moringa pterygosperma* or *M. oleifera)*

1. Moringa hairy caterpillar

Eupterote mollifera

Order - Lepidoptera

Family - Bombicidae

Nature of damage. The pest is prevalent wherever moringa is grown. During the daytime, the caterpillars aggregate in large numbers and take shelter in-between the places where the stems branch or in depressions and hollow places in the stem or in cracks, crevices and such concealed places and remain without any movement. After dusk, they come out of the hiding places in swarms and feed voraciously on the leaves and within a few days all the leaves in the trees are completely eaten away. Under the places where the caterpillars hide during the daytime and under the infested trees, the faecal matter of the caterpillars is found in large quantities as dark pellets. The caterpillars scrape and feed on the bark of the stem portion also. The pest also attacks *Thespesia populnia* trees (Fig. 51).

Life cycle of the pest. The female moth lays eggs on the leaves and young shoots in clusters. The young caterpillars, which hatch out of the eggs become full-grown in 10-15 days. The grown-up caterpillar is stout, cylindrical in shape, grayish-black in color and the body covered with black, thin, long hairs. The hairs cause severe irritation to the human skin. The grown-up larva pupates under the soil and emerges as an adult moth at the time of the next flush, when there is abundant food supply for the hatching larvae. The large-sized moth is yellowish-brown in color, with yellowish-brown wings. The wings have dull, thin, wavy lines and markings (Fig. 51).

Fig. 51. Moringa hairy caterpillar-*Eupterote mollifera*

1. Eggs 2. Caterpillars feeding on the leaves 3. Caterpillar 4. Resting caterpillars in groups 5. Pupa in earthen cocoon 6. Moth

Control measures

Physical method. Burning with a flame torch can destroy the assemblage of caterpillars.

Chemical control. The young caterpillars can be destroyed by dusting with carbaryl 10 D or methyl parathion 2 D on the caterpillars, which shelter in large numbers during the daytime. Grown-up caterpillars can be destroyed by spraying with methyl parathion-0.05 % or dichlorvos-0.076 %.

2. Moringa bud worm

Noorda moringae

Order - Lepidoptera

Family - Pyraustidae

Nature of damage. The caterpillars bore into the flower buds and feed on the inner content, as a result the flower buds fall off. In case of severe infestation, up to 75 per cent of the flower buds are affected and shed causing severe loss in yield.

Life cycle of the pest. The female moth lays oval-shaped, light yellow-colored eggs singly or in small clusters on the flower buds. The young caterpillars hatch out of the eggs in 3-4 days and become full-grown in 8-16 days. The grown-up caterpillar is 1.1-1.4 cm. in length, dirty brown in color with a black head, a distinct line on the dorsal surface of the body and a black, thickened, shield-like skin on the first thoracic segment. When the flower buds drop due to the pest attack, the caterpillars inside the buds also fall down along with the buds and pupate under the soil. Only one caterpillar is found in each of the bud. The adult moth emerges from pupa in 6-10 days. The moth is small-sized, with dark brown forewings and white hindwings. The outer margin of the hindwings is dark brown in color.

Control measures

Chemical control. Spraying with malathion-0.1% or profenofos-0.05% when the trees come to flowering controls the pest.

3. Moringa leaf webber

Noorda blitealis

Order - Lepidoptera

Family - Pyraustidae

Nature of damage. The larvae of the pest web the leaflets together with fine silken thread, remain inside the web and feed on the green matter of the leaflets, as a result the leaflets become papery, dry and fall off. In case of severe attack, the whole tree is defoliated quickly.

Life cycle of the pest. The female moth lays creamy white, oval eggs in clusters, each having 34-96 eggs on the leaves. One insect lays up to 230 eggs. The eggs hatch in about 3 days and the larval period lasts for 7-15 days. The larva is 1.4-2.0 cm. long, with a brown head and devoid of a prothoracic shield. The full-grown larva pupates in the soil and emerges as an adult moth in 6-9 days. The moth appears very much similar to that of *N. moringae* but slightly bigger in size.

Control measures

Chemical control. Spraying with malathion-0.1% or profenofos-0.05% controls the pest.

Pests of minor importance. The leaf weevils-*Myllocerus viridanus* and *M. maculosus* bite the leaves from the margin and feed on the tissues; the nymphs and adults of the cotton aphid-*Aphis gossypii* suck and feed on the sap from the leaves and young shoots; the American boll worm-*Helicoverpa (Heliothis) armigera* feeds on the leaves; the caterpillars of the bark borer pest-*Indarbela tetraonis* construct long galleries on the stem portion with silken thread, frass and dry bark tissues, remain inside, scrape and feed on the bark; the maggots of the moringa bud midge-*Stictodiplosis moringae* bore into the flower buds, feed on the inner contents and cause shedding; the maggots of the fly-*Gitona distigma* infest the fruits and cause drying from the tips; the leaf caterpillar-*Pericallia ricini* and the caterpillars of the big-sized moon moth-*Actias selene* feed on the leaves; the grubs of the stem borer beetle-*Coptops aedificator* bore into the stems and the affected stems are killed.

CURRY LEAF *(Murraya koenigii)*

Many of the pests that infest citrus plants are found to infest curry leaf plants also. The nymphs and adults of the citrus psyllid bug-*Diaphorina citri* congregate in large numbers on the young leaves and tender shoots, suck and feed on the sap resulting in tip drying; the lemon butterfly pest-*Papilio demoleus* feeds voraciously on the leaves; the larvae of *Tonica zizyphi* roll the leaflets, remain inside, scrape and feed on the leaf tissues; the bark borer-*Indarbela tetraonis* scrapes and feeds on the bark of the stem; the leaf miner-*Phyllocnistis citrella* mines the leaves, feeds on the inner tissues and cause damage.

AMARANTHUS *(Amaranthus viridis)*

1. Amaranthus weevil

Hypolixus truncatulus

Order - Coleoptera

Family - Curculionidae

Nature of damage. The common leaf vegetable is attacked by the amaranthus weevil, which is almost a specific pest of the crop. The grub of the weevil bores into the top shoots and stem of the plants and produces gall-like swellings. The adult weevils bite and feed on the leaves. The infested plants become stunted in growth.

Life cycle of the pest. The life cycle of the pest is quite similar to that of the cotton shoot weevil-*Alcidodes affabar.* The small grub hatching out from the egg bores into the top shoot, feeds on the inner tissues and grow. The grub is whitish in color and lacks legs. It pupates inside the gall-like swelling and emerges as an adult weevil. The weevil is ashy dark gray in color with a prominent snout (Fig. 52).

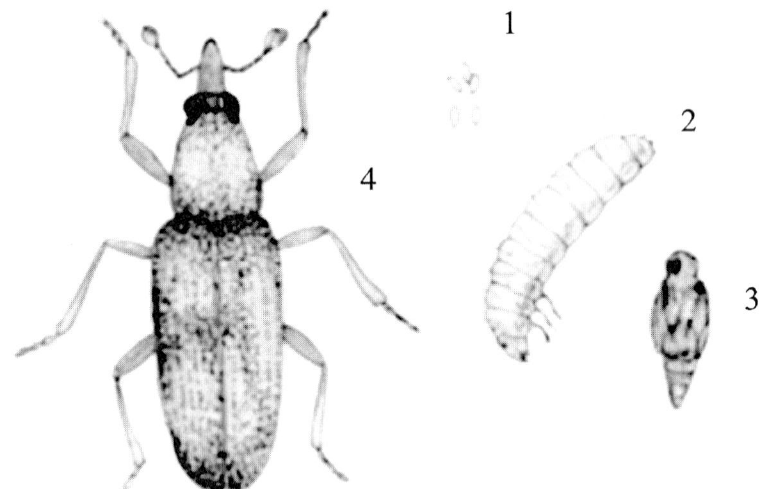

Fig. 52. Amaranthus weevil-*Hypolixus truncatulus*

1. Eggs 2. Grub 3. Pupa 4. Beatle

Control measures

Chemical control. Spraying the crop with endosulfan-500 ml. or carbaryl 50 WP-500 gm. or chlorpyrifos-500 ml. in 250 liters of water per acre with a high volume sprayer controls the pest.

2. Amaranthus leaf caterpillar

Hymenia recurvalis

Order - Lepidoptera

Family - Pyraustidae

Nature of damage. The pest attacks all amaranthus species, beet root, soybean etc. The insect breeds on grasses and during daytime the moths fly about among grasses and low vegetation. The larvae fold and web the leaves and top shoots, remain inside, scrape and feed on the green matter. In case of severe infestation, the leaves are completely skeletonized and eventually they dry.

Life cycle of the pest. The female moth lays about 156 eggs in masses on the leaves. The eggs hatch in 3-4 days. The larvae emerging from the eggs fold and web the leaves and top shoots, feed on the green matter and grow. The larval period lasts for 12-15 days. The caterpillar is slender and greenish in color with a brown head. The full-grown larva pupates in an earthen cocoon and emerges as an adult moth in 8-11 days. The moth has dark wings with wavy, white markings (Fig. 53).

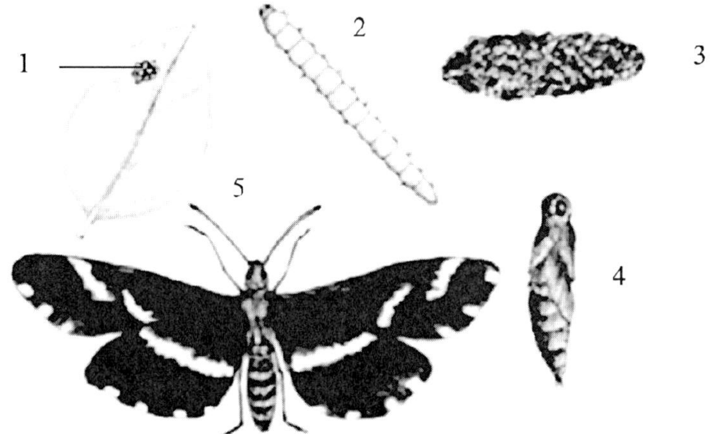

Fig. 53. Amaranthus leaf caterpillar-*Hymenia recurvalis*

1. Eggs 2. Caterpillar 3. Pupa inside earthen cocoon 4. Pupa 5. Moth

Control measures

Physical methods. The leaf folds may be collected and destroyed

Chemical control. Foliar spraying with endosulfan-500 ml. or phosalone-500 ml. or profenofos-500 ml. in 250 liters of water per acre with a high volume sprayer is effective in controlling the pest.

Other pests. The larvae of *Spodoptera exigua* and *Plusia eriosoma* feed gregariously on the leaves and defoliate the plants; the nymphs and adults of the thrips-*Haplothrips ceylonicus* infest the inflorescence and feed on the plant sap.

PESTS OF COMMERCIAL OR CASH CROPS

SUGARCANE *(Saccharum officinarum)*

1. Sugarcane early shoot borer

Chilo (Chilotraea) infuscatellus

Order - Lepidoptera

Family - Crambidae

Nature of damage. The pest occurs worldwide and infests only 1-3 months old, young crops. The caterpillar burrows the stem, enters into the stem region, and feeds on the soft inner tissues of the central shoot and severes it. The severed shoot rots, turns brown and dries, causing the characteristic **'dead heart'** symptom. Due to the infestation, 10-15 per cent of the young plants may die. The 'dead hearts' if pulled slightly comes off easily from the inside and the rotten tissues inside emit a foul smell. Even after rotting of the central shoot, the other leaves may remain green for many more days before the plant dies ultimately. After the central shoot rots and when there is no food for the caterpillar, it may move to the adjacent plant and likewise a single caterpillar may damage many young plants. When the temperature is high and the humidity is low, the infestation is more severe. After a few heavy showers, the pest incidence is very much reduced (Fig. 54).

Life cycle of the pest. The female moth lays whitish, scale-like, flattish, overlapping eggs in rows of 8-60 on the underside of the leaf sheaths. There may be 3-5 rows in each cluster. One female lays 300-500 eggs within a period of one week. The eggs hatch in 4-6 days. The young caterpillar is white in color, with a dark brown head and fringed with thin, long hairs and brown dots on the body. The young caterpillars separate themselves, find different plants and bore into the stem just above the ground level. After entering into the stem, the caterpillar starts feeding on the soft, inner tissues of the central shoot and in this process severes the shoot from inside causing 'dead heart'. The young caterpillar becomes full-grown in 4-5 weeks. The grown-up caterpillar is white in color, with a brownish head, 2.5-3.5 cm. in length, the body mottled with dark brownish markings and has setae-like hairs and tubercles. There are 5 dorsal and subdorsal, deep violet colored lines on the body. The grown-up larva pupates inside the stem and emerges as an adult moth in 8-10 days. There may be 5-6 overlapping broods in one season. The moth is small-sized, about 3.5 cm. in size, dull gray in color, with brown-colored markings on the forewings and whitish hindwings (Fig. 54).

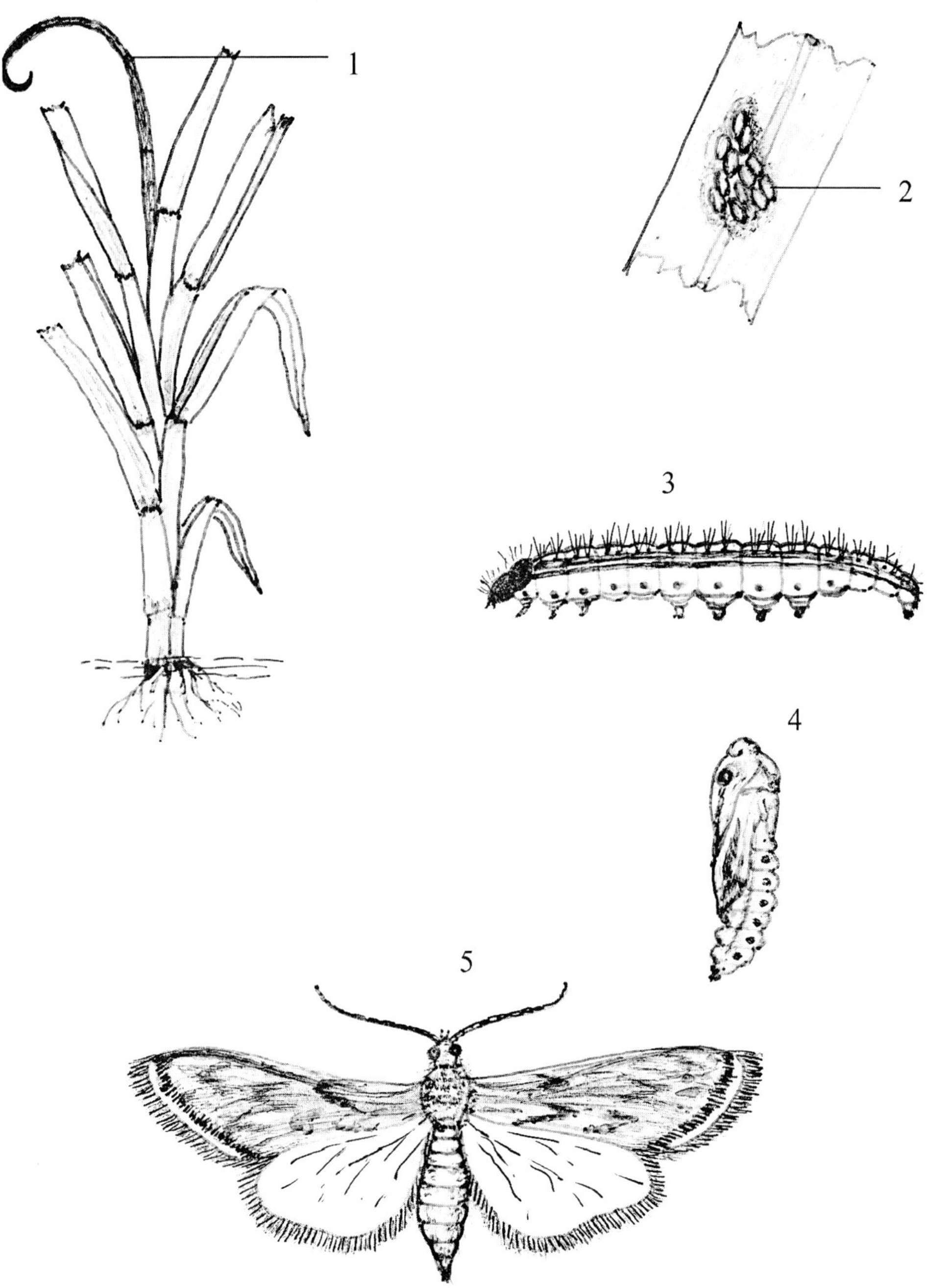

Fig. 54. Sugarcane early shoot borer-*Chilo infuscatellus*

1. Dead heart 2. Egg mass 3. Larva 4. Pupa 5. Moth

Control measures

Physical and cultural methods (i) Late planting (April-May), when the temperature is high and the humidity is low may be avoided (ii) The infested young plants should be removed and destroyed (iii) Earthing up one month after planting, followed by a second earthing up after one month, prevents the moth from laying eggs on the lower leaves and also prevents the caterpillars reaching and boring into the stem (iv) Trash mulching up to a height of 10-15 cm. on the third day after planting also prevents the moths from laying eggs on the leaves and also prevents the caterpillars from reaching the stem. Trash mulching increases the humidity of the environment around the crop canopy, which is also unsuitable for the pest infestation (v) The dead hearts may be removed and through the hole a long, thin iron needle is poked and the caterpillar may be killed.

Chemical control (i) Application of sevidol 4 G-10 kg. or carbofuran 3 G-13 kg. or lindane 10 G-5 kg., mixed with 10 kg. of sand along the planting furrows at the time of planting and irrigating the field helps to prevent the attack by the pest (ii) Spraying monocrotophos-300 ml. or endosulfan-300 ml. or phosalone-600 ml. or chlorpyrifos-600 ml. in 300 liters of water per acre with a high volume sprayer, twice, 15 and 45 days after planting, so as to cover the stem portion thoroughly is also effective in controlling the pest (iii) Application of sevidol 4 G-10 kg. or carbofuran 3 G-13 kg, or quinalphos 5 G-8 kg. or chlorpyrifos 10 G-4 kg., mixed with 10 kg. of sand per acre in the leaf whorls twice, 30 and 60 days after planting controls the pest.

Biological control (i) The egg parasite-*Trichogramma australicum* parasitizes the eggs and destroys the eggs of the pest (ii) The formulation of the bacterial parasite-*Bacillus thuringiensis* containing 25 x 10^6 spores/mg., commercially available as Thuricide may be sprayed to infect and destroy the caterpillars (iii) The virus parasite-Granulosis virus may be sprayed at 100 larval equivalent (100 LE), mixed with Teepol-0.01 % per acre, 30 and 50 days after planting to destroy the caterpillars. Hand operated sprayers are preferable to low volume sprayers to spray the parasites and the parasites should be sprayed during the evening hours (iv) Releasing 50 gravid females of the larval parasitoid-*Sturmiopsis inferens* per acre 45 days after planting helps to eliminate the larvae of the pest.

2. Sugarcane internode borer

Chilo (Chilotraea) sacchariphagus indicus

Order - Lepidoptera

Family - Crambidae

Nature of damage. The pest, which occurs in all sugarcane growing states of India, generally infests the crop at the later stages of the crop and continues till harvest. Low temperature and high humidity favor infestation by this pest. The young caterpillars, which hatch out from the eggs start feeding by scraping the young, unopened leaves. Later they bore into the internode region of the tender cane tops. They make long tunnels in the stem region and feed on the inner tissues. The tissues around the tunnelled areas turn red. One larva may traverse 1-3 internodes and the attack is more in the top five internodes. Many caterpillars may be found within a single stem. The excreta of the caterpillars are pushed out through the entry holes. Dead-heart symptoms may appear generally in the infested canes. In case of severe infestation, the yield loss may go up to 25 per cent, at the same time the sugar content of the infested canes is also very much reduced. High tillering varieties are more prone to attack by the pest (Fig. 55).

Life cycle of the pest. The female moth lays oval-shaped, whitish, scale-like, flattish eggs in two rows on the upper surface of leaves. There may be 10-15 eggs in each row. Likewise one moth may lay up to 400 eggs. The eggs hatch in 3-5 days. The larva, which becomes full-grown in 30-40 days, is almost similar to the early shoot borer, but slightly larger in size, about 3.5-4.0 cm. in length and whitish in color, with a brown head. There are four thin, violet-colored lateral stripes on the body and four dark brown spots across each of the segment. The full-grown caterpillar makes an exit hole in the stem, covers it with a fine, silken mesh, pupates within the stem and emerges as an adult moth in 10-15 days. The moth is slightly bigger than the early shoot borer moth and pale brown in color. The pale brown forewings of the moth have small, golden spots near the lateral margins and the hindwings are whitish (Fig. 55).

Control measures

Physical and cultural methods (i) The setts for planting should be selected from canes not infested by the borer pest (ii) The egg masses can be collected and destroyed (iii) The crops should be detrashed 150 and 210 days after planting to destroy the egg masses on the leaves and to prevent the larvae from burrowing the internode region (iv) Application of excessive doses of nitrogenous fertilizer should be avoided.

Biological control (i) The egg parasitoid-*Trichogramma chilonis* parasitizes and destroys the eggs. The parasitoids should be released at the rate of 20,000 numbers (1.0 cc. of parasitized eggs) per acre per release and are released 6 times at 15 days intervals commencing from the fourth month of planting (ii) A few species of *Telenomus* also parasitize and destroy the eggs.

Resistant varieties. A few varieties viz., Co.975, Co.62175, Co.6506 and Co.77-1 are found to have resistance to this pest.

Chemical control. No chemical control measures are recommended to eradicate the pest.

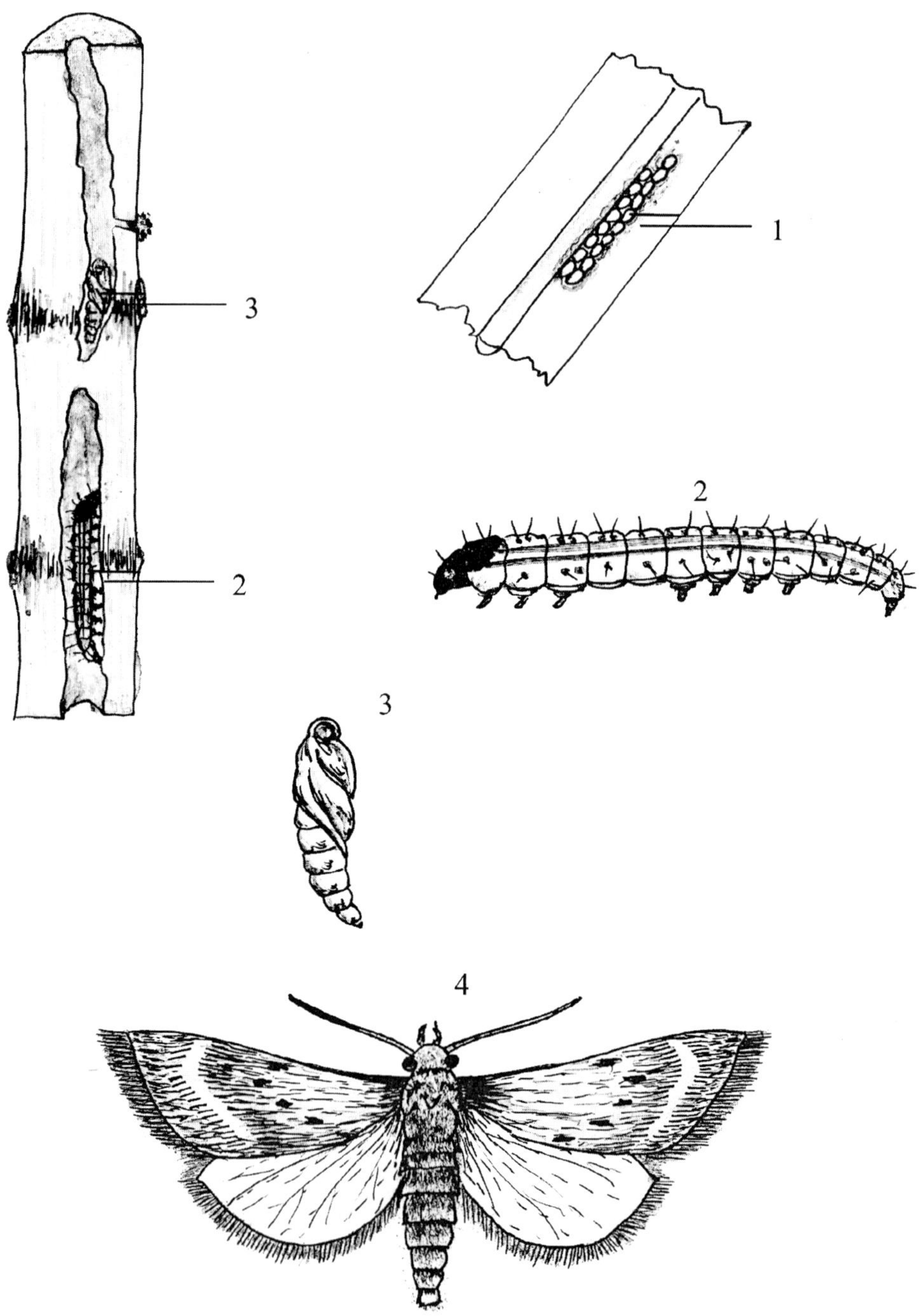

Fig. 55. Sugarcane internode borer-*Chilo sacchariphagus indicus*

1. Eggs 2. Larva 3. Pupa 4. Moth

3. Sugarcane top borer

Scirpophaga excerptalis (nivella)

Order - Lepidoptera

Family - Pyraustidae

Nature of damage. The pest is prevalent in all sugarcane growing tracts of India. Though the pest occurs generally in the later stages of crop and extends up to the harvest stage, it is found to infest the crop at all growth stages. The infested canes come to maturity much earlier and thus may result in a yield loss of up to 20-30 percent. Further, the quality of the infested canes becomes poor and the sugar content in the juice is also very much reduced. The young caterpillar, which hatches out of the egg burrows through the mid rib of the young leaf downwards, reaches the central shoot, bores through it and enters the stem region, as a result when the young leaves unfurl, small holes are seen on either side of the mid rib and in the mid rib also. This is followed by the formation of a reddish-brown dead heart. Through the burrow in the central shoot, excreta of the larva is pushed out. After the death of the central shoot, the growth of the buds at the top portion of the cane is stimulated and they grow as small branches and the whole top of the cane presents a characteristic bushy appearance. The growth of the infested plants is also very much stunted (Fig. 56).

Life cycle of the pest. The moths are very active and fly about and mate during the nighttime. The female moth lays whitish, scale-like, flattish, oval-shaped, overlapping eggs near the mid rib of young leaves in clusters and covers them with orange-brown anal hairs. One insect may lay up to 500 eggs in batches of 60-70 eggs in each. The eggs hatch in 7-10 days. The young caterpillar after emerging from the egg, rests for a short while, then starts burrowing the mid rib, reaches the central shoot and bores through it into the stem, as a result the central shoot dries up forming a dead heart. The caterpillar tunnels the stem portion downwards, feeds on the inner tissues and becomes full-grown in 25-40 days. The caterpillar may go down up to 4-6 nodes. The full-grown larva is 2.5-3.0 cm. in length and pale yellowish in color. It burrows an exit hole in the stem, covers it with a fine silken mesh, constructs a silken cocoon and pupates inside it within the stem. The pupa is shiny and brownish-yellow in color. The moth emerges from the pupa in 15-20 days. The moth is silvery white in color and about 3.0 cm. in size. The female moth has a stout abdomen, with orange-brown tuft of anal hairs at the crimson-colored tip of the abdomen. There may be 5-6 generations in a year (Fig. 56).

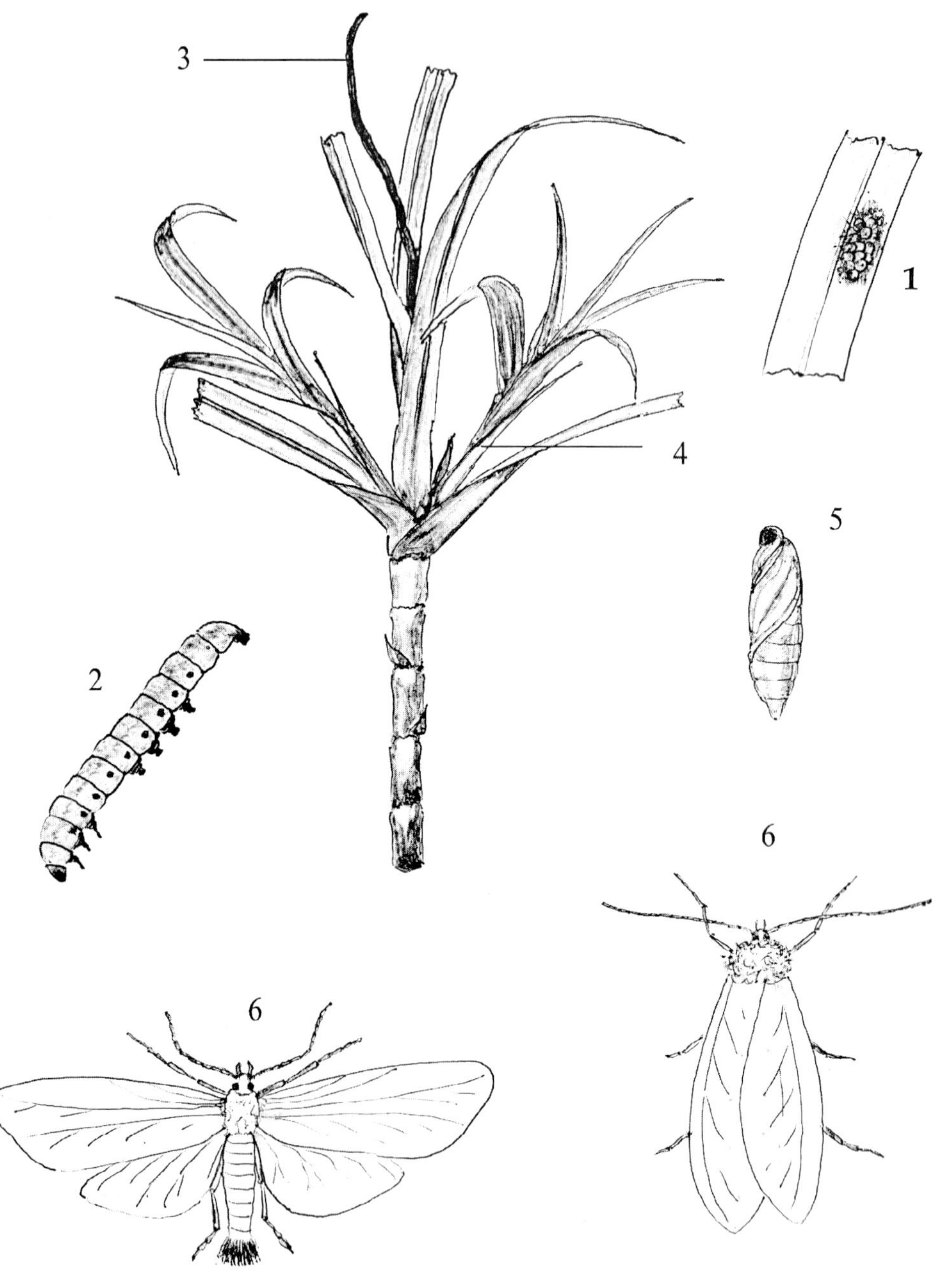

Fig. 56. Sugarcane top borer-*Scirpophaga nivella*

1. Egg mass 2. Larva 3. Dead heart 4. Infested plant showing sprouting of terminal buds 5. pupa 6. Moth

Control measures

Physical and cultural methods (i) The egg masses, which are clearly visible in the sunlight can be collected and destroyed (ii) The top burrowed portion of the infested canes may be cut off and destroyed along with the caterpillars inside (iii) The canes should be detrashed 150 and 210 days after planting, which may prevent the moths from laying eggs on the leaves.

Biological control (i) The Ichneumonid wasp-*Isotima javensis* parasitizes and destroys the caterpillars (ii) The egg parasitoids-*Trichogramma minutum* and *Trichogramma intermedium* parasitize and destroy the eggs of the pest

Resistant varieties. A few sugarcane varieties such as Co.859, Co.1158, Co.7224 and CoS.767 are less prone to attack by this pest.

Chemical control. No chemical control measures have been recommended for the control of this pest.

4. Sugarcane leaf hopper

Pyrilla perpusilla

Order	-	Hemiptera
Sub order	-	Homoptera
Family	-	Fulgoridae

Nature of damage. The pest is prevalent in the states of Uttar Pradesh, Bihar, Maharashtra and Tamil Nadu. In some tracts of Tamil Nadu, the pest appears in large numbers during the months of October-November after the monsoon rains. In the infested fields, both the nymphs and adults are found on the young and older leaves in swarms of thousands and the sucking up and continued drainage of sap from the leaves causes fading and drying up of leaves. Severely infested plants present a sickly and blighted appearance. When such plants are disturbed, hundreds of nymphs and adults hop about on all sides. The yield of infested crop is reduced and the quality of cane juice is also poor. The honeydew secreted by the nymphs attracts many other insects and ants, which come to feed on the sugary secretion. Sooty mould, a fungal disease appears on the leaves covered by the honeydew. Besides sugarcane, the pest attacks pearl millet, sorghum, maize, rice, wheat, barley, oats etc.

Life cycle of the pest. The female hopper lays pale greenish-yellow eggs in clusters on the under surface of leaves near the mid rib or inside the leaf sheaths and covers them with a white, fluffy, filamentous, waxy material, which is secreted by it. One female lays up to 750 eggs in batches of 20-60 numbers. The eggs hatch in about 7 days. The young nymphs, which hatch out of the eggs are pale brown, active, hopping insects with a pair of characteristic long, wax covered, bunch of hair-like anal processes. The nymphs are also covered with a white, waxy coating. During their growth period, the nymphs moult 5-6 times and become adult hoppers in 40-60 days. The adult hopper is very active, straw-colored, with the head prominently drawn forward as a sort of rostrum. The wings are folded on the back like a roof. They are densely knitted with veins and are transparent. The females have a pair of wax covered, dense hairy anal processes. There may be 3-4 generations in a year (Fig. 57).

Control measures

Physical and cultural methods (i) The egg masses, which are quite visible, can be collected and destroyed (ii) The nymphs and adults can be collected by bagging and destroyed (iii) Application of excessive doses of nitrogenous fertilizer should be avoided (iv) The crop should be detrashed 150 and 210 days after planting to destroy the egg masses on the leaves and also to reduce the humidity within the plant canopy, as high humidity favors large scale multiplication of the pest (v) Light traps can be set up to attract and destroy the adult hoppers.

Chemical control (i) Dusting the foliage with carbaryl 10 D at 10-15 kg./ acre controls the pest (ii) Spraying on the foliage with endosulfan-800 ml. or fenitrothion-400 ml. or malathion-800 ml. in 400 liters of water per acre with a high volume sprayer is effective in controlling the pest.

Biological control (i) The predatory *Coccinella* beetles-*Coccinella septum punctata* and *Coccinella undicempunctata* feed on the nymphs of the pest (ii) The fungal parasites-*Mucor heimalis* and *Metarrhizium anisopliae* infect and destroy the nymphs.

5. Sugarcane white fly

Aleurolobus barodensis

Order	-	Hemiptera
Sub order	-	Homoptera
Family	-	Aleyrodidae

Nature of damage. Both the nymphs and adults harbor on the under surface of young leaves in large numbers, suck and feed on the sap from the leaves. The infested leaves gradually turn yellow and pinkish and eventually dry up. Continuous drainage of sap from the leaves affects the vigor of the canes resulting in stunted growth and poor quality juice. The pest appears in a severe form in low-lying, water inundated fields, as well as in the ratoon crops. In recent years, the pest has assumed serious proportions in the states

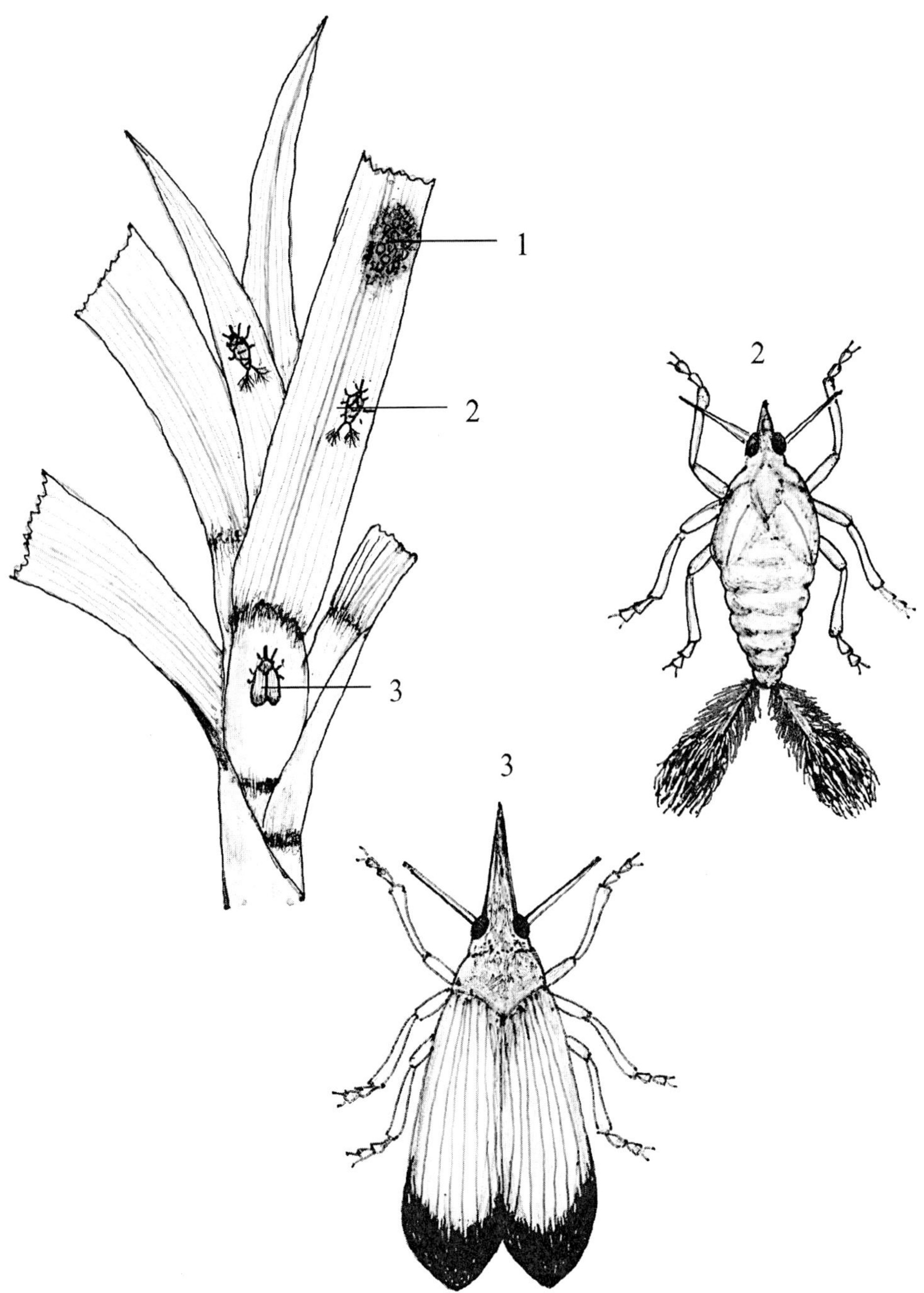

Fig. 57. Sugarcane leaf hopper-*Pyrilla perpusilla*

1. Eggs 2. Nymph 3. Adult hopper

of Bihar, Gujarat, Maharashtra, Haryana, Karnataka, Punjab, Tamil Nadu, Uttar Pradesh and Telungu Desam.

Life cycle of the pest. The female white fly lays very small, dark brown, shiny eggs in rows of 3-50 on the under surface of top, tender leaves, generally along the mid rib. The eggs hatch and the nymphs emerge in about 7 days. The young nymph is very small, oval in shape, light brown in color, with its body covered with white, powdery, waxy coating. The outer margin of the body is also fringed with white, waxy filaments. They are more or less stationary. There are three nymphal instars and one pupal instar, after which they become winged, full-grown adult flies. The entire life cycle is over in about 30-40 days. The adult white flies are very small, delicate, pale yellow in color, with black eyes and white wings. Their body and wings are covered with white, powdery, waxy coating. They are active, able to fly short distances and thus distribute themselves. The females are stouter than the males. There may be about 9 generations in a year.

Control measures

Cultural methods (i) Proper drainage facilities should be provided to the fields. In water inundated fields ratooning should be avoided (ii) Application of nitrogenous fertilizer in excess of the recommended dose should be avoided (iii) In tracts where the pest occurs regularly, ratooning should be avoided (iv) The crop should be detrashed 150 and 210 days after planting and maintained properly.

Chemical control. Foliar spraying with fenitrothion-400 ml. or methyl parathion-400 ml. or monocrotophos-400 ml. or endosulfan-800 ml. or malathion-800 ml. in 400 liters of water per acre with a high volume sprayer controls the pest.

Biological control. The fungal parasite-*Fusarium subglutinens* is found to infect and destroy the nymphs during highly humid weather conditions.

6. Sugarcane scale insect

Melanaspis glomerata

Order - Hemiptera
Sub order - Homoptera
Family - Diaspididae

Nature of damage. The pest is prevalent in the states of Tamil Nadu, Telungu Desam, Gujarat and Maharashtra. Both the nymphs and adult females stick on to the nodal regions and inside the leaf sheaths within an oval-shaped, grayish-black, hard, scale-like protective covering, suck and feed on the sap, as a result the leaves turn yellow and the plants become stunted. In due course the insects form a thick encrustation on the entire stem. The nymphs and the female insects are sedentary once they fix themselves in a particular place and do not move about.

Life cycle of the pest. The female insect lays 5-500 eggs inside the leaf sheaths or on the soil under the canopy of the plants. The nymphs, which hatch out of the eggs, crawl about and fix themselves in a particular place in the nodal region or inside the leaf sheaths and form a hard, grayish-black, oval, scale-like protective covering, stay inside, suck and feed on the sap and grow. There are five nymphal instars as they grow and they remain sedentary all the time. The full-grown nymph pupates inside the scale-like covering and emerges as an adult insect. Only the males have wings and can fly short distances. The adult male scale insect is very small, like an ant in general appearance, with a black body and thin, whitish, papery wings. There may be 8-9 generations in a year. Varieties having high stomatal density are more prone to attack by the pest, as the stylets are pushed through the stomata while feeding.

Control measures

Cultural methods (i) Setts free from scale insect infestation should be selected for planting (ii) Adequate drainage facilities should be provided to drain off excess water from the fields (iii) The crop should be detrashed 150 and 210 days after planting to destroy the egg masses as well as the sedentary insects.

Chemical control (i) Spraying the crop with malathion-400 ml. or methyl parathion-400 ml. or dimethoate-800 ml. or monocrotophos-400 ml. or methyl demeton-800 ml. in 400 liters of water per acre with a high volume sprayer, so as to cover the stem portion and leaf sheaths is effective in controlling the pest The sprayings may be given at 30 days interval from the month of June (ii) Soil application of carbofuran 3 G-13 kg./ acre when the crop is 4-6 months old also controls the pest.

7. Sugarcane white grub

Holotrichia consanguinea

Order	-	Coleoptera
Family	-	Melolonthidae

Nature of damage. The pest occurs in the states of Tamil Nadu, Bihar, Uttar pradesh, Punjab and Kerala sporadically and sometimes causes severe damage to the crop. The grubs of the pest remain under the soil, cut, bite and feed on the roots, rootlets and underground stalk. Rarely the grubs bore into the stem of canes and feed on the inner tissues. Severely affected canes come off easily, when pulled slightly. The pest is also found to infest finger millet

Life cycle of the pest. The beetles become active with the onset of rains during May-June. They come out of the soil during dusk and mate. The female beetle lays eggs on the moist soil below the plant canopy. The young grubs hatching out of the eggs, burrow into the soil, reach the roots, bite and feed on the roots and root hairs and grow. The full-grown grub is whitish in color, long, smooth, shining, fleshy with a dark head, wrinkled body segments and curved ventrally like a 'C' in its natural position. It pupates under the soil and emerges as an adult beetle. The beetle is cylindrical in shape and black in color. There is only one generation in a year (Fig. 58).

Control measures

Chemical control (i) During the months of May-June, when the beetles start coming out of the soil in large numbers, carbaryl 10 D or lindane 0.65 D is incorporated into the soil at 20 kg. per acre to destroy the beetles (ii) Application of fensulfothion (Dasanit) 5 G or quinalphos 5 G at 10 kg. per acre uniformly over the soil, followed by irrigation is effective in controlling the grubs.

8. Termites

Odontotermes obesus

Order - Isoptera
Family - Termitidae

Nature of damage. The pest occurs in serious proportions in the states of Tamil Nadu, Telungu Desam and Uttar Pradesh. The infestation starts from the planting time. The insects start feeding on both the cut ends and buds of the planted setts initially and the affected setts fail to germinate leading to gappiness wherever the setts are destroyed. In due course, the entire inner portion of the setts, except the hard outer rind are completely eaten away by the pest. After germination, when the young plants are attacked, the outermost leaves start drying initially and ultimately the entire infested plants completely dry and die. Such plants when pulled slightly come off easily from the bottom. When grown-up plants are attacked, the lower portion of the stem up to the sixth internodal region from the base are affected. The inner tissues are completely eaten and mud galleries constructed by the white ants are seen inside the stem portion. In case of severe infestation, the pest may damage 40-50 per cent of the canes. Light sandy soils and drought conditions favor large-scale infestation.

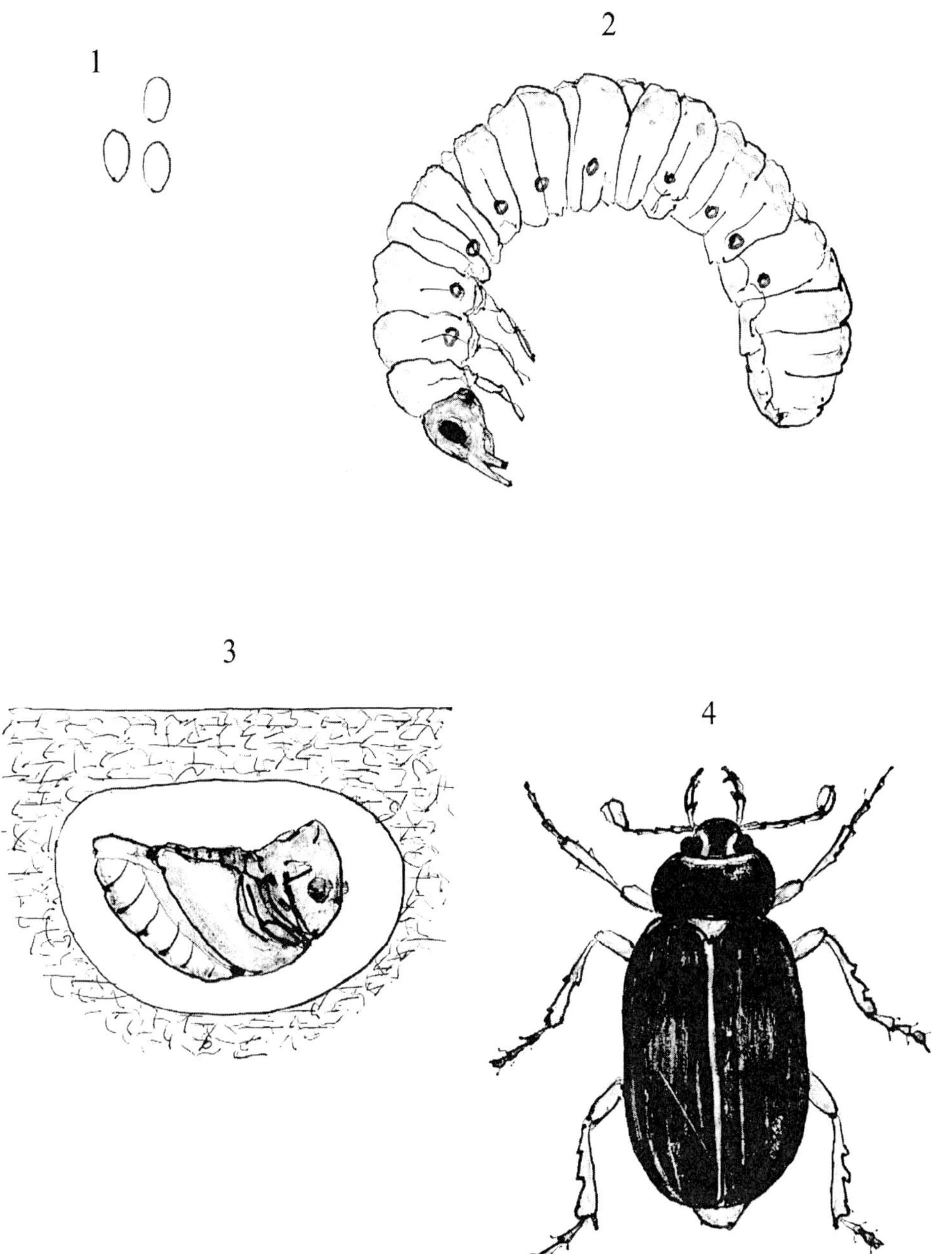

Fig. 58. Sugarcane white grub-*Holotrichia consanguinea*

1. Egg 2. Grub 3. Pupa inside soil 4. Adult beetle

Life cycle of the pest. The winged kings and queens, which are the reproductive forms that come out of a old colony after the first monsoon showers are responsible for building up new termite colonies. After flying about for some time they cast off their wings. Afterwards, a male and a female seek a suitable place in the ground, excavate a small chamber in which they mate frequently several times. Soon after copulation, the female starts laying eggs. A queen may lay thousands of eggs daily and a colony comprising of over a million individuals is formed. The earlier broods of nymphs are mostly sterile castes of workers and soldiers and they attend on the king and queen. After the death of the primary queen, a supplementary queen may develop. The life of termites is very long, ranging from 15-50 years. In a termite colony, the workers do the entire work, such as cleaning the nest, procuring food, feeding and caring of the queen and the young ones etc. The workers are responsible for causing damage to sugarcane crop (Fig. 102).

Control measures

Chemical control (i) Application of carbaryl 50 WP-0.1% or lindane 20 EC-0.4% at 1000 liters per acre uniformly over the planting furrows before planting controls the pest (ii) Application of carbaryl 10 D or chlordane 5 D or heptachlor 5 D or aldrin 5 D at 20 kg./acre uniformly over the setts after planting and then covering with soil is also effective in controlling the pest.

Pests of minor importance. Besides the pests discussed above in detail, several other pests are also known to attack sugarcane, but to a lesser extent. However, some of them may assume serious proportions when environmental conditions become favorable to them. The grasshoppers-*Hieroglyphus banian* (Fig. 59) and *H. nigrorepletus* cut, notch and feed on the leaves; the pink borer pest-*Sesamia inferens*, a polyphagous pest bores into the tender shoots of young plants, feeds on the inner tissues and causes damage; the larvae of the Gurdaspur borer-*Acigona (Bissetia) steniellus* bore into the shoots, tunnel upwards and feed on the inner tissues; *Odontotermes assumthi*, another termite species also feeds on the planted setts and growing canes and causes gappiness in the field; the sluggish caterpillars of the root borer-*Emmalocera depressella*, a specific pest of sugarcane, bore into the underground portions of canes, resulting in dead hearts in the early stages of growth; the nymphs and adults of the black leafhopper-*Proutista moesta* suck and feed on the sap from the leaves. The pest is also the vector of **'grassy shoot disease'** caused by a mycoplasma; the nymphs and adults of the black bug-*Cavelerius (Macropes) excavatus* attack the top whorl of leaves; the grubs of the hispa-*Asamangulia cuspidata* feed on the tissues in-between the epidermal layers of leaves producing brown blisters; the mealy bugs-*Ripersia sacchari* and *Saccharicoccus sacchari* suck and feed on the sap near the nodes and leaf sheaths; the nymphs and adults of the aphid pest-*Rhopalosiphum maidis* suck the sap from the leaves

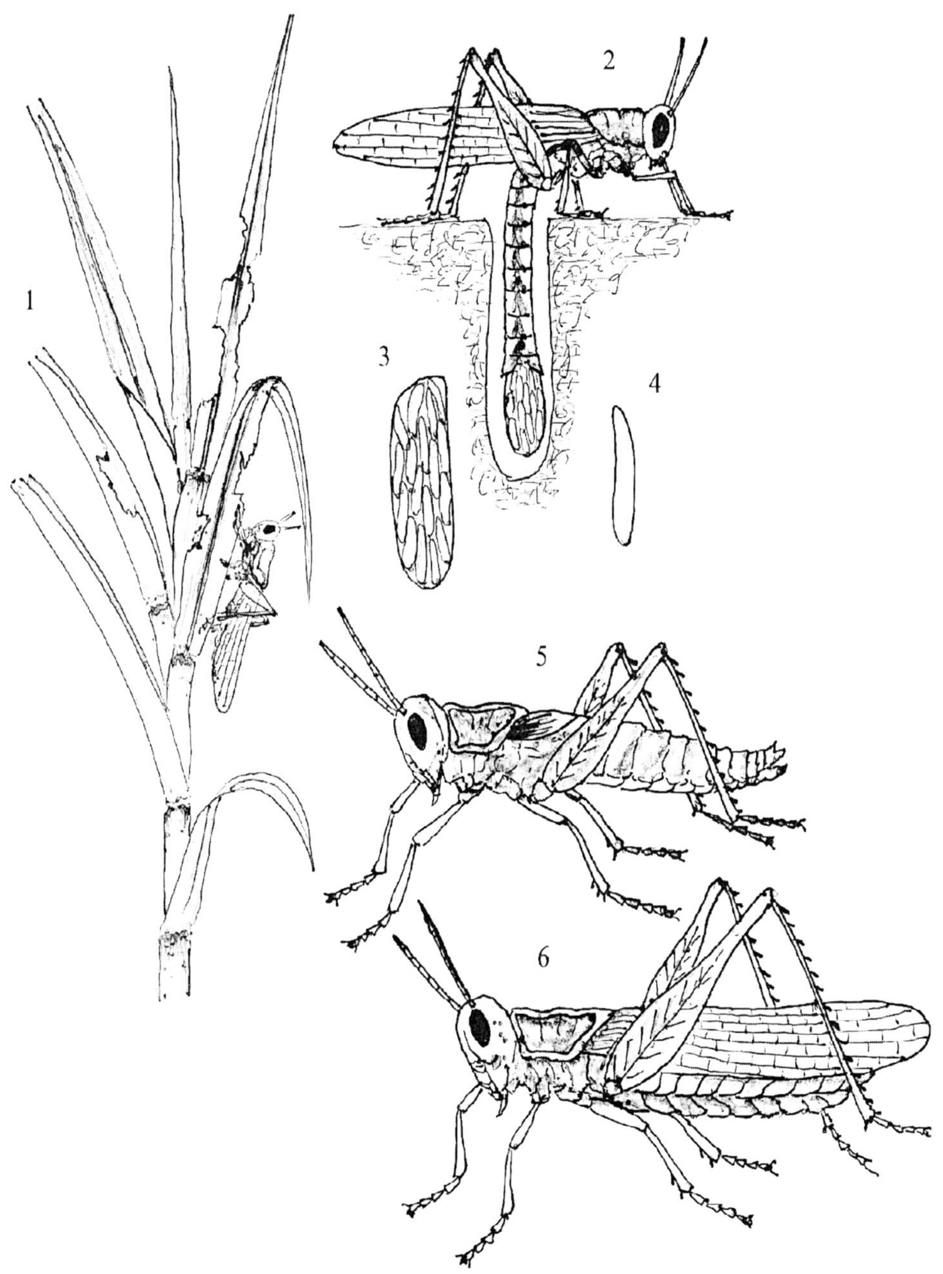

Fig. 59 : Rice grasshopper-*Hieroglyphus banian*

1. Insect feeding on the leaf 2. Female laying eggs inside soil 3. Egg pod 4. Egg 5. Nymph 6. Adult

and inflorescence and also serve as the vector of **'mosaic disease'**; the mites-*Oligonychus indicus* and *Schizotetranychus andropogoni* web fine nests with silken threads on the under surface of leaves, remain inside, suck and feed on the sap; the root aphid-*Tetraneura javensis* infests roots of sugarcane; the thrips-*Stenchaetothrips saccharicidus, Sorghothrips jonnaphilus* and *Anephothrips sudanensis* lacerate, suck and feed on the sap from the leaves.

COTTON *(Gossypium* species)

1. Cotton spotted boll worms

(i) *Earias insulana* and (ii) *Earias vitella (fabia)*

Order - Lepidoptera

Family - Cymbidae

Nature of damage. These are the most serious of cotton pests in South India, though they are prevalent all over India and are commonly known as **'the shoot and boll borers'**. The infestation starts when the plants are about 6 weeks of age and continues till full maturity of the crop. The caterpillars bore into the young shoots, feed on the inner tissues, as a result the shoots wilt, droop and die. In the wilted shoots, small burrows made by the caterpillars are visible, plugged with the faecal matter of the larvae. When young plants are attacked, their growth is very much stunted and they may even die. In the grown-up plants, the larvae burrow the flower buds, squares and bolls. The flower buds and squares are shed in large numbers. The infested young bolls shrink, dry and fall down. The matured bolls when attacked do not fall down, but open prematurely and the cotton from such bolls are stained, blackish and of poor quality. A single caterpillar burrows and destroys many bolls and one boll may have many burrows and all the seeds are eaten away and the burrows are plugged with the excreta of the larvae. Besides cotton, the pest infests lady's finger, hollyhock, shoe flower and Deccan hemp (Fig.60)

Life cycle of the pest. The pest occurs throughout the year. The female moth lays spherical, bluish-green, shining eggs with minute, parallel longitudinal ridges, which project at the top to give the appearance of a crown. The eggs are laid during the nighttime on the young shoots, flower buds, flowers, bracts and young leaves singly or in batches of two to three. One female lays up to 385 eggs. The eggs hatch in 3-4 days and the young caterpillars, which emerge from the eggs are pale brownish in color, with black head and about 2.0 mm. in length. If the eggs are laid on a young plant, the hatching larvae bore down the central shoots and if laid on a plant with bolls, the caterpillars bite a hole through the tender bolls and enter them. The larvae become full-grown in 10-12 days and they measure 1.8-2.0 cm. in length. The larva of *Earias insulana* is dull greenish in color with a dark brown head

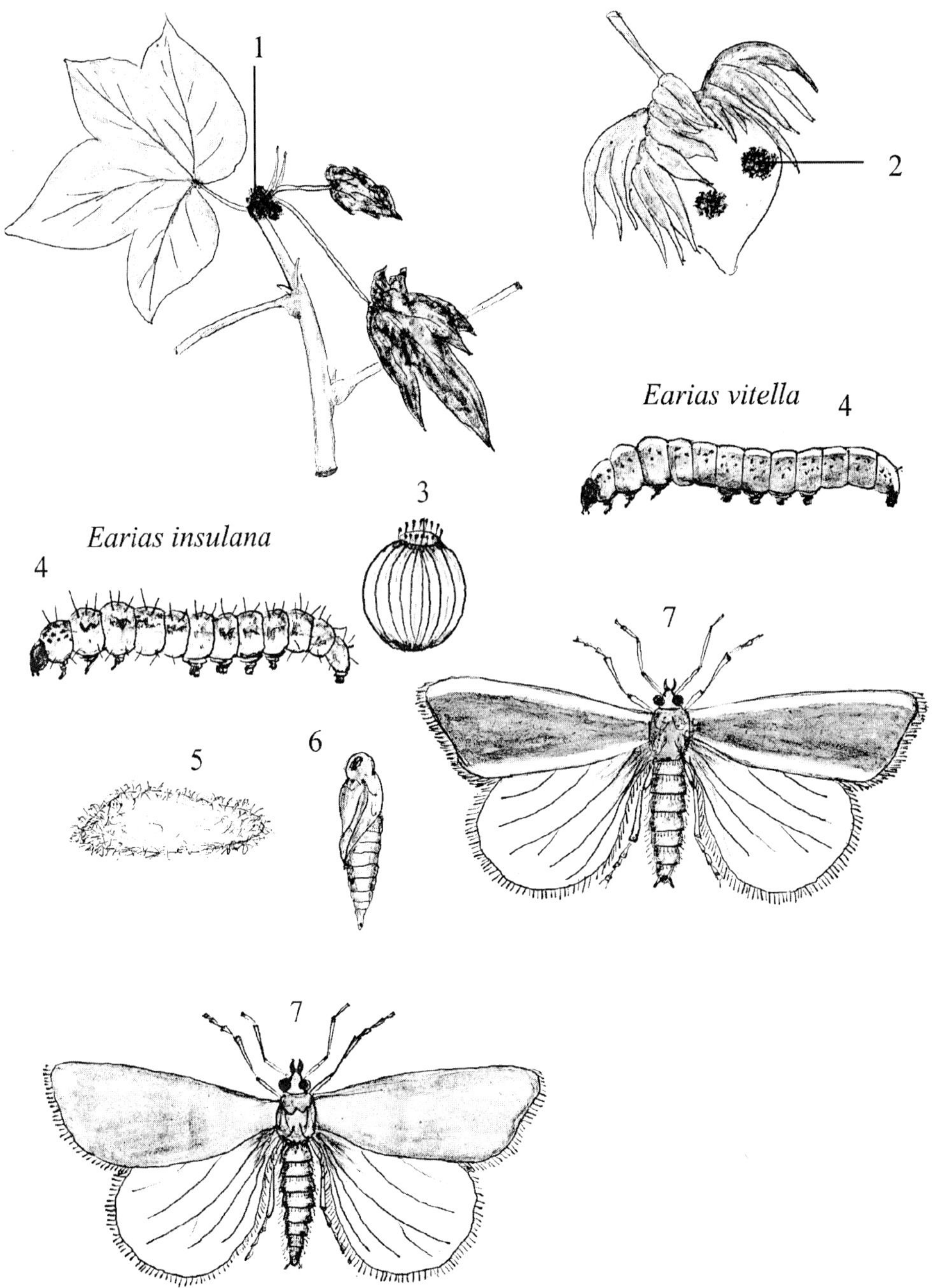

Fig. 60. Cotton spotted boll worms-*Earias species*

1. Infested shoot 2. Infested boll 3. Egg 4. Larva 5. Cocoon 6. Pupa 7. moth

and a number of dark markings on the body, finger-like short hairs and orange-colored spots on the prothorax. The larva of *Earias vitella* is light brownish in color with a black head, a median longitudinal whitish streak and dark spots on the body. Before pupation, the larva comes out, constructs a boat-shaped cocoon with dirty white and pale brown silken threads and pupates inside it. The pupa is purplish-brown in color. The cocoons are found on the surface of bolls, bracts and stems or on the ground among the shed leaves and flowers. The pupal stage lasts for 8-10 days in warm weather, about 3 weeks in autumn and 6-12 weeks during the winter season. The moths measure about 2.5 cm. across the wings. In *Earias insulana,* the forewings are completely green, whereas in *Earias vitella,* the forewings are pale white with a broad greenish band in the central portion from the base up to the lateral margin. The hindwings are white in color in both the species (Fig. 60).

Control measures

Cultural methods (i) The infested shoots should be removed and destroyed (ii) The flower buds, flowers, squares and young bolls, which have fallen down should be collected and destroyed periodically (iii) The crop should be ploughed or uprooted immediately after completion of harvest.

Chemical control (i) If the pest appears at the flowering time, foliar spraying may be given with endosulfan-800 ml. in 400 liters of water using a high volume sprayer (ii) If the pest occurs during the time of square formation, spraying with phosalone-800 ml. or quinalphos-800 ml. or carbaryl 50 WP-800 gm. or monocrotophos-400 ml. or chlorpyrifos-800 ml. or the synthetic pyrethroids, fenvalerate-150 ml. or permethrin-160 ml. or cypermethrin-100 ml. or decamethrin-200 ml. in 400 liters of water per acre may be taken up with a high volume sprayer. The sprayings should be continued at 15 days interval, if necessary. Use of different insecticide for each spraying affords better control of the pest.

When sucking pests such as white flies, aphids or spider mites are found along with the borer pests, use of synthetic pyrethroids should be restricted to one or two sprayings and a spraying with an organo phosphorus insecticide should follow this. Insecticides such as quinalphos, fenthion, methyl parathion and synthetic pyrethroids may induce resurgence in the case of sucking pests.

Biological control (i) *Trichogramma chilonis* and *T. evanescens* parasitize the eggs and destroy them. *Trichogramma chilonis* should be released at the rate of 50,000 parasitoids per acre per release and three releases should be made at an interval of 15 days commencing from 45 days after sowing (ii) The larval parasitoids-*Microbracon lefroyi, M. greeni, M. hebetor* and *M. brevicornis* parasitize and destroy the larvae (iii) The pupal parasitoid-*Chelonus rufus* attacks and destroys the pupae (iv) The mud wasp collects the larvae and uses them as food for its developing grubs.

2. Cotton pink boll worm

Platyedra (Pectinophora) gossypiella

Order - Lepidoptera

Family - Gelechiidae

Nature of damage. The pest has a very wide distribution all over the world and is considered to be one of the worst pests of cotton. In India, it is distributed in all the cotton growing tracts and is most destructive particularly in the states of Uttar Pradesh, Punjab, Tamil Nadu, Telungu Desam and Maharashtra. The caterpillars bore into the flower buds, flowers, squares and bolls, feed on the inner contents and cause severe damage. But the pest does not generally attack the shoots. The infested flower buds do not open normally and present a characteristic rosetted appearance. The caterpillar bores through the tip of young bolls and when the boll develops, the borehole is naturally closed and it is impossible to know whether the boll has been attacked by the pest. Inside the boll the caterpillar feeds on the immature seeds and lint. The larva burrows through the septa of the boll and moves from one locule to the other and feeds on all the seeds. The attacked young boll may fall down. But the developed bolls remain in the plant and burst prematurely. The cotton lint is stained, discolored and the quality is also poor. The matured seeds are also burrowed, the contents eaten and hollowed out. In case of severe incidence, 40-85 per cent of the bolls are affected and destroyed. The oil content and germination capacity of the seeds and the quality of the lint are badly affected. Each boll may have up to 10 larvae inside. Besides cotton, the pest infests lady's finger, hollyhock, *Thespesia populnia* and few other host plants (Fig. 61).

Life cycle of the pest. The female moth lays whitish, elongated, flattish eggs singly or in batches of 2-10 on the flower buds, flowers, bracts and young bolls. One insect lays up to 125 eggs. The eggs hatch in 4-25 days. The young caterpillar is pale pinkish in color with a brown head. It becomes full-grown in 25-35 days. The grown-up larva is 1.5-1.6 cm. in length, uniformly pink in color with a brownish head and has a thick, brownish, shield-like skin on the prothorax. The grown-up caterpillar pupates inside the boll or often inside the hollowed out seed and is enclosed within a loosely constructed silken cocoon. The pupal period lasts for 8-20 days. The moth is small, about 1.25 cm. across the wings and dark brownish in color. The brown-colored forewings have many wavy lines and black spots. The hindwings are gray in color. The lateral wing margins are fringed with long, brown hairs. The moth is active after dusk and is attracted by light (Fig. 61).

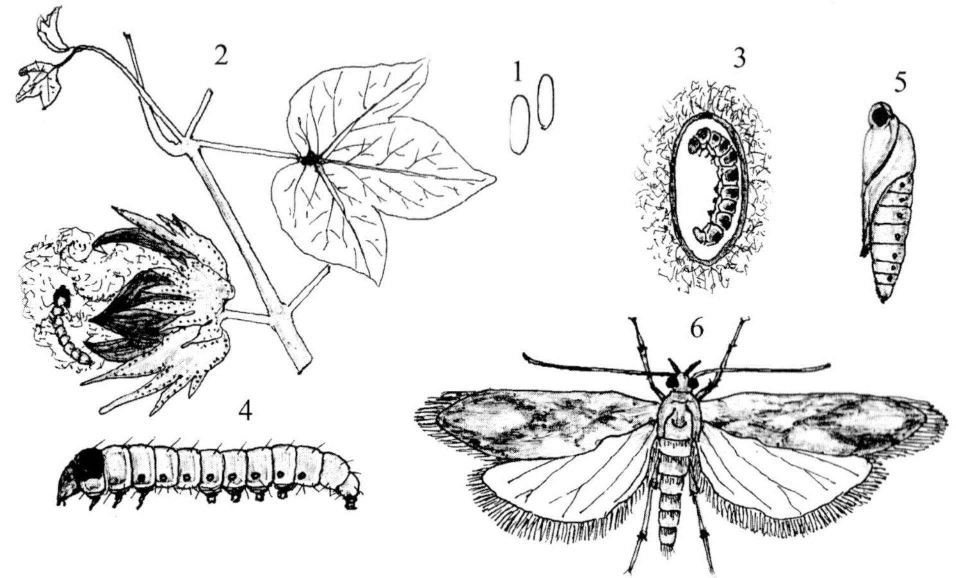

Fig. 61. Cotton pink boll worm-*Pectinophora gossypiella*

1. Eggs 2. Larva inside a boll 3. Larva inside a seed 4. Larva 5. Pupa 6. Moth

Control measures

Cultural methods (i) The infested flower buds, flowers, squares, young bolls etc., which have fallen down should be collected and destroyed periodically (ii) The cotton plants should be removed and disposed off as soon as the harvest is completed (iii) Ratooning should be avoided (iv) Light traps may be set up to attract and destroy the adult moths (v) Pheromone traps with 'Gossyplure' as a lure may be set up at the rate of 5 per acre to attract and destroy the male moths.

Seed treatment (i) The cotton seeds should be thoroughly dried in sunlight or treated in hot air at 60°C for a few minutes before storing to destroy the larvae and pupae that may be present inside the seeds (ii) Seeds intended for sowing purposes may be subjected to acid delinting treatment with concentrated sulfuric acid (iii) Fumigation of seeds with Methyl bromide at 70-90 ml./cu. ft for 48 hours or Aluminium phosphide at the rate of one 3.0 gm. tablet for every 1,000 kg. of seeds for 5 days is also effective in eliminating the larvae present inside the seeds.

Chemical control (i) If the pest occurrence is noticed at the time of flowering, spraying with endosulfan-800 ml. in 400 liters of water per acre may be taken up with a high volume sprayer (ii) If the pest occurs at square formation stage, spraying with phosalone-800 ml. or quinalphos-800 ml. or carbaryl 50 WP-800 gm. or monocrotophos-400 ml. or chlorpyrifos-800 ml. or synthetic pyrethroid, fenvalerate-150 ml. or permethrin-160 ml. or cypermethrin-100 ml. or decamethrin-200 ml. in 400 liters of water per acre may be taken up with a high volume sprayer. The sprayings should be repeated at 15 days interval. Different insecticides may be used for every spraying. Care should be taken in the selection of insecticides, whenever sucking pests are also present.

Biological control (i) The egg parasitoid-*Trichogramma chilonis* parasitizes the eggs and destroys them. The parasitoids should be released at the rate of 50,000 parasitoids per acre per release. A total of 3 releases should be made at 15 days interval, commencing from the time of flowering (ii) The larval parasitoids-*Bracon kitcheneri, Microbracon lefroyi, M. hebetor* and *M. greeni* parasitize and destroy the larvae (iii) The bacterial parasite-*Bacillus cereus* affects the larvae and destroys them.

3. American cotton boll worm

Helicoverpa (Heliothis) armigera

Order	-	Lepidoptera
Family	-	Noctuidae

Nature of damage. This polyphagous pest is also known as **'Gram green caterpillar'** and **'Gram pod borer'**, as gram is also a favorite host for the pest. Because of its large-scale attack of cotton crop in the United States of America, it is commonly called as the **'American boll worm'**. It is widely distributed throughout the world. The pest attacks and destroys gram plants and pods, cotton shoots and bolls, red gram pods and top shoots, soybean shoots and pods, maize cobs, tomato fruits, sorghum seeds, lady's finger fruits, chilli fruits, tobacco leaves, black gram and green gram shoots and pods, potato shoots and leaves etc.

The young caterpillars, which hatch out of the eggs, initially scrape and feed on the tender as well as slightly older leaves. Later they bore and feed on the flower buds, flowers, squares and tender bolls. A single caterpillar damages several flower buds. The infested flower buds, flowers, squares and tender bolls drop off. The growing caterpillar bites a big hole near the tip of the boll, puts its head inside and feeds on the immature seeds, while the rest of the body remains outside the boll. The larva never enters the boll completely. The faeces of the larva are found below the infected bolls in large quantities as pellets. The infested matured bolls do not fall down, but remain in the plant, rot or dry.

Life cycle of the pest. The female moth lays spherical, light yellow-colored eggs with longitudinal parallel lines singly on the flower buds, flowers and tender bolls. One moth lays 300-1,000 eggs. The eggs hatch in 4-12 days. The young caterpillar hatching out of the egg is whitish-yellow in color, with a dark brown head, minute black dots on the body and thin hairs on the dots. It becomes full-grown in 15-25 days. The grown-up larva is large in size, 4.0-5.0 cm. in length, cylindrical, and green in color with two lateral yellowish lines and a gray dorsal line, and thin hairs on the body. The full-grown larva pupates under the soil in an earthen cocoon. The pupa is brown in color and 1.2-1.3 cm. in length. The adult moth emerges from the pupa in 10-25 days and comes out of the soil through the burrow made by the larva before pupation. The moth is large in size, stout, 3.5-3.75 cm. across the wing span and light brownish in color. The greenish-gray forewings have gray-colored wavy bands and black spots of varying sizes. The hindwings are whitish, with prominent black veins and a dark band along the margin (Fig.42).

Control measures

Cultural methods (i) The field should be ploughed well after harvest of the crop to bring out and destroy the pupae present underneath the soil (ii) Light traps should be set up to attract and destroy the adult moths (iii) Pheromone traps should be set up at the rate of 5 traps per acre to lure and destroy the male moths (v) The cotton plants should be uprooted and disposed off soon after completion of the harvest (vi) Ratooning should be avoided.

Chemical control (i) The chemical control methods suggested for the control of spotted boll worms may be adopted. However, the pest has developed resistance to synthetic pyrethroids and so, the use of such insecticides against the pest should be restricted.

Biological control (i) The egg parasitoids-*Trichogramma confusum* and *T. chilonis* are capable of parasitizing and destroying the eggs of the pest. *Trichogramma chilonis* may be released at the rate of 50,000 parasitoids per acre per release and 3 releases should be made at an interval of 15 days, commencing from 45 days after sowing (ii) *Bracon kitcheneri, Chelonus narayani* and a few other larval parasitoids, parasitize and destroy the larvae (iii) Nuclear polyhedrosis virus can control the larvae, especially the early instar larvae effectively. The viral spray fluid containing 200 larval equivalent (200 LE) of virus particles and Teepol-0.01% in 400 liters of water per acre should be sprayed on the foliage during the evening hours with a high volume sprayer (iv) Spraying a combination of 200 LE of Nuclear polyhedrosis virus particles, cotton seed kernel powder extract-1.0 kg., crude sugar-1.0 kg. and endosulfan-400 ml. in 400 liters of water per acre controls the pest very effectively.

4. Tobacco caterpillar or Cut worm

Spodoptera (Prodenia) litura

Order - Lepidoptera

Family - Noctuiidae

Nature of damage. The caterpillars hide in cracks and crevices in the field or under mud clods during the daytime. They come out in swarms of thousands after dusk and destroy the crop. The caterpillars bite and feed on the leaves of seedlings and cut the stem portion from the bottom, leading to death of the seedlings. In the grown-up plants, the larvae appear in large numbers and feed on the entire foliage leaving the plants barren. During the matured stage of the crop, the caterpillars bore into the bolls, feed on the inner contents and cause severe damage. Besides cotton, the polyphagous pest attacks tobacco, groundnut, castor, gingelly, maize, soybean, sunflower, tomato, cabbage, sweet potato, banana, chillies and many other crops.

Life cycle of the pest. The female moth lays small, spherical eggs on the leaves of host plants or weed plants in clusters of 200-300 and covers them with gray-colored anal hairs. The eggs hatch in 4-5 days. The young caterpillars emerging from the eggs are greenish in color. They become full-grown in 14-21 days. During the growth period of the larva its color continues to change. The grown-up larva is 3.5-4.0 cm. in length, stout, cylindrical, smooth, and generally pale greenish-brown in color with dorsal and subdorsal longitudinal black lines. There may be lots of variations in the color pattern of the larvae. The full-grown larva pupates under the soil in an earthen cocoon and emerges as an adult moth in 10-15 days. The moth is large-sized, stout, 3.5-5.0 cm. across the wing span and brownish in color. The blackish forewings have yellow-colored cross bands and markings. The hindwings are whitish in color with a dark brown band along the margin (Fig.68).

Control measures

Cultural methods (i) Castor may be grown as a trap crop along the border of the field or as an intercrop. The caterpillars coming out in large numbers after dusk and attacking the taller castor crop may be easily collected and destroyed (ii) The egg masses on the leaves of the host plants may be collected and destroyed (iii) Trenches may be dug around heavily infested fields and water is let in to stagnate in the trenches. A small quantity of kerosene oil or crude oil or insecticide emulsion is added to the water. The caterpillars, which try to cross over to the adjoining fields, fall into trenches and are killed (iv) Pieces of blue cloth may be spread in the field in different places in the evening. The larvae, which hide under the blue cloth may be collected during the next morning and destroyed.

Setting up traps (i) Pheromone traps with **'Pherodin SL'** as a sex lure may be set up in the field to attract and destroy the male moths. The septa impregnated with the pheromone should be replaced once in 3 weeks (ii) Poison baits may be used to attract and destroy the caterpillars. To prepare poison bait, rice bran-5.0 kg., molasses or sugar syrup-1.0 kg. and carbaryl 50 WP-500 gm. are mixed with enough water to form a thick dough and then it is made into several small balls. These balls are placed along the bunds and inside the field in many places at random to cover 1.0 acre of field during the evening hours. The caterpillars, which are attracted to the bait, eat them and are killed.

Chemical control (i) The young caterpillars up to the third instar can be controlled by spray application of fenitrothion-400 ml. or dichlorvos-400 ml. or quinalphos-400 ml. or endosulfan-400 ml. in 400 liters of water per acre with a high volume sprayer (ii) The fourth to sixth instar grown-up caterpillars can be controlled by spraying with chlorpyrifos-800 ml. or dichlorvos-400 ml. or phenthoate-800 ml. in 400 liters of water per acre. The pest has developed resistance to carbaryl.

Biological control (i) The egg parasites-*Trichogramma evanescens* and *T. minutum* parasitize and destroy the eggs (ii) The mud wasps are predatory on the larvae in the sense, they use the larvae for feeding their young ones (iii) Nuclear polyhedrosis virus capable of infecting and killing *Helicoverpa* larvae may be used to destroy the larvae. The virus spray fluid containing 100 larval equivalent of virus particles (100 LE) and crude sugar-1.0 kg. is mixed with 300 liters of water and sprayed on the foliage with a high volume sprayer during the evening hours.

5. Red cotton bug

Dysdercus koenigii

Order - Hemiptera
Suborder - Heteroptera
Family - Pyrrhocoridae

Nature of damage. It is a serious pest of cotton in Tamil Nadu, Telungu Desam, Uttar Pradesh, Bihar and Maharashtra. Both the nymphs and adults pierce, suck and feed on the sap from young, immature bolls. Such infested bolls fail to mature normally and the cotton lint is stained and of poor quality. The seeds from such bolls are unfit for sowing, for oil extraction or for use as cattle feed. The nymphs sit on the cotton of opened bolls and stain the lint with their excreta. Through the punctures made by the insects, a few saprophytic fungi and bacteria also enter into the bolls and further spoil the lint (Fig. 62).

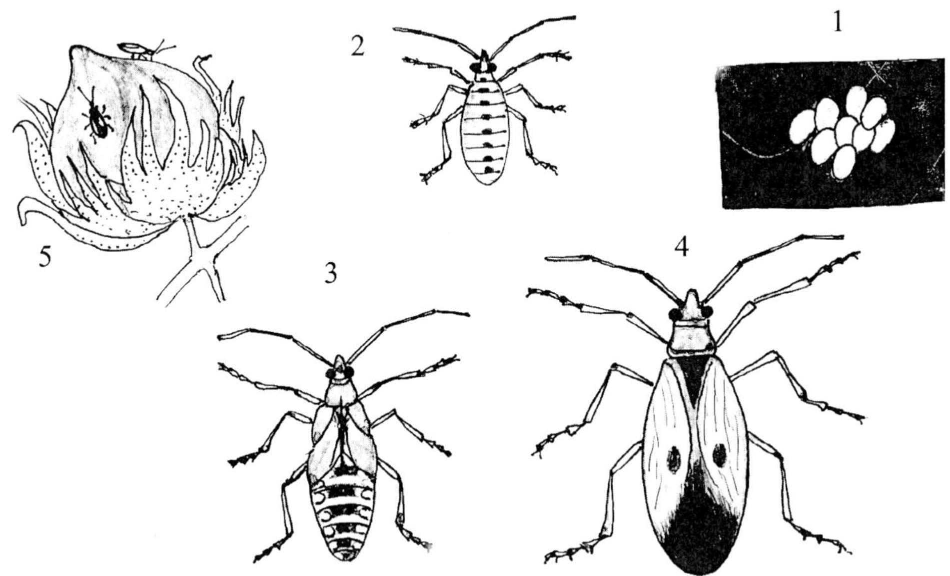

Fig. 62. Red cotton bug-*Dysdercus koenigii*

1. Eggs 2,3. Nymph 4. Full grown adult bug 5. Bugs feeding on the boll

Life cycle of the pest. The female bug lays whitish-yellow, oval-shaped eggs in clusters of 70-80 eggs inside cracks and crevices in the soil. The eggs hatch in about 7 days. The young nymph is pinkish in color with a row of black spots on the median dorsal surface and a row of white spots on both the lateral margins of the abdomen and is wingless. They are very active and hop about sucking and feeding on the sap from young, tender bolls. During their growth, the nymphs moult 5 times and reach the winged, adult stage in 50-90 days. The full-grown adult bug is about 1.5 cm. in length, red-colored, with black eyes and antennae. A black, inverted triangular spot covers the thoracic region and the tip of the forewings is blackish in color. The thoracic and abdominal segments have a row of small, white bands on both the lateral margins. The pest occurrence is more severe during the months of August-October. The pest attacks lady's finger crop also (Fig.62).

Control measures

Cultural methods (i) The field may be ploughed well to bring out the eggs from inside the soil and to destroy them (ii) The nymphs and adults may be caught by using hand nets and destroyed.

Chemical control (i) Dusting with carbaryl 10 D at 10 kg/acre, so as to give a thorough covering of the bolls is effective in controlling the pest

(ii) Spraying with endosulfan-800 ml or fenitrothion-800 ml. or phosphamidon-220 ml. in 400 liters of water per acre with a high volume sprayer, so as to give a thorough covering of the bolls is more effective in controlling the pest

6. Cotton white fly

Bemisia tabaci

Order - Hemiptera

Suborder - Homoptera

Family - Aleyrodidae

Nature of damage. In India, this pest is prevalent in all cotton growing states and sometimes causes severe damage to the crop. The pest attacks the crop at all growth stages however, when the young plants are infested, the damage caused is much more severe. Both the nymphs and adults harbor on the under surface of leaves in large numbers, suck and feed on the sap continuously, as a result the leaves turn yellow and dry. In case of severe infestation, the leaves, flowers and squares drop down in large numbers.

Besides cotton, the polyphagous pest attacks several other crops such as, lady's finger, hollyhock, tobacco, sunflower, beans, tapioca, soybean etc. In tobacco, tapioca, cotton and soybean, the pest is the vector of the virus that causes **'leaf curl disease'**, while in lady's finger, it is the vector of the virus that causes **'vein clearing disease'**. The honeydew secreted by the pest invites several other insects and ants, which come to feed on the sugary secretion. The fungal disease, **'sooty mould'** occurs on the surface of the leaves covered with the honeydew.

Life cycle of the pest. The mother insect lays minute eggs in rows of 3-50 eggs in each row on the under surface of tender leaves. The eggs hatch in 3-5 days during the summer months and in 9-14 days during the cold season. The young nymphs are very small, delicate, oval in shape, wingless and their body covered with a white, powdery, waxy coating. The nymphs become full-grown in 5-33 days in summer and 17-73 days in the winter season. During their growth, the nymphs moult 3 times. After the third moult, the nymphs loose their legs and antennae and attain a pupal stage and after 2-6 days the adult white flies emerge from the pupae. The life cycle is completed in 20-107 days. The adult white fly is very small in size, whitish in color with whitish wings and the body and the forewings are covered with a whitish, powdery, waxy coating (Fig. 63).

Control measures

Cultural methods (i) The cotton plants should be uprooted and disposed off soon after the harvest is completed (ii) Ratooning should be avoided (iii)

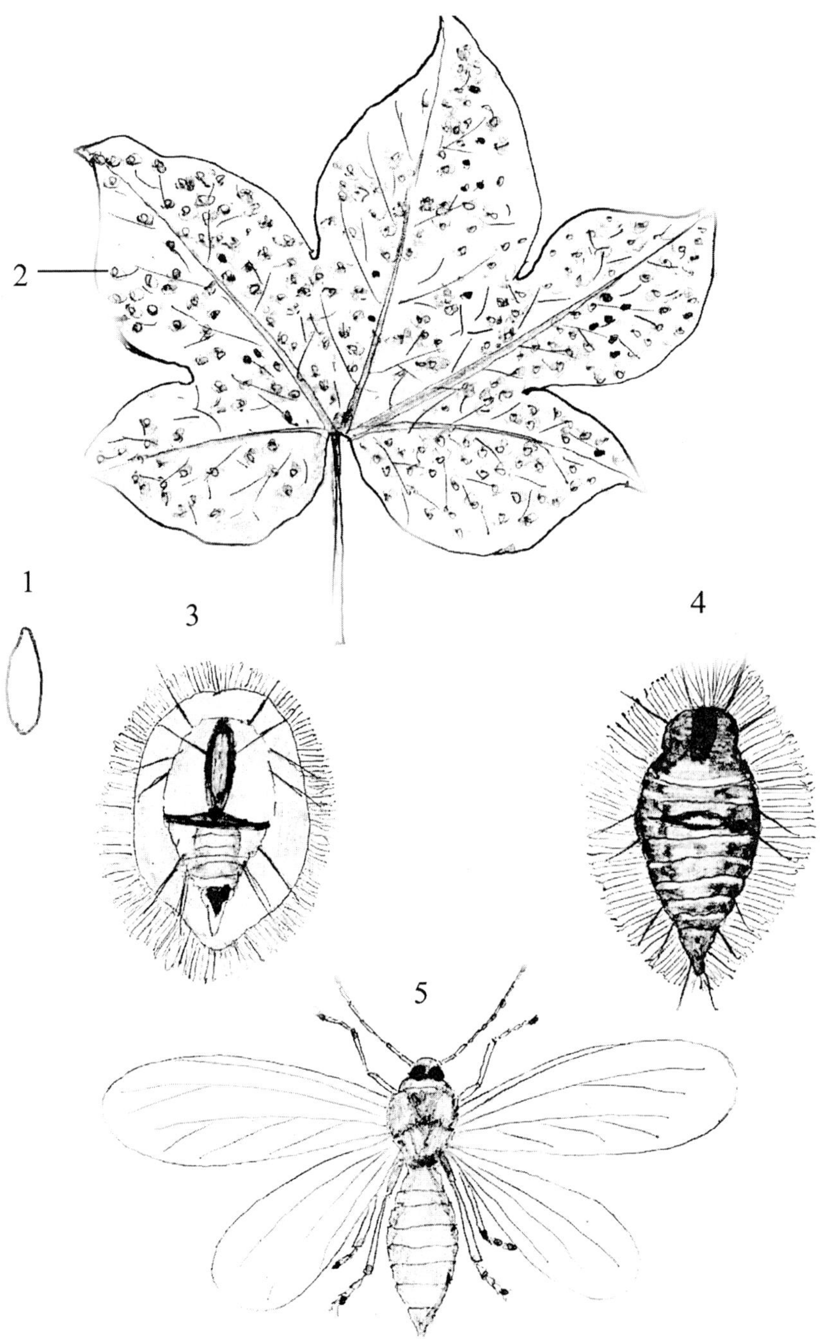

Fig. 63. Cotton white fly-*Bemisia tabaci*
1. Egg 2. Nymphs and adults feeding on the leaves 3. Nymph 4. Pupa 5. Adult white fly

Yellow sticky traps should be set up in the field at the rate of 10 traps per acre (iv) Application of nitrogenous fertilizer in excess of the recommended dose should be avoided.

Chemical control. Spraying the crop with methyl demeton-400 ml. or dimethoate-400 ml. or phosphamidon-220 ml. or acephate-800 gm. or ethion-400 ml. or neem oil-2% or neem seed kernel extract-5% in 400 liters of water per acre with a high volume sprayer, so as to cover the under surface of leaves thoroughly controls the pest effectively.

In fields infested by white flies, repeated sprayings with quinalphos, fenthion or synthetic pyrethroids should be avoided, as these insecticides may induce resurgence of the pest.

7. Cotton aphid

Aphis gossypii

Order	-	Hemiptera
Suborder	-	Homoptera
Family	-	Aphididae

Nature of damage. The pest is found in all the cotton-growing countries of the world. Both the nymphs and adults infest the tender portions of the plants in colonies and with their well-developed proboscis pierce, suck and feed on the plant sap. The continuous drainage of sap caused by thousands of the insects at varying stages of development produces a very unhealthy effect on the plants. The leaves become curled up, the young shoot fades gradually and the whole plant becomes very much stunted. The insects also attack the flower buds and flowers, suck and feed on the sap. The insect secretes honeydew through the two honey tubes, which project from the posterior end of the abdomen, which attracts many other insects that come to feed on the sugary solution. **'Sooty mould'**, a fungal disease occurs on the leaf surface bolls covered with the honeydew. The pest also infests lady's finger, eggplant, chillies, guava, gingelly and a few other hosts.

Life cycle of the pest. The grown-up insects are brownish-green in color, very small, soft-bodied and delicate. The males are winged, whereas in the females both winged and wingless forms are found. The females are bigger in size than the males and have a much bigger abdomen. The females are parthenogenetic and viviparous. One female lays 8-22 young ones and they become full-grown adults capable of reproducing in 7-9 days and thus the pest multiplies very rapidly (Fig. 64).

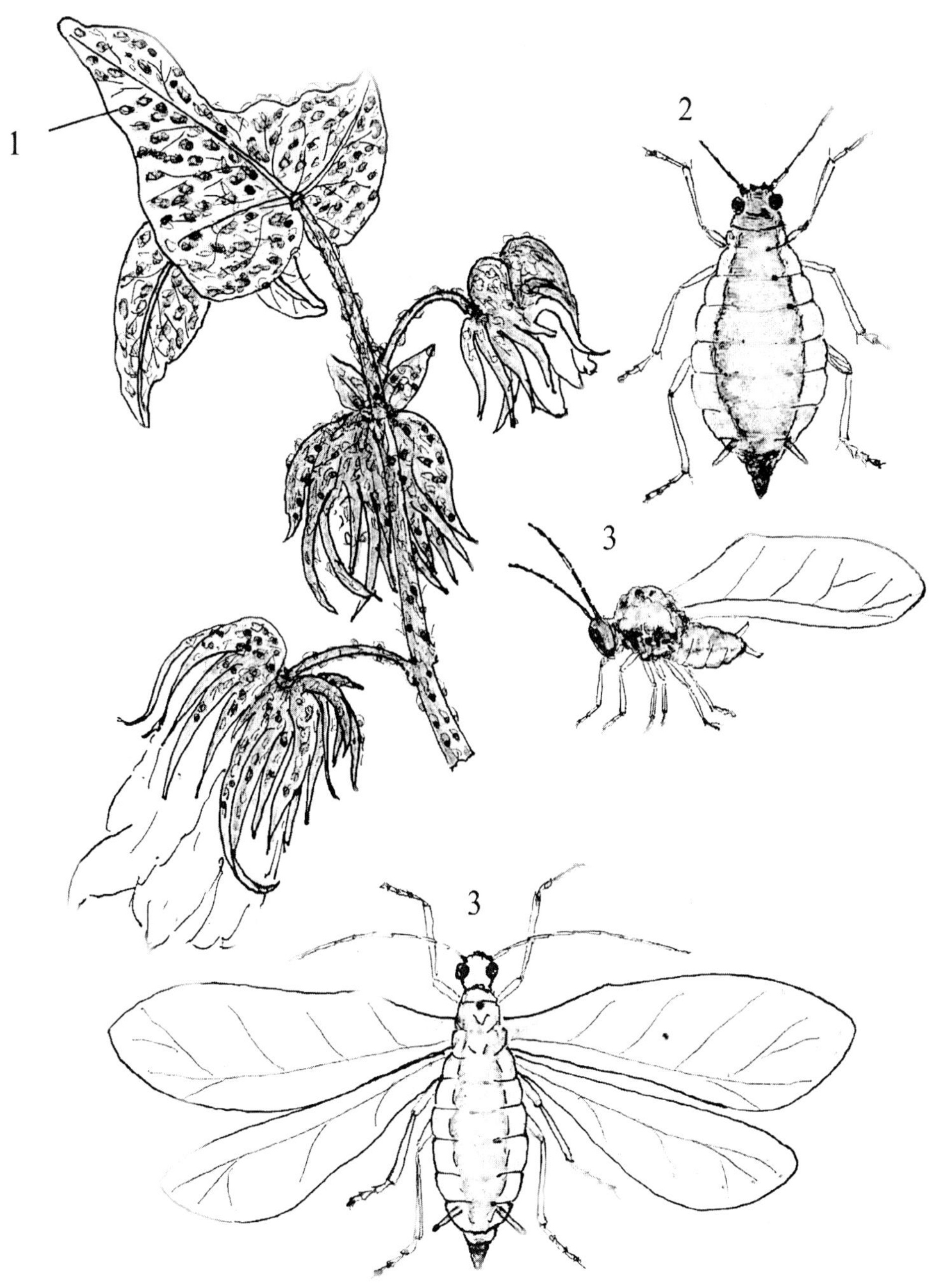

Fig. 64. Cotton aphid-*Aphis gossypii*

1. Nymphs and adults feeding on the shoot 2. Nymph 3. Full grown aphid

Control measures

Cultural methods. Severely affected parts of the plants may be removed along with the insects and destroyed

Chemical control. Spraying the foliage with methyl parathion-400 ml. or phosalone-800 ml. or methyl demeton-400 ml. or monocrotophos-400 ml. or phosphamidon-220 ml. or dimethoate-400 ml. or neem oil-2% in 400 liters of water per acre with a high volume sprayer controls the pest.

Biological control. The lady bird beetle-*Coccinella septum punctata* is predatory on the pest.

8. Cotton stem weevil

Pempherulus affinis

Order - Coleoptera

Family - Curculionidiae

Nature of damage. The pest generally appears in the summer season and is common in all cotton-growing tracts of South India. Only the grubs are capable of causing injury to the plants. The young grub burrows into the stem, especially between the bark and stem and feed on the soft, inner tissues. When 15-30 days old young plants are attacked they generally die. But the older plants may survive the pest attack, though they loose their vigor. External symptoms are visible as characteristic nodular and irregular gall-like swellings on the stem generally at the lower portion of the stem just above ground level. During strong winds, severely affected plants may lodge and break at the region where the swellings are formed. The pest also attacks lady's finger and shoe flower plants (Fig. 65).

Life cycle of the pest. The female weevil deposits small, oval, shining, whitish, smooth eggs singly under the surface of the bark of the stem just above the ground level in a small cavity excavated by its snout. Similarly 7-8 eggs may be laid in a single stem. One female lays about 50 eggs. The eggs hatch in 6-10 days. The small, pale whitish-yellow grub hatching out of the egg, bites through into the region between the bark and the main stem, makes irregular galleries in the stem and feed on the soft inner tissues. The infestation causes abnormal growth of the tissues of the attacked stem portion, resulting in the formation of irregular, gall-like swellings. Several grubs may be found in an infested stem. The grub becomes full-grown in 35-57 days and pupates inside the gallery inside the stem. The adult weevil emerges from the pupa in 25-30 days. The weevil is small-sized, about 0.3 cm. in length and dirty brown in color, with a conspicuous, long, downward-curved snout. The weevil lives for more than 30 days. There may be 3 generations during the period from October to April (Fig. 65).

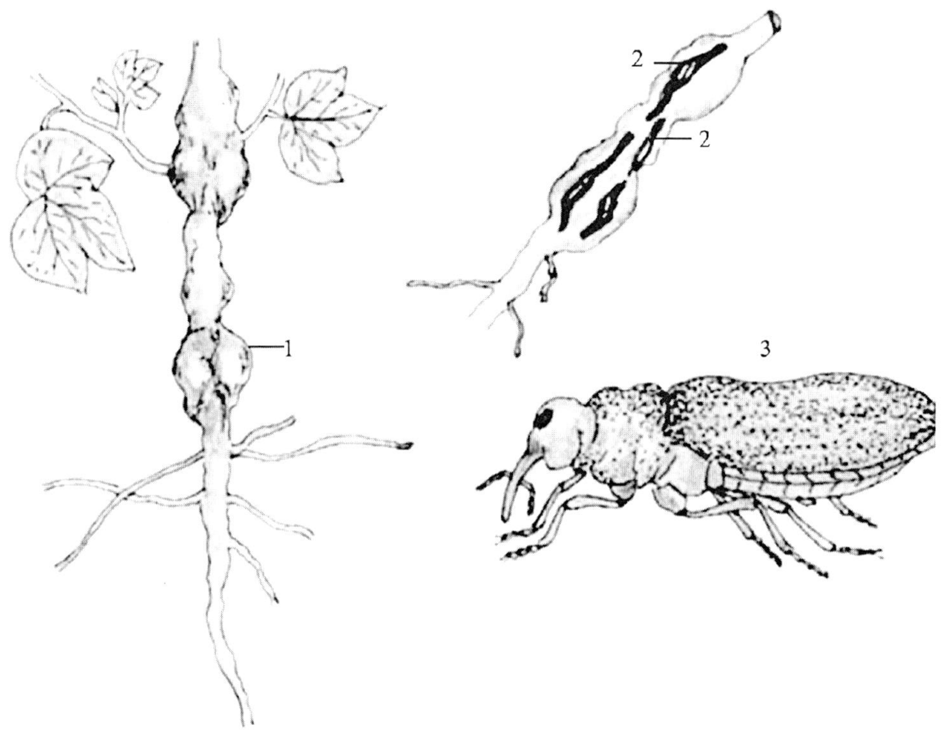

Fig. 65. Cotton stem weevil-*Pempherulus affinis*

1. Gall-like swellings 2. Larva and pupa inside the gall 3. Adult weevil

Control measures

Cultural methods. Infested plants should be removed and destroyed.

Chemical control (i) Spraying with dieldrin-800 ml. or methyl parathion-400 ml. or carbaryl 50 WP-800 gm. or chlorpyrifos-800 ml. in 400 liters of water per acre with a high volume sprayer, so as to give a thorough covering of the stem portion, especially near the ground level is effective in controlling the pest (ii) Application of carbofuran 3 G-13 kg. or aldicarb 10 G-4.0 kg., mixed with 10 kg. of sand per acre uniformly on the soil followed by irrigation also controls the pest.

9. Cotton leaf hopper or Cotton jassid

Amrasca biguttula biguttula

Order - Hemiptera, Suborder-Homoptera,

Family - Cicadellidae

Nature of damage. The nymphs, as well as the adult hoppers occupy the under surface of leaves in large numbers, pierce, suck and feed on the sap from the leaves. The infested leaves turn yellow and start curling downwards. In severe cases of infestation, the leaves turn chocolate-brown in color, present a burnt-up appearance and crumple. This is commonly known as **'hopper burn injury'**. The growth of the infested plants are also very much affected. The insects, while feeding on the sap, inject some toxic substances into the plants along with the saliva, which also leads to yellowing, cupping or curling downwards of leaves, reddening and drying of leaves in patches, which are characteristic of hopper burn injury. The pest infests other crops such as, lady's finger, hollyhock, eggplant, potato, sunflower etc.

Life cycle of the pest. The female hopper inserts eggs in rows of about 18 eggs in the tender stem portion or in the midrib of leaves. One hopper lays 200-300 eggs. The eggs hatch in 4-11 days. The young nymphs moult 4 times during their growth period and reach the winged adult stage in 10-21 days. The adult hopper is small, thin and greenish in color.

Control measures. The control measures suggested for the control of cotton aphids may be adopted.

10. Cotton leaf roller

Sylepta derogata

Order - Lepidoptera

Family - Pyraustidae

Nature of damage. The pest is found in all the cotton-growing tracts of South India. Infested plants can be easily spotted by the presence of numerous characteristic leaf rolls on the plants and when such a leaf roll is opened, the caterpillar or the pupa is found inside. The young caterpillar emerging from the egg starts scraping and feeding on the green matter of tender leaves. Later, it webs a nest around it with silken thread, stays inside it, scrapes and feeds on the leaf tissues. Then it rolls and webs the leaf like a funnel, stays inside, bites holes and feeds on the leaf tissues. Sometimes the adjacent leaves are also joined and spun together to form a nest and from within, the larva feeds on the leaf tissues. Only one larva is found inside each leaf roll (Fig.66)

Life cycle of the pest. The female moth lays smooth, flattish, pale white eggs singly on the under surface of leaves near the midrib. The eggs hatch in 4-6 days. The small, greenish-yellow caterpillars, which emerge from the eggs become full-grown in 15-18 days. The grown-up larva is about 2.5 cm. in length, glistening green in color with a dark brown head and prothorax. It pupates within the leaf roll or in-between fallen leaves on the ground. The

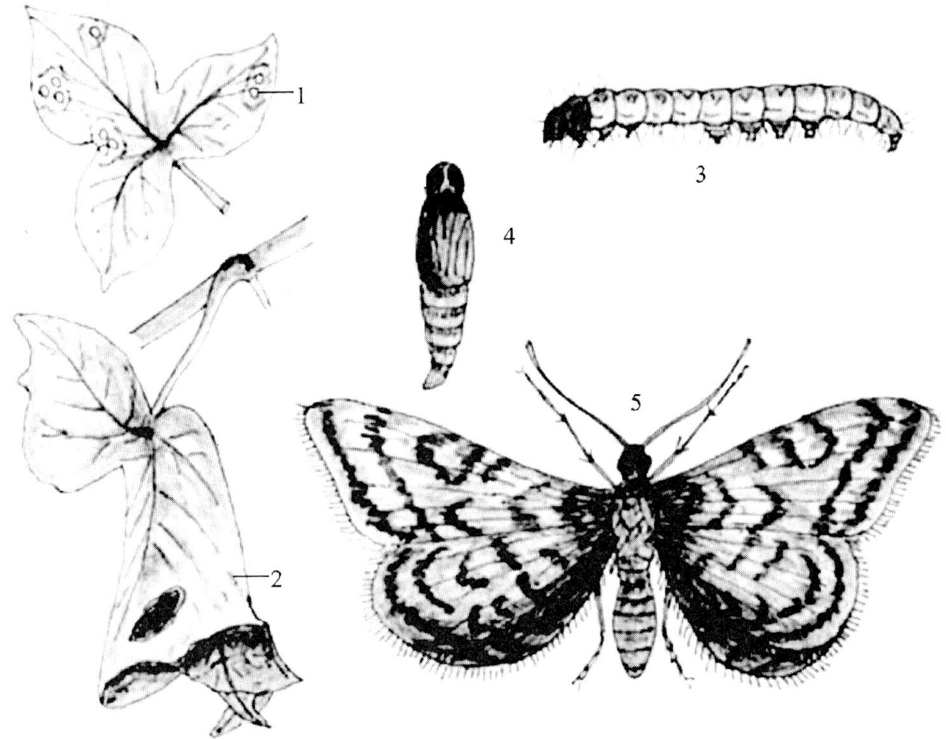

Fig. 66. Cotton leaf roller-*Sylepta derogata*

1. Eggs 2. Larva inside the leaf roll 3. Larva 4. Pupa 5. Adult moth

pupa is reddish-brown in color. The pupal stage lasts for 7-8 days. The adult moth is medium-sized, 3.0-3.5 cm. across the wing span and whitish-yellow in color. The yellowish wings have many dark brownish, thin, wavy markings across. There may be many generations in a year. The pest also infests lady's finger, hollyhock and a few other hosts belonging to Malvaceae (Fig.66)

Control measures

Cultural methods (i) Growing alternate hosts near about cotton fields should be avoided (ii) The leaf rolls attached to the plants may be collected and destroyed (iii) The leaf rolls and other fallen leaves fallen on the ground may be collected and destroyed periodically.

Chemical control. Foliar spraying with endosulfan-400 ml. or fenitrothion-400 ml. or methyl parathion-400 ml. or phosalone-800 ml. in 400 liters of water per acre with a high volume sprayer is effective in controlling the pest.

11. Ash weevil

Myllocerus maculosus

Order - Coleoptera

Family - Curculionidae

Nature of damage. Both the grubs and adult weevils cause damage. The grubs cut and feed on the roots, while the weevils feed on the margin of leaves, flower buds and occasionally the flowers. Infested plants may wilt and dry up. The polyphagous pest attacks red gram, finger millet, sorghum, pearl millet, maize, lady's finger, sugarcane etc.

Life cycle of the pest. The female weevil lays pale yellowish, oval eggs in small groups in a hole in the soil, about 8.0 cm. deep, dug by the weevil with its snout. One insect may lay up to 361 eggs. The eggs hatch in 3-11 days. On emergence from the eggs, the grubs start feeding on the roots. Full-grown grub pupates in the soil in an earthen cocoon and emerges as an adult weevil. The entire life cycle is completed in 42-54 days.

Control measures

Chemical control (i) Incorporating quinalphos 4 D or carbaryl 5 D or lindane 0.65 D at 20 kg./acre into the soil before sowing controls the pest (ii) Drenching the soil with a solution of chlorpyrifos-2,000 ml. or lindane-2,000 ml. in 1,000 liters of water per acre in the standing crop is also effective in controlling the pest.

12. Stem boring jewel beetle

Sphenoptera gossypii

Order - Coleoptera

Family - Buprestidae

Nature of damage. The grubs of this pest alone are responsible for causing injury to the plants. The grubs bore into the stem and feed on the soft inner tissues, as a result the plant starts wilting and in many cases the infested plants are killed. The insect belongs to the family of jewel beetles. The pest does more damage during years of drought.

Life cycle of the pest. The female beetle lays eggs singly on the bark of the tender stem region. The grubs emerging from the eggs bore into the stem and tunnel through the stem. The grub is pale whitish in color, elongated and with a flattish anterior region. The full-grown grub pupates inside the gallery. After the pupation period, the adult beetle emerges through a hole already made by the grub before pupation. The beetle is coppery-brown in color, shining and about 8.0 mm. in length.

Control measures

Cultural methods. The plants showing symptoms of infestation may be removed and destroyed.

Pests of minor importance. Besides the pests described above in detail, several other pests are also known to infest cotton, but they are of lesser importance. Cotton crop is infested by the flower thrips-*Thrips tabaci,* which lacerates, sucks and feeds on the sap from the floral parts leading to flower shedding; the leaf thrips-*Scirtothrips dorsalis* infest the under surface of young leaves, lacerate, pierce, suck and feed on the sap from the leaves and may cause severe damage to the plants; the hairy caterpillars-*Amsacta albistriga, Euproctis fraterna* and *Pericallia ricini* bite and feed on the foliage; the nymphs and adults of the dusky cotton bug-*Oxycarenus hyalinipennis (lactus)* feed gregariously on the sap of immature and partially matured bolls and seeds and impart a yellow tinge to the lint; the surface grass hoppers-*Chrotogonus trachypterus* and *C. oxypterus* cut the seedlings at the ground level resulting in the death of seedlings and later feed on the leaves of grown-up plants; the bud worm -*Phycita infusella* damages the buds; the grubs of the shoot weevil-*Alcidodes affaber* bore into the shoots and leaf axils; the semilooper-*Anomis flava* feeds on the leaves; the mealy bug-*Ferrisia virgata* sucks and feed on the sap from the leaves and young shoots.

TOBACCO *(Nicotiana tobacum)*

1. Tobacco aphid or Green peach aphid

Myzus persicae

Order - Hemiptera

Suborder - Homoptera

Family - Aphididae

Nature of damage. The pest occurs in all tobacco-growing countries of the world. In Tamil Nadu, the pest occurs in a severe form in the major tobacco-growing districts of Coimbatore, Thanjavur and Madurai. The nymphs as well as the adults harbor in crowded colonies on the young leaves and tender shoots, suck and feed on the plant sap. The infested leaves curl upwards, droop and turn brown. The infested plants are stunted in growth and look pale and unhealthy. The leaves become unfit for curing purposes and loose their economic value. The insects secrete honeydew through the two honey tubes protruding out from the posterior end of the abdomen, which attracts several other insects and ants that come to feed on the sugary secretion. The fungal disease, **'sooty mould'** may also appear on the leaves covered with honeydew, which further deteriorates the quality of the leaves for commercial

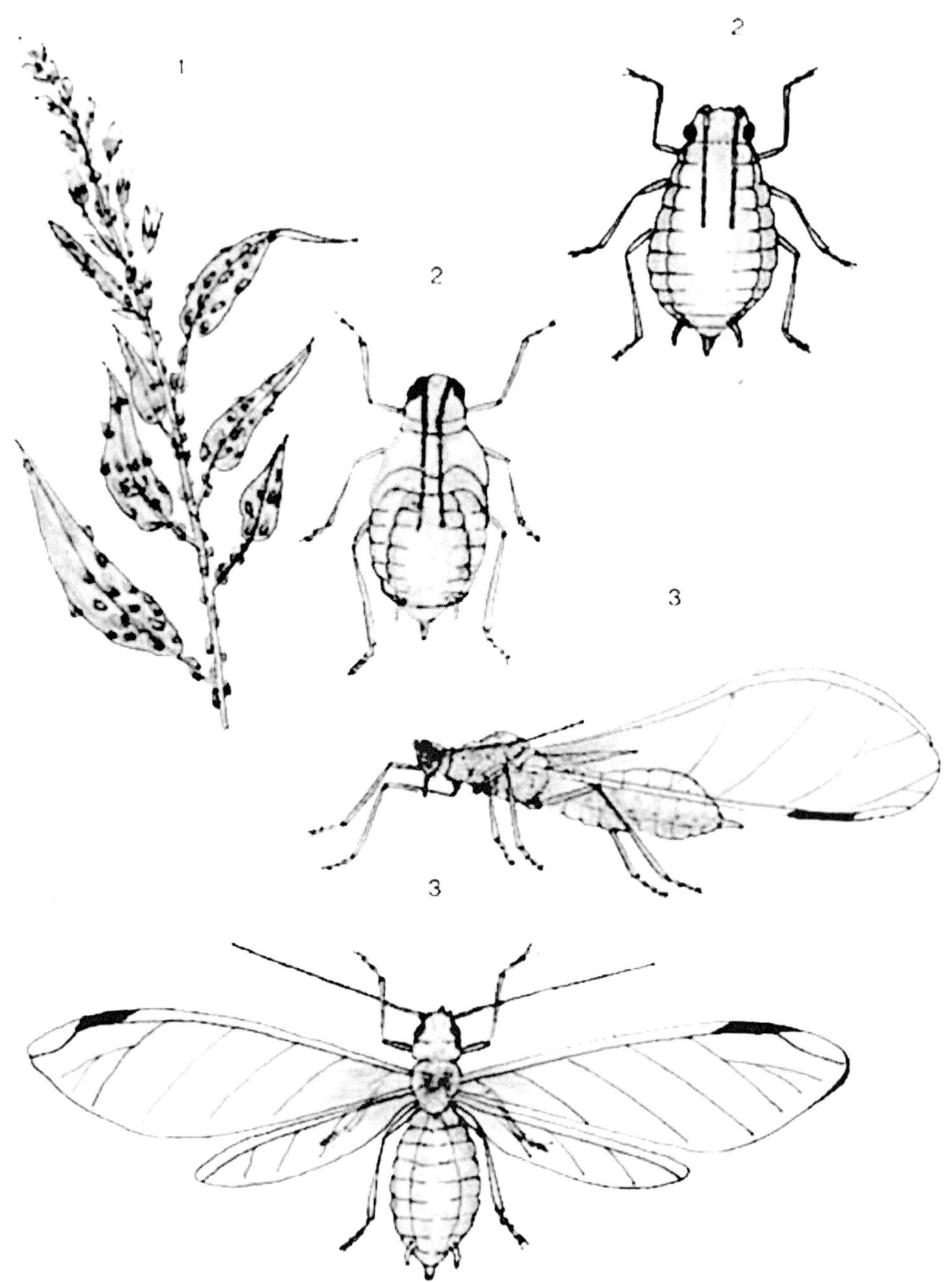

Fig. 67. Tobacco aphid-*Myzus persicae*

1. Nymphs and adults feeding on the shoot of mustard plant 2. Wingless adult 3. Winged adult

purposes. Besides causing direct damage to the crop, the pest is the vector of the virus, which causes **'leaf curl disease'**. The pest also infests potato, wheat, mustard and a few other Crucifers.

Life cycle of the pest. The females are capable of reproducing parthenogenetically and are viviparous. The adult female lays many young nymphs one by one, which are yellowish-green in color. Within a few days, the nymphs reach the adult stage and start reproduction as a result, within a short time the pest multiplies enormously. The adult aphids are green or brownish-green in color, very small and delicate insects. The males are winged and smaller than the females. The bigger females have a much larger abdomen than the males. Both winged and wingless forms are found among the females (Fig. 67).

Control measures

Physical methods. Severely affected plant parts may be removed along with the insects and destroyed.

Chemical control. Foliar spraying with methyl demeton-250 ml. or dimethoate-250 ml. or phosphamidon-120 ml. in 250 liters of water per acre with a high volume sprayer controls the pest.

2. Tobacco caterpillar

Spodoptera (Prodenia) litura

Order - Lepidoptera

Family - Noctuidae

Nature of damage. This is the most serious and common caterpillar pest of tobacco, which causes very serious damage to the seedlings in the nursery and also to the growing plants in the main field and hence it is commonly known as the **'tobacco caterpillar'**. The larvae hide in cracks and crevices in the soil and under mud clods during the daytime and come out of their hiding places after dusk in swarms, bite big holes in the leaves, feed on the leaves gregariously and within a few days all the leaves are completely eaten away. In the nursery, the caterpillars besides feeding on the leaves, cut the stem of seedlings from the bottom and cause extensive damage (Fig. 68).

Life cycle of the pest and **control measures.** Refer pests of cotton (Page 161).

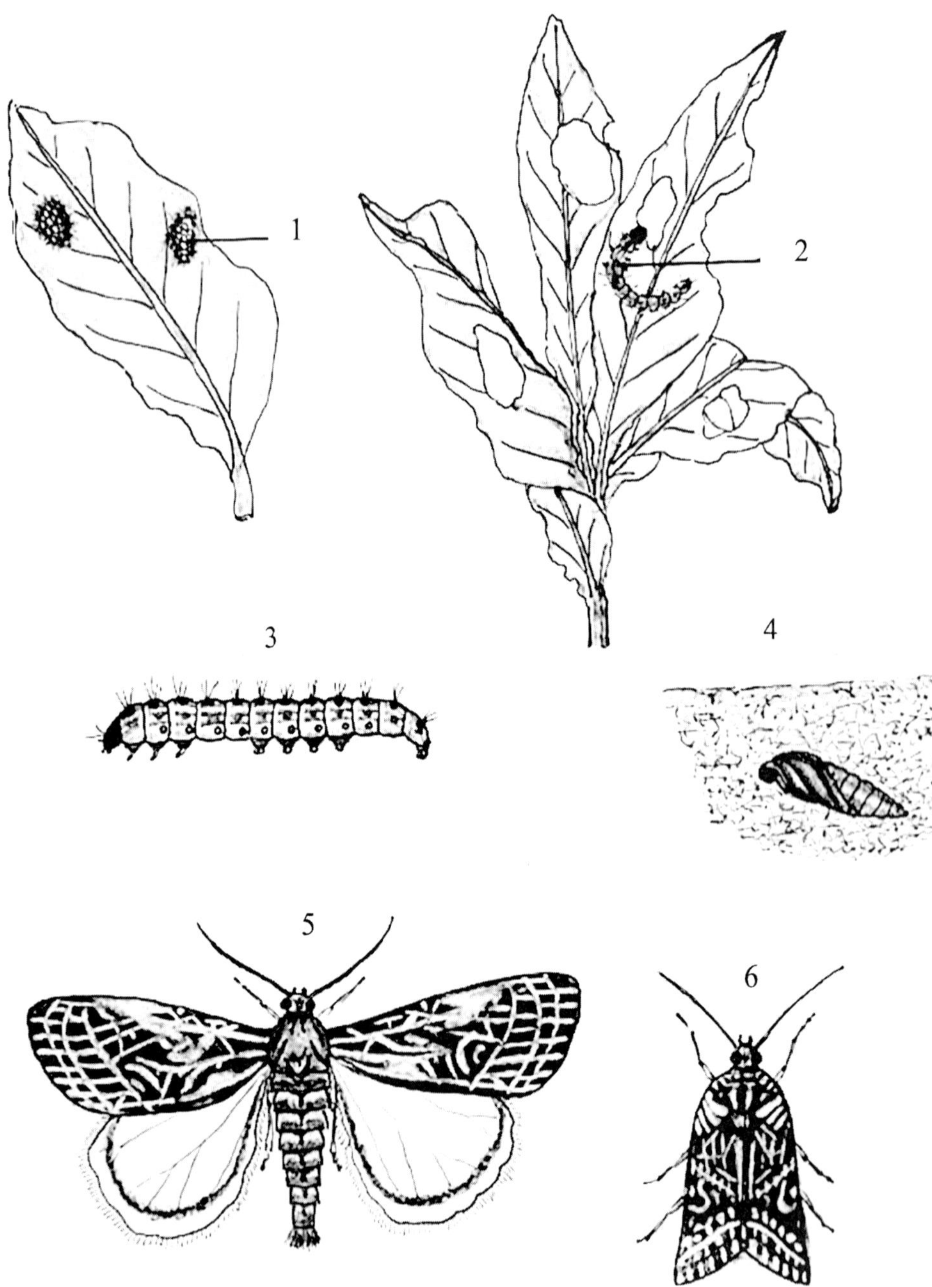

Fig. 68. Tobacco caterpillar-*Spodoptera (Prodenia) litura*

1. Egg mass 2. Larva feeding on the leaf 3. Larva 4. Pupa inside the soil 5. Moth

3. Tobacco stem borer

Gnorimoschema (Phthorimaea) heliopa

Order - Lepidoptera

Family - Gelechiidae

Nature of damage. The caterpillars mine through the midrib and leaf sheath, and ultimately bore into the stem. The irritation caused as a result of infestation leads to formation of gall-like swellings due to proliferation of cells around the site of attack. The infested plants become stunted in growth and may produce unusual branching.

Life cycle of the pest. The female moth lays eggs generally on the leaf stalks. The tiny caterpillar emerging from the egg burrows along the leaf stalk, enters the stem, feeds on the inner tissues and grows. The caterpillar is whitish-brown in color and the full-grown caterpillar is 1.2-1.3 cm. long. The grown-up caterpillar pupates inside the larval gallery inside the stem. The pupa is small and reddish-brown in color. The moth is small, very active, brown in color and closely allied to the potato tuber moth. The entire life cycle is completed in about 6-7 weeks. The pest is found to breed in the tobacco stubble.

Control measures

Cultural methods (i) Infested seedlings should be removed and destroyed and they should not be used for transplanting in the main field (ii) In the main field, plants showing symptoms of infestation may be removed and destroyed (iii) Soon after harvest, the stubble that may harbor the larvae and pupae should be pulled out and destroyed by burning or buried deep in the soil.

Other pests. Tobacco is also attacked by a number of pests other than the ones described above in detail. The nymphs and adults of the white fly-*Bemisia tabaci,* besides causing direct damage by sucking and feeding on the plant sap, also transmit **'tobacco mosaic'** and **'leaf curl'** virus diseases; the onion thrips-*Thrips tabaci* lacerate, suck and feed on the sap from the leaves, as a result the leaves curl up and dry; the caterpillar pest-*Helicoverpa (Heliothis) armigera* and the semilooper-*Plusia signata* cut and feed on the foliage gregariously; the hairy caterpillars-*Amsacta* spp. and the cut worm-*Agrotis ipsilon* feed on the foliage voraciously and may cause severe damage; the lucerne caterpillar-*Laphygma exigua* also feeds on the foliage; the nymphs and adults of the tobacco grasshopper-*Atractomorpha crenulata* and the surface grasshopper-*Chrotogonus robertsoni* notch, cut and feed on the leaves; the grubs of the beetle pest-*Holotrichia serrata* attack the roots of transplanted crop.

TURMERIC *(Curcuma longa)*

1. Turmeric thrips

Panchaetothrips indicus

Order - Thysanoptera

Family - Thripidae

Nature of damage. This is mostly a specific pest of turmeric, but infests banana plants also. Both the nymphs and adults congregate in large numbers on the under surface of leaves, lacerate, pierce, suck and feed on the sap from the leaves. The infected leaves roll and dry from the tip downwards.

Life cycle of the pest. The nymphs, which hatch out from the eggs are minute, yellowish-brown in color and are wingless. The life cycle is completed in 30-40 days. The adult insect is minute, fragile, reddish-brown or brown in color and has fringed wings like a comb (Fig.94).

Control measures

Chemical control. Spraying the foliage with phosalone-400 ml. or monocrotophos-200 ml. or phosphamidon-110 ml. is effective in controlling the pest.

2. Turmeric stem borer

Conogethes (Dichocrocis) punctiferalis

Order - Lepidoptera

Family - Pyraustidae

Nature of damage. The caterpillars bore into the central shoots, feed on the inner tissues and severe the growing shoot from inside, leading to dead hearts. Later, the caterpillars burrow into the developing rhizomes, feed on the inner contents and cause severe damage.

Life cycle of the pest and **control measures.** Refer pests of castor (Fig. 31, Page 81).

Minor pests. Besides the pests mentioned above, the banana lace wing bug pest-*Stephanitis typicus* pierces, sucks and feeds on the sap from the leaves causing whitish patches in the attacked portions; the green, stout, black-headed caterpillars of the turmeric skipper butterfly-*Udaspes folus* fold the leaves, remain securely inside the folds and feed on the leaf tissues; colonies of the gray, hard scale insect-*Aspidiotus hartii* infests turmeric haulms.

GINGER *(Zingiber officinale)*

The caterpillars of the stem borer-*Conogethes (Dichocrocis) punctiferalis* bore into the pseudostem and feed on the inner tissues causing dead hearts; the caterpillars of the leaf roller-*Udaspes folus* feed on the leaves; the maggots of the fly-*Chalcidomyia atricornis* bore into the rhizomes and cause damage.

BETELVINE *(Piper betle)*

1. Betelvine bug

Disphinctus politus

Order - Hemiptera

Suborder - Heteroptera

Family - Miridae

Nature of damage. This is an important pest of betelvine and causes much damage to the leaves. Both the nymphs and adults cause damage to the young leaves. They puncture the leaves, suck and feed on the sap, as a result brownish, blighted areas are formed around the punctures, the leaves shrivel and in severe cases of infestation dry completely. The pest is active from June-October, when the climate is warm and humid, which is conducive for the multiplication of the pest (Fig. 69).

Life cycle of the pest. The female bug thrusts the eggs into the tender plant tissues by means of its sharp ovipositor. The long, somewhat cylindrical egg has two slender hair-like processes, which project outside the tissues. The nymphs emerging out of the eggs are very active and are quite similar to that of the adult bugs, but they do not have wings and are smaller in size. The adult bug is slender, about 1.25 cm. in length, reddish-brown in color and very active (Fig. 69).

Control measures

Chemical control. Because betel leaves are eaten raw, application of highly poisonous insecticides should be avoided, as the toxic residues in the leaf tissues may cause serious consequences to the consumer. Only when absolutely necessary insecticides should be used and that too only less toxic chemicals should be used. Before insecticide application, the leaves should be harvested and after insecticide application, no harvest should be done for at least 7-10 days (i) The plants may be sprayed with Nicotine sulfate-0.02% (0.5 ml. per liter of water) or with tobacco decoction. These plant products are non-toxic to human beings (ii) Spraying with less toxic chemicals viz., malathion-0.1% (2.0 ml. per liter of water) or endosulfan-0.07% (2.0 ml. per liter of water) may be advocated.

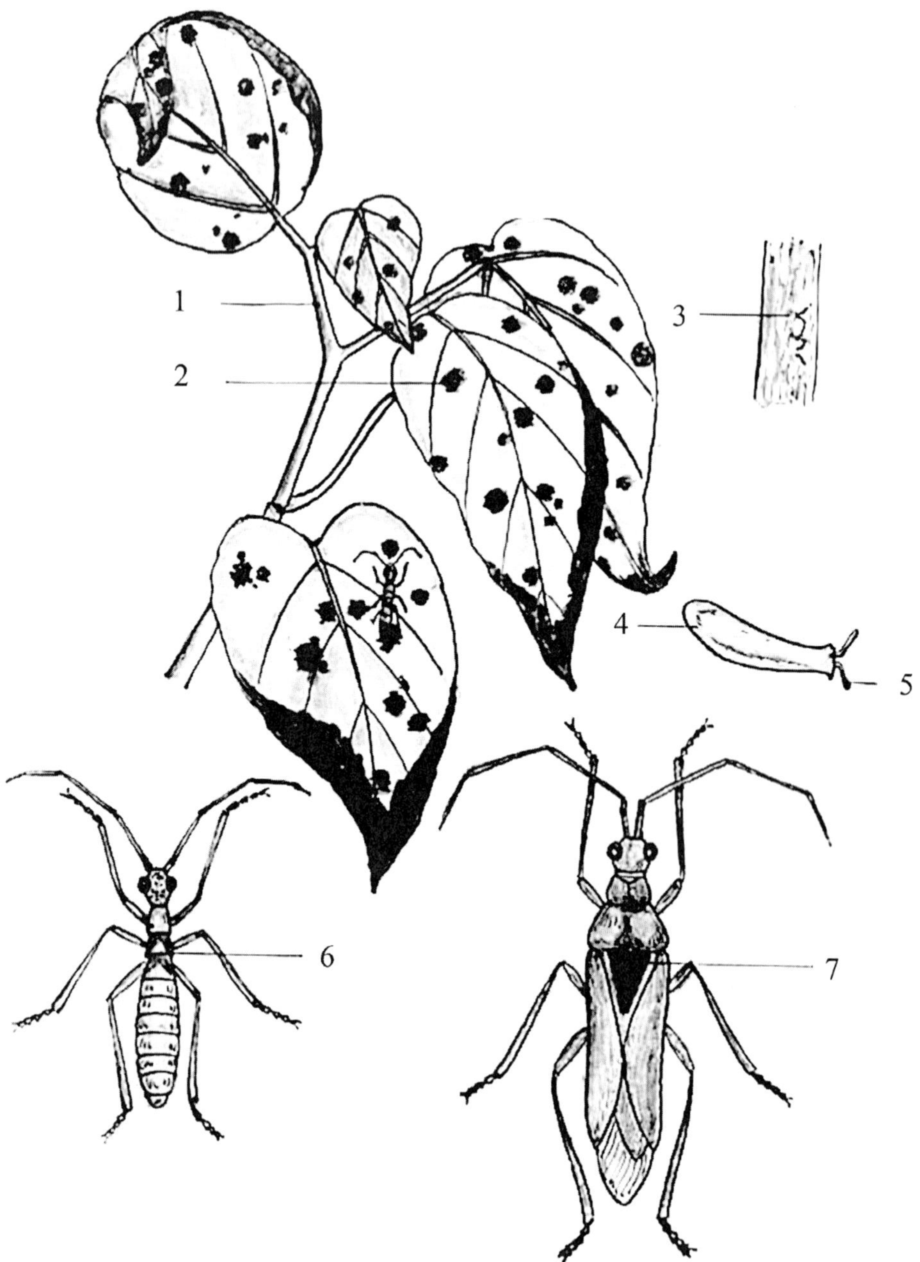

Fig. 69. Betelvine bug-*Disphinctus politus*

1. Infested shoot 2. Dried patches around punctures made by the insects 3. Eggs inserted into the tissues 4. Single egg 5. Hair-like processes 6. Nymph 7. Adult bug

Pests of minor importance. Besides the above pest, the nymphs and adults of the mealy bugs-*Ferrisiana virgata* and *Pseudococcus* spp. suck and feed on the plant sap from the tender shoots, petioles and leaves and cause severe damage. Nicotine sulfate-0.02% or tobacco decoction may be sprayed to control these pests; the white fly-*Dialeurodes pallida* infests the leaves; the root knot nematodes-*Meloidogyne* spp. may also attack the roots, cause root galls and damage the crop.

PESTS OF TUBER AND BULB CROPS

POTATO *(Solanum tuberosum)*

1. Potato tuber moth

Gnorimoschema (Phthorimaea) operculella

Order - Lepidoptera

Family - Gelechiidae

Nature of damage. This is the only borer pest, which causes very severe damage to potato plants in the field, as well as potato tubers that are exposed in the field and potato tubers in storage. The pest was introduced to India from Italy about 70 years back and has since established itself as a serious pest of potato. The caterpillars mine the leaves and bore into the young shoots, feed on the inner tissues and cause drooping, wilting and dying of the infested shoots. At the time of tuber formation, the caterpillars bore into the exposed tubers in the field, feed on the inner content and cause damage. Further, the caterpillars burrow the potato tubers in the godowns and store houses and cause damage. In case of severe infestation, the loss of the produce may go up to 30-70 per cent. The infested tubers, when cut open show tunnels and galleries made by the caterpillars. Many caterpillars may be found inside a single tuber. The excreta of the caterpillars are pushed out through the burrows made by the larvae. The adult moths are seen flying inside the godowns where potato tubers are stored. The infested tubers are later subjected to attack by saprophytic microorganisms and rot. Besides potato, the pest attacks crops such as, eggplant, tomato and tobacco (Fig. 70).

Life cycle of the pest. The female moth lays whitish eggs singly on the tender shoots of plants or on the eyes of tubers or on aberrations or cracks on the outer skin of the tubers at the time of harvest and transit or on the tubers in the godowns. One insect lays 100-200 eggs. The eggs hatch in 5-7 days. In the field, the young caterpillar emerging out of the egg, mines into the leaf or bores into the tender shoot of the plant or burrows into the exposed tuber in the field or the stored potato tuber in the godown, feeds on the inner content and becomes full-grown in 15-18 days. The grown-up larva is about 1.25 cm.

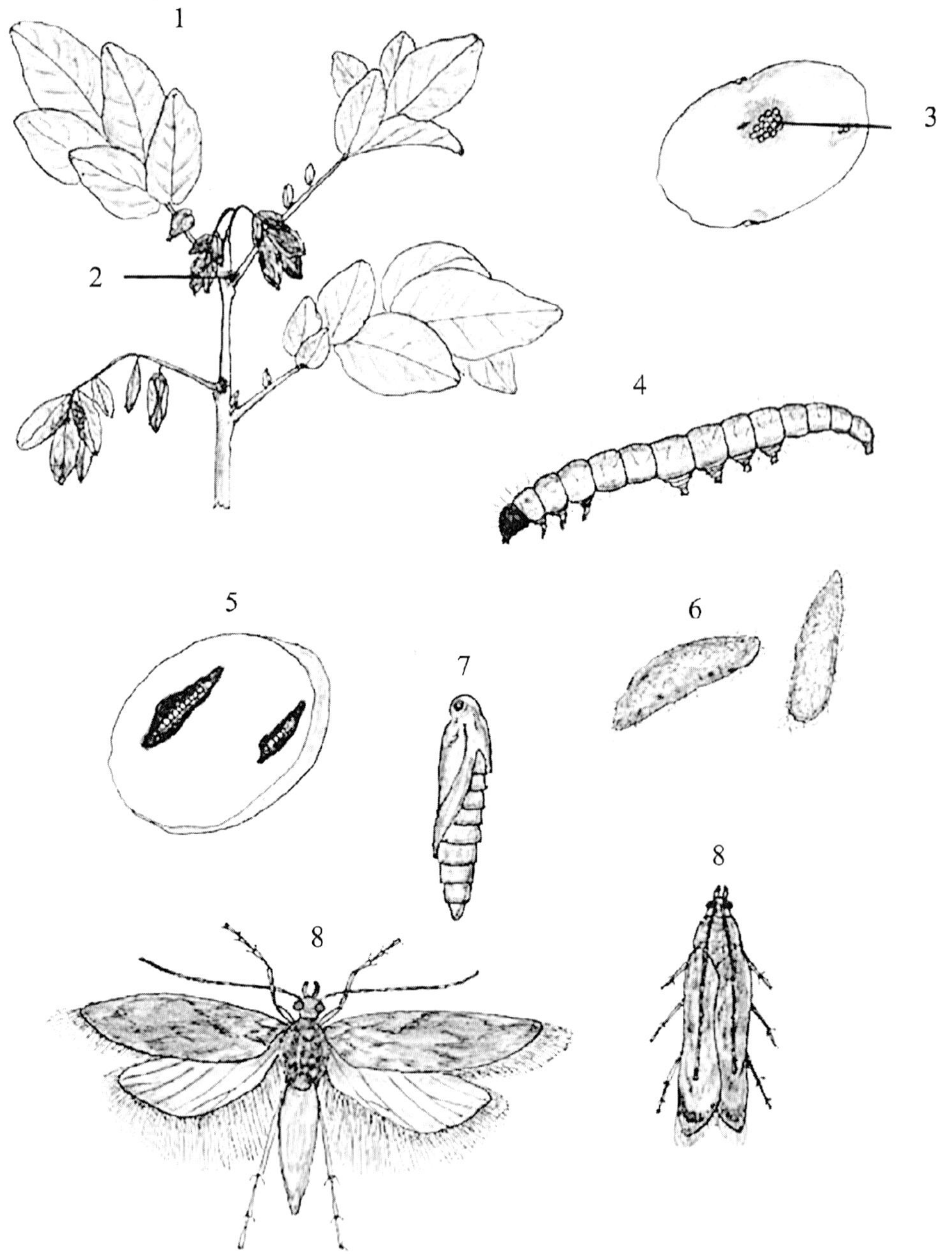

Fig. 70. Potato tuber moth-*Gnorimoschema operculella*

1. Infested shoot 2. Point of entry of the larva 3. Egg mass 4. Larva 5. Larva inside the tuber 6. Cocoon 7. Pupa 8. Moth

in length, yellowish in color with a brown head. Before pupation, the larva comes out of the plant or tuber, constructs a white cocoon with silken thread and pupates in-between mud clods or crevices or in the interspaces between the tubers or cracks in the walls or floors of the godown or between folds of gunny bags. The pupal stage lasts for 7-8 days and after that period, the moth emerges out from the pupa. The moth is small-sized, dark brown in color and the grayish-brown forewings have yellow bands and minute dark markings. The hindwings have thin, dense, long hairs around the margin. Several generations may occur in a season (Fig. 70).

Control measures

Cultural methods (i) The godowns should be kept clean, well ventilated and should be free from any cracks on the walls or on the flooring (ii) Before storing the tubers, damaged and infested tubers should be sorted out and removed (iii) The tubers may be spread out uniformly on the floor of the godown and covered with 2.5 cm. layer of sand to prevent the moths from laying eggs on the tubers and also to prevent the moths from emerging out from the tubers.

Chemical control (i) Malathion 5 D or carbaryl 10 D should be dusted uniformly on the gunny bags in which tubers are stored for seed purposes (ii) If the tubers are stored for consumption purposes, the godown is fumigated with Methyl bromide at the rate of 2.5-5.0 kg. per 1,000 cubic meter (700-900 ml. per 1,000 cu. ft.) space and kept under air tight condition for 3 hours.

2. Gram cutworm

Agrotis ipsilon

Nature of damage, life cycle of the pest and control measures. Refer pests of gram (Fig. 19, Page 55).

3. Tobacco aphid

Myzus persicae

Nature of damage. Both the nymphs and adults occupy mostly the under surface of young leaves and tender shoots in crowded colonies, suck and feed on the sap from the plants. The infested leaves curl upwards, droop and turn brown. The infested plants are stunted in growth and look pale and unhealthy. Besides causing such direct damage to the crop, the pest serves as a vector that transmits the viruses causing **'leaf crinkle'** and **'yellow mosaic'** virus diseases.

Life cycle of the pest and control measures. Refer pests of tobacco (Fig. 67, Page 173).

Minor pests. Besides the pests described above, several other pests are also known to infest potato and may cause serious damage to the crop during some seasons. The larvae of the white grub-*Holotrichia conferta* cut and feed on the roots of potato plants. The grubs also bore into the newly formed tubers, feed on the inner content and cause damage to the tubers; the lucerne caterpillar-*Laphygma exigua* feeds on the foliage; the leaf weevil-*Myllocerus subfasciatus* cuts the leaves from the margin and feeds on the leaf tissues; the spotted beetle-*Henosepilachna (Epilachna) vigintioctopunctata* scrapes and feeds on the green matter of leaves; the bug pest-*Nezara viridula* and the leaf hopper-*Amrasca biguttula biguttula* suck and feed on the sap from the leaves and tender shoots; the aphids-*Aphis gossypii* and *Macrosiphum euphorbiae* and the cotton white fly-*Bemisia tabaci* suck and feed the sap from the leaves and tender shoots; the mite-*Hemitarsonemus latus*, a sucking non-insect pest feeds on the sap from the leaves and tender shoots.

SWEET POTATO *(Ipomoea batatas)*

1. Sweet potato weevil

Cylas formicarius

Order - Coleoptera

Family - Apionidae

Nature of damage. The pest causes damage to the crop in the field, as well as to the tubers in the storehouses during storage. The grubs bore into the vines, feed on the inner tissues, as a result the vines wilt, droop and finally die. The weevils also bite holes in the leaves, bore into the vines or burrow into the tubers, feed on the tissues and cause damage. In the godowns, the grubs burrow the tubers and feed on the inner content. A foul smell emanates from the infested tubers and the tubers become unfit for consumption. Several holes and tunnels may be found in a single tuber.

Life cycle of the pest. The female weevil bites a small hole in the vine or in the tuber and lays eggs singly. The eggs are small, oval, whitish in color and shining. One insect may lay about 200 eggs. The eggs hatch in 3-5 days. The young grubs hatching out of the eggs bore into the vines or into the tubers, feed on the inner content and become full-grown in 25-30 days. The grown-up grub is whitish in color, footless and has a brown head. It pupates either inside the vine or inside the tuber and emerges as an adult weevil in 7-10 days. The weevil is thin, with a long body and looks very much like an ant. It is shiny black in color with reddish thoracic region and legs. It has a long snout, curved downwards.

Control measures

Cultural methods (i) Planting of infested vines should be avoided (ii) The infested vines and tubers should be removed and destroyed (iii) In the godown, the damaged and infested tubers should be sorted out and stored separately (iv) In the godown, the tubers may be spread on the floor uniformly and covered with a 2.5 cm. thick layer of sand, so as to prevent the weevils from laying eggs on the tubers and also to prevent the pest from attacking the tubers.

Chemical control (i) Spraying the crop with fenthion-0.1 % or carbaryl 50 WP-0.1 % or fenitrothion-0.05 %, when the crop is 45 days old controls the pest. This should be followed by 2 more sprayings at an interval of 3 weeks (ii) Fumigating the godowns with Methyl bromide at the rate of 2.5-5.0 kg. per 1,000 cubic meter (700-900 ml./cu.ft.) of space in an air-tight condition for 3 hours is effective in eliminating the grubs and weevils present in the tubers.

Minor pests. Besides the above-mentioned pest, a few other pests also attack sweet potato. Sweet potato stem borer-*Omphisa anastomosalis* bores into the vine at the basal portion and causes damage to the crop; the greenish, stout caterpillars of the sphinx hawk moth-*Herse convolvuli* feed on the foliage; sometimes the tobacco caterpillar-*Spodoptera (Prodenia) litura* appears in large numbers, feeds voraciously on the foliage and causes severe damage; the Bihar caterpillar-*Spilosoma (Diacrisia) obliqua* and the hairy caterpillar-*Pericallia ricini* also infest sweet potato crop; the sweet potato rust mite-*Oxypleurites convolvuli* sucks the plant sap from the leaves and tender shoot portions leading to formation of rusty spots on the infested areas.

TAPIOCA *(Manihot utilissima)*

Only a few pests attack tapioca plants. The cassava scale insects-*Aonidomytilus albus* adhere to the stem and petioles in large numbers, suck and feed on the plant sap; the white fly-*Bemisia tabaci* also infests tapioca crop.

CARROT *(Daucus carota)*, **RADISH** *(Raphanus sativus)* and **TURNIP** *(Brassica rapa)*

Carrot, radish and turnip are subjected to attack by several pests, which attack some of the other important crops. The cabbage borer-*Hellula undalis* burrows the shoots and feeds on the inner tissues; the pea semi-looper-*Plusia orichalcea* feeds on the foliage; the lucerne caterpillar-*Spodoptera exigua* feeds voraciously on the foliage; the leaf webbing caterpillar-*Crocidolomia binotalis* webs the leaves together, remains inside the web securely, scrapes and feeds on the leaf tissues; the Bihar caterpillar-*Diacrisia obliqua* feeds on the foliage gregariously; the grubs of the mustard saw fly-*Athalia lugens proxima* bore into the shoots, feed on the inner tissues and cause damage; the caterpillars of the diamond back moth-*Plutella maculipennis* also infest the foliage; the nymphs and adults of the painted bug-*Bagrada cruciferarum* suck and feed on the plant sap; the nymphs and adults of the onion thrips-*Thrips tabaci* aggregate in large numbers on the under surface of leaves and tender shoots, lacerate, suck and feed on the plant sap; the nymphs and adults of the aphid pest-*Lipaphis erysimi* also suck and feed on the sap from the leaves and tender shoots.

Colocasia (*Colocasia antiquorum*)

The leaf caterpillars-*Spodoptera (Prodenia) litura* and *Pericallia ricini* feed on the foliage; the horned caterpillars of the hawk moth-*Hippotion oldenlandiae* and *Agrius convolvuli* feed gregariously on the leaf tissues; the grubs of the lace wing bug-*Stephanitis typicus* feed on the foliage; colonies of nymphs and adults of the aphid-*Pentalonia nigronervosa* congregate on the tender parts of the plants, suck and feed on the plant sap; the black thrips-*Caliothrips indicus* and *Heliothrips haemorrhoidalis* suck and feed on the sap from the leaves; the leaf flea beetle-*Monolepta signata* bores holes in the leaves and feed on the tissues.

Elephant-foot yam (*Amorphophallus campanulatus*)

Only a few pests are known to attack elephant-foot yam. The scale insect-*Aspidiotus hartii* adheres to the stem portion in large numbers and feeds on the plant sap; the polyphagous horned caterpillars of the hawk moth, with a prominent horn at the anal end-*Hippotion celerio* feed gregariously on the foliage; the larvae of the leaf beetle-*Galerucida bicolor* feed on the leaves.

BEET ROOT *(Beta vulgaris)*

Many of the pests that infest cole crops are found to infest beet root also. The cabbage borer-*Hellula undalis* feeds on the foliage and bores into the roots and causes severe damage to the crop; the lucerne caterpillar-*Laphygma exigua* and the tobacco caterpillar-*Spodoptera (Prodenia) litura* feed on the foliage gregariously; the caterpillars of the hawk moth-*Hippotion celerio* also feed on the leaves; the leaf roller pests-*Hymenia fascialis* and *Marasmia trapezalis* roll the leaves, remain securely inside the folds, scrape and feed on the leaf tissues; colonies of nymphs and adults of the beet leaf hopper-*Eutettix tenellus* suck and feed on the sap from the foliage and also acts as a vector of virus diseases; the peach green aphid-*Myzus persicae,* a sucking pest also infests beet root; the flea beetle-*Monolepta signata* and the leaf weevil-*Tanymecus indicus* bite holes in the leaves and feed on the leaf tissues.

ONION *(Allium cepa)* and GARLIC *(Allium sativum)*

1. Onion thrips

Thrips tabaci

Order - Thysanoptera

Family - Thripidae

Nature of damage. The pest occurs in all the countries where the crops are cultivated and causes considerable damage to the crop and yield loss. The nymphs and adults are found in large numbers on the young leaf bases and in-between the leaf whorls, lacerating the leaf tissues, sucking and feeding on the sap. The affected parts of the leaves turn into whitish patches. In case of severe infestation, the leaves start drying from the tip downwards and the leaf tips become twisted and shrivelled and the plants appear blighted.

Life cycle of the pest. The females reproduce parthenogenetically. It inserts the eggs inside the tissues of young leaves. One insect lays 45-50 eggs. The eggs hatch in 4-5 days and the young nymphs emerging from the eggs are pale yellowish in color and are wingless. The life cycle is completed in 11-21 days and thus, the pest multiplies very rapidly. The adult thrips are minute, fragile, dark brown in color and the wings are fringed with fine, long hairs. Male thrips are rarely found or not al all found (Fig.94).

Control measures

Chemical control. The pest can be controlled effectively by foliar spraying with monocrotophos-200 ml. or phosalone-400 ml. or dimethoate-400 ml. or phosphamidon-110 ml. in 200 liters of water per acre with a high volume sprayer.

Other pests. Besides the pest described above, onion and garlic are also infested by the polyphagous leaf caterpillar-*Laphygma exigua* and the cosmopolitan tobacco caterpillar-*Spodoptera (Prodenia) litura*, which feed on the foliage gregariously and defoliate the plants.

Chapter-2

Pests of Horticultural Crops

Mango (*Mangifera indica*)

1. Mango hoppers

(i) ***Amritodes atkinsoni*** *(ii)* ***Idioscopus niveosparsus*** and *(iii)* ***I. clypealis***

Order	-	Hemiptera
Suborder	-	Homoptera
Family	-	Cicadellidae

Nature of damage. Three different species of hoppers are found to cause damage to mango trees. During some years, especially during the months of November-February, when the mango trees come to flowering, the pest appears in large numbers. During the other months, the pests are seen in much lesser numbers in an active form, while the other insects hibernate under the bark of the trees. Even in the peak period of pest occurrence, after a few heavy showers, the number of insects is considerably reduced.

During the cold season, when the trees put out the flower spikes, thousands of hoppers, young and adults appear in large numbers, pierce, suck and feed on the sap from the tender shoots, flower stalks, flowers and buds. This drain of the sap causes the buds and flowers to dry, wither and they are shed in large numbers, as a result fruit production may go down by

25-60 per cent. When a severely infested tree is approached, thousands of hoppers hop or fly about and make a rustling noise, as they dash against the foliage. The hoppers exude a sweet, sticky fluid excretion and this fluid thrown out by myriads of these insects after feeding on the tree sap, wets the soil underneath the trees and on the surface of the leaves of the trees. In case of severe infestation, the fluid may fall down in drops. The secretion attracts scores of flies, bees and other insects, which come to feed on the secretion. The infested tree presents a sickly appearance. The trees get devoid of buds and blossoms. The leaves appear shiny and are covered by **'sooty mould'**, a fungal disease. Thousands of caste skins of the nymphs are found on the shoots and leaves. The continuous damage caused to the trees year after year by the pest makes them loose their vigor gradually and eventually leads to severe reduction in their yielding capacity.

Life cycle of the pest. The female hopper inserts spherical, pale whitish eggs singly into the plant tissues through slits made in the young shoots, flower stalks and flower spikes. The eggs hatch in 4-6 days and the young nymph hatching out of the egg looks quite similar to the adult. The nymphs are very active and move about rapidly along the flower spikes and shoots, suck and feed on the plant sap. They are smaller in size, wingless and unable to fly or jump. The young ones moult their skin a few times during their growth and become adults in 10-15 days. The adult hoppers are small insects, about 3.0 mm. in length with a wedge-shaped body. The head is broader and the body gradually narrowing backwards. The legs are well developed, especially the hind legs. The general color is light greenish-brown. They are extremely active and can fly and hop around. They have a tubular, sucking mouthpart by which they puncture the tender plant parts and floral parts. The adults of *Idioscopus clypealis* are smaller, with two black spots on the prothorax and a few black spots on the front portion of the head, while those of *Idioscopus niveoparsus* are slightly bigger with three black spots on the prothorax and whitish, distinct cross bands on the light brownish forewings. The adults of *Amritodes atkinsoni* are bigger than the other two species, about 5.0 mm. in length, with two black spots on the prothorax. Two or more generations may appear during the flowering season of mango trees (Fig.71).

Control measures

Cultural methods. The mango trees should be pruned properly, so that there is good aeration and sunlight in the garden.

Chemical control (i) Spraying with fish oil-rosin soap affords good control of the pest (ii) Foliar spraying with phosalone-0.07% or carbaryl 50 WP-0.1% or Phosphamidon-0.05% is also effective in controlling the pest. Chemical spraying, when the trees are in full bloom should be avoided, as it may interfere with pollination and may kill the pollinating insects.

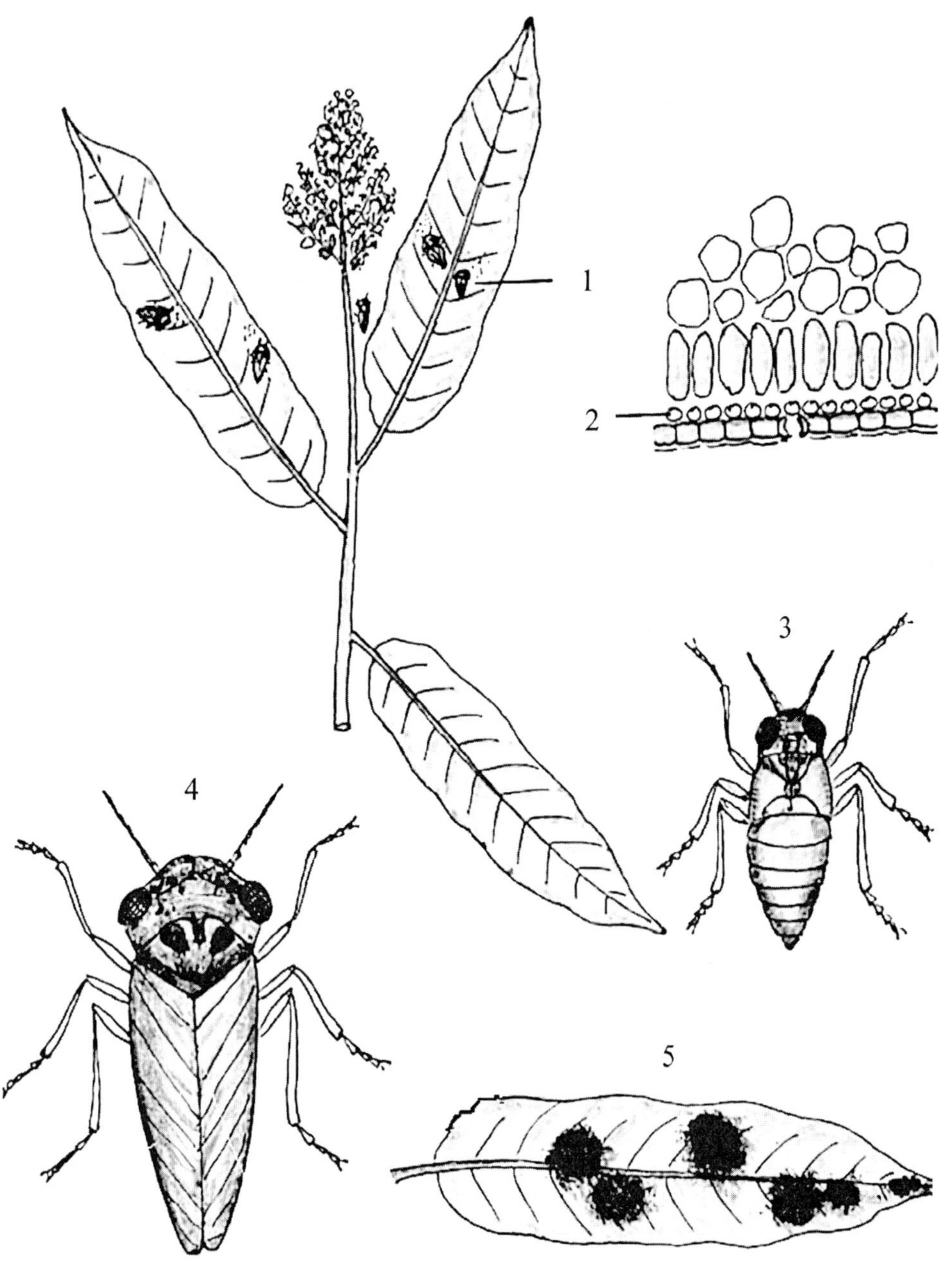

Fig. 71. Mango hopper-*Amritodes (Idiocerus) atkinsoni*

1. Nymphs and adults feeding on the leaves and floral parts 2. Eggs inside leaf tissue 3. Nymph 4. Adult hopper 5. Sooty mould on the infested area.

2. Mango stem borer

Batocera rufomaculata

Order - Coleoptera

Family - Cerambycidae

Nature of damage. The young grub hatching out of the egg burrows into the stem, mostly through the tender portion, where the stem branches. After entering into the stem, the grub starts feeding on the inner tissues by making tunnels. Through the boreholes, mass of faecal matter, gummy sap and plant tissues come out. Gradually the borehole is closed naturally. Due to the tunnelling and feeding of the inner tissues, the vascular system of the tree is damaged leading to shedding of the leaves of the attacked branch and sudden collapse and drying of the entire branch. When several branches are attacked like this, the whole tree may die ultimately (Fig. 72).

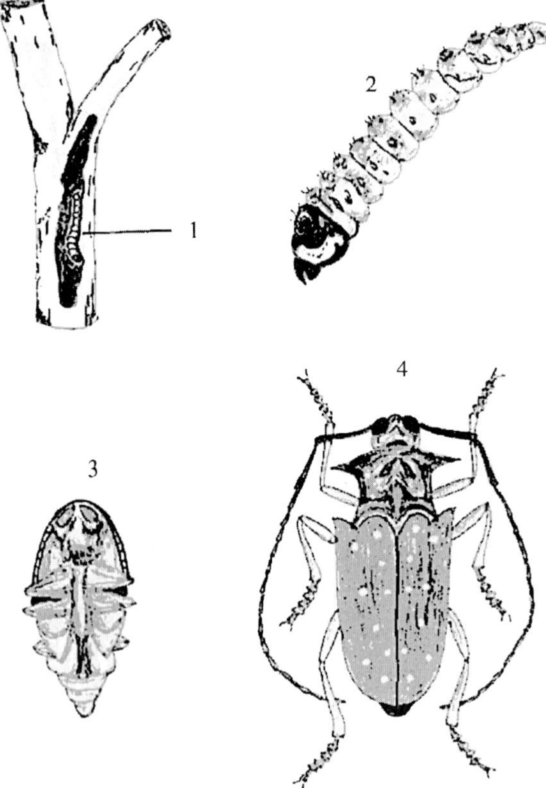

Fig. 72. Mango stem borer-*Batocera rufomaculata*

1. Larva inside the stem 2. Larva 3. Pupa 4. Adult beetle

Life cycle of the pest. The whitish, cylindrical eggs are inserted singly under loose bark, in wounded or diseased portions of the stem or branches or in-between the branching stem portion. The young grub that hatches out of the egg in 7-14 days, which has very strong, biting mouthparts gradually burrows into the stem, generally through the tender portion, where the stem branches. After penetrating into the stem, the grub continues to burrow inwards and makes tunnels. Maximum damage is done during the growing period of the grub. The grub becomes full-grown in 7-8 months. Full-grown grub is stout, yellowish-white in color, fleshy, measuring about 10.0 cm. in length and 1.25-2.00 cm. in breadth, with the anterior portion broader. It is footless and the head is dark brown in color. The grown-up grub pupates inside a chamber inside the stem. The pupal period lasts for 3-6 months depending upon the weather conditions. The adult is a large, longicorn beetle, measuring 4.0-6.0 cm. in length, grayish-brown in color, with long legs and antennae. The beetle has a few yellowish or orange spots on the prothorax and forewings. The prothorax has two, horn-like, pointed projections on the lateral sides. The forewings or elytra are hard and strong. The beetles fly about freely during the nighttime, scrape and feed on the bark tissues of trees. Besides mango, the pest attacks jack, mulberry, fig, rubber, eucalyptus etc. (Fig. 72).

Control measures

Physical methods. The infested and dried branches should be cut off and destroyed

Chemical control (i) The bored holes are cleaned thoroughly and Petrol or Ether or Chloroform or Carbon disulfide is poured into the hole and plugged with clay mixed with a fungicide such as, copper oxychloride. If the holes are too small, then the holes may be made a little wider with a hand drill or drilling machine to expose fresh tissues and then any one of the chemicals may be poured (ii) Monocrotophos-10 ml. or dichlorvos-10 ml. or monocrotophos-5.0 ml. + dichlorvos-5.0 ml. is poured into the hole and then the hole plugged with clay mixed with a fungicide (iii) A few crystals of Paradichlorobenzene may be put into the hole and then the hole plugged.

3. Mango fruit fly

Bactrocera (Dacus) dorsalis and *Bactrocera (Dacus) zonatus*

Order - Diptera

Family - Tephritidae

Nature of damage. The cosmopolitan, polyphagous pest attacks semi-ripe mango fruits during June-August. The young maggot that hatches out of the egg, bores into the fruit, feeds on the inner fleshy portions of the fruit and causes damage. The infested fruits show brownish, circular patches on the outer skin around the boreholes. Severely affected fruits rot and a brownish fluid exudes through the holes and they drop off prematurely. The maggots attack mostly unripe, but matured fruits.

Life cycle of the pest. The female fly inserts shiny white, long, cylindrical eggs into the tissues of the outer skin of matured, unripe fruits or partially ripened fruits in batches of 5-10. The latex, which comes out of the injured portion of the fruit, covers the eggs completely. One female lays 50-200 eggs. The young maggot, which comes out of the egg, bores into the fruit, starts feeding on the fleshy pulp of the fruit and becomes full-grown in 7-28 days. The grown-up maggot is whitish in color, long, 0.8-0.9 cm. in length, footless, with the head region broader and the rear portion narrow. The full-grown maggot comes out of the fallen, rotten fruit, enter into the soil and pupates at a depth of 7.0-15.0 cm. The pupation period lasts for about 7 days during the hot season and about 42 days during the colder months. The adult flies are small, slightly bigger than the housefly, orange or brownish-red in color, with black and yellow markings. The thin, transparent wings have red markings near the anterior margin of the forewings. The hindwings are reduced into halteres. The abdomen is broader at the anterior end and narrow at the posterior end. Besides mango, the pest also infests guava, pomegranate, apple, banana and orange fruits. The pest is active throughout the year in one or the other hosts. There may be many overlapping generations in a year.

Control measures

Cultural methods (i) The infested fruits that have fallen down to the ground should be removed and destroyed (ii) The mango garden, especially the area around and below the canopy of the trees should be ploughed well to bring out and destroy the pupae from under the soil.

Chemical control. If possible, the trees may be sprayed with malathion-0.1% or fenitrothion-0.05%, when the fruits are young.

Poison bait (i) A mixture of malathion 50 EC or trichlorphon 50 EC-1.0 liter as poison and Protein hydrolysate-1.0 kg. as bait are mixed in 10 liters of water and the fluid poured in small containers and are hung from the branches of trees. The flies, which are attracted by the smell of the bait, fall into the poisoned bait and die (ii) Poison bait with Methyl eugenol as a bait can also be used.

4. Mango nut weevil or Stone weevil

Sternochetus mangiferae

Order - Coleoptera

Family - Curculionidae

Nature of damage. This is a specific pest of mango. The young grub immediately after emerging from the egg, penetrates the skin of the young fruit, makes irregular, zigzag tunnels in the fleshy portion and finally bores into the nut. When the fruit develops, the bore-hole is sealed naturally and it is impossible to detect the infested fruit. The grubs feed on the cotyledons and kernel and destroy them. The adults emerging from the pupae also feed on the kernel of the seed. The infested fruits mature and ripen early. The pest is partial to certain varieties of mango and in susceptible varieties there may be up to 7 weevils inside a single nut.

Life cycle of the pest. The adult female makes a small incision in the skin of the young fruit and lays shiny, white eggs singly just under the skin and covers it with latex, which comes out of the incision. On each fruit, 12-20 eggs are laid. The incision in which the egg is laid heals up and for all outward appearance the fruit appears quite normal. The young grub, which hatches out of the egg in 5-7 days, penetrates the outer skin, burrows through the developing fleshy pulp, bores into the nut, feeds on the kernel and becomes full-grown in 25-30 days. The grown-up grub is stout, fleshy, light yellow in color, footless and with a dark brown head. It pupates inside the nut and emerges as an adult weevil after 10-12 days. The weevil is short, stout, ovoid, dark brown in color, about 6.0 mm. in length and with a long snout, which is curved downwards. Sometimes the weevils, which come out of the fruits, hibernate under the bark of the stem for a long period of time.

Control measures

Cultural methods. The infested fruits, which had fallen prematurely should be collected periodically and destroyed.

Chemical control. The pest can be controlled by spraying with fenthion-0.05% or malathion-0.1% or deltamethrin-0.025%, when the fruits are young. This should be followed by 2-3 more sprayings at 15 days interval.

5. Mango flower webber

Eublemma versicolor

Order - Lepidoptera

Family - Noctuidae

Nature of damage. The caterpillars web the flowers in the inflorescence together with fine silken thread, make galleries, remain inside the galleries, bite and feed on the floral parts and cause extensive damage. The caterpillars also burrow the peduncles of the inflorescence and feed on the inner tissues. On the trees infested with the pest, numerous webbed and dried top shoots with frass and dry plant tissues can be seen sticking on to the galleries conspicuously. The infested flowers fail to produce fruits.

Life cycle of the pest. The female moth lays hemispherical, red-colored eggs in batches of 8-10 on the petals of flowers. The eggs hatch in 3-4 days. The young larvae coming out of the eggs become full-grown in 18-20 days. The grown-up caterpillar is greenish-yellow in color, about 2.0 cm. in length, smooth, with a light brown head and brownish shield-like, thick, brownish skin on the prothorax. The full-grown caterpillar constructs a cocoon with fine, silken thread, pupates inside it and emerges as an adult moth in 8-9 days. The female moth has grayish-violet wings with black markings near the lateral margin. The male moth has light orange-colored wings.

Control measures

Chemical control. Wherever possible, the trees can be sprayed with carbaryl 50 WP-0.1% or endosulfan-0.035% or malathion-0.01%, so as to cover the inflorescence thoroughly.

6. Mango mealy bug

Drosicha mangiferae

Order - Hemiptera

Suborder - Homoptera

Family - Coccidae

Nature of damage. The nymphs of both the sexes and the wingless female adults only cause the damage. They congregate in large numbers on the tender parts of the plant, pierce, suck and feed on the plant sap from the green twigs, shoots and flowers, as a result they dry up and fruit formation is drastically affected. Whenever fruits are formed in the affected fruit bearing twigs, they become very weak and the fruits fall down easily. The insects secrete a thick, sugary liquid from their body, which falls on the leaves and inflorescence leading to infection by a fungal disease known as **'sooty mould'**. The adult males have wings, but no mouthparts and so they do not feed. Besides mango, the pest attacks many other trees such as, jack, guava, citrus, pomegranate, papaya, castor, plum, peach, apricot, fig, litchi, shoe flower etc.

Life cycle of the pest. During early summer, between the months of March to middle of May, the female bug lays 400-500 eggs, enclosed in sac-like pouches at the end of white, silken, thread-like material at a depth of 5.0-15.0 cm. in the soil, mostly in cracks and crevices near the base of the infested trees. The clusters of eggs in such pouches are deposited by the female in instalments over a period of 7-15 days, after which it dies. The individual egg is oval in shape and pink in color. The eggs remain in the soil till the coming January, when they hatch. The dirty brown, tiny nymphs on emergence crawl up the host trees and fix themselves on tender twigs and terminal shoots. When flowering starts, they migrate to the inflorescence. The female nymphs become full- grown, reach the sexually matured adult stage in about 3 months, after moulting 3 times during their growth period. In the case of the males, a short pupal stage of a few days duration intervenes before they attain the adult stage. The nymphs of both the sexes and the females have a flattish, plump body, pink-colored and are covered all over the body with a white, mealy, powdery coating. The adult male is a winged insect, with dark red body, measuring about 1.25 cm. across the wing span. The wings are smoky in color and the antennae are very long. The female lives for about a month. The male immediately after emergence from the pupa mates and then dies. There is only one generation in a year (Fig. 73).

Control measures

Cultural methods (i) The soil below and around the infested trees may be stirred or ploughed during July-November, so as to expose and destroy the eggs in pouches under the soil (ii) Severely infested branches may be pruned and destroyed

Chemical control. Spraying the crop with monocrotophos-0.04% or dimethoate-0.06% or malathion-0.1%, when the crop is in full flush during the Spring season and also at the flowering stage of the tree controls the pest.

Other pests. Mango is subjected to attack by several other pests and some of them may cause serious damage to the crop and yield loss under conditions favorable for them. The grubs of the medium-sized, reddish-brown, leaf-twisting weevil with a long neck-*Apoderus tranquebaricus,* have the habit of cutting and twisting the mango leaves into a knot, remain inside it, bite and feed on the leaves (Fig. 74); the maggots of the mango flower gall midge (gall flies)-*Procystiphora mangiferae* and *Erosomya indica* infest the floral parts, form galls and damage the flowers, as a result the flowers are shed; the grubs of the leaf weevil-*Rhynchaenus mangiferae* mine the leaves and feed on the inner tissues; the nymphs and adults of the guava scale insect-*Chloropulvinaria (Pulvinaria) psidii* and the scale insect-*Aspidiotus destructor* suck and feed on the plant sap from the tender shoot portions; the polyphagous bark-eating

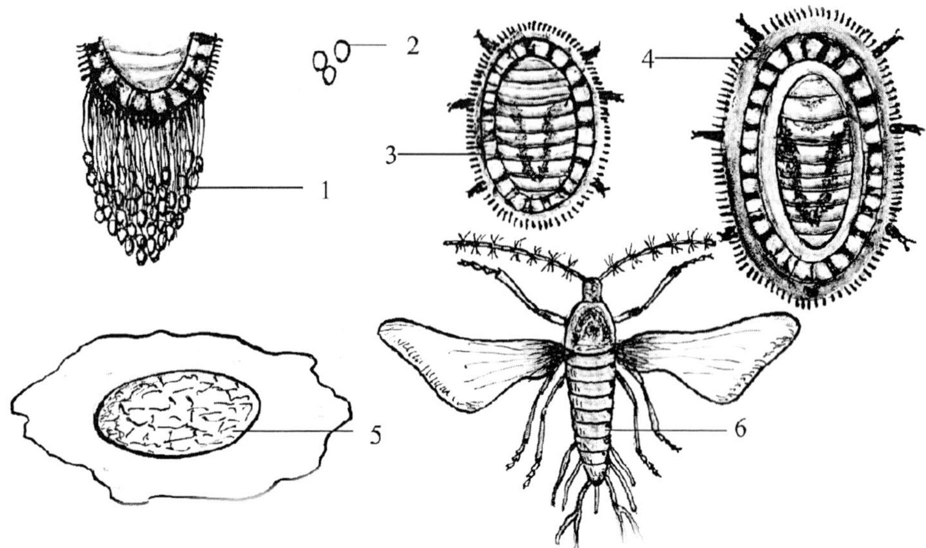

Fig. 73. Mango mealy bug-*Drosicha mangiferae*

1. Egg clusters in sac like pouches 2. Individual eggs 3. Nymph 4. Female wingless adult 5. Pupa 6. Male winged adult

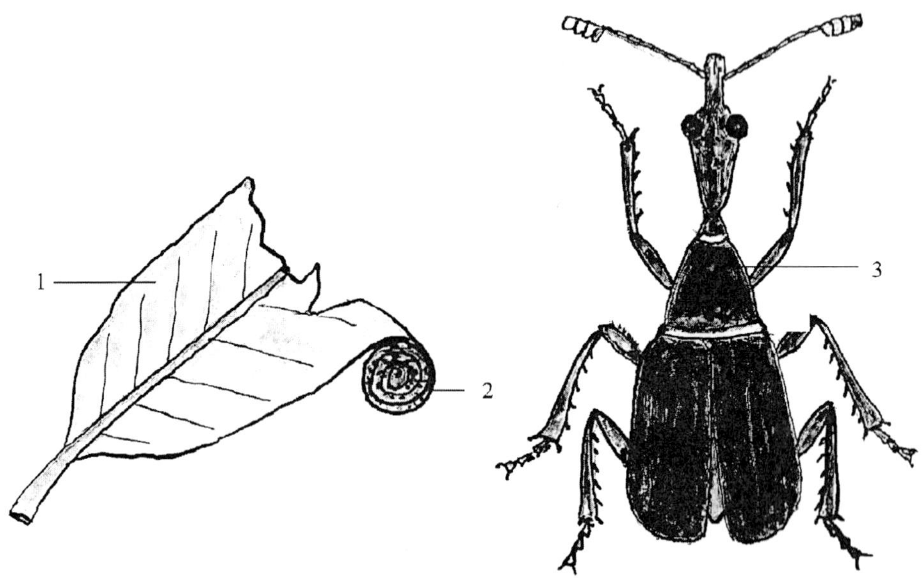

Fig. 74. Mango Leaf twisting weevil-*Apoderus tranquebaricus*

1. Infested leaf 2. Leaf, cut and twisted into a knot 3. Adult weevil

caterpillars-*Indarbela tetraonis* and *I. quadrinotata* bore the bark of the stem and branches, construct long, ribbon-like galleries with silken thread, frass and bark tissue, remain inside the galleries, scrape and feed on the bark tissues. Later they bore into the trunk or branches, make zigzag tunnels and feed on the inner tissues; the leaf eating caterpillar-*Parasa lepida* feeds on the leaves; the caterpillar pest-*Euproctis fraterna* feeds on the flowers; the maggots of the leaf gall flies-*Alassomyia tenuspatha* and *Amradiplosis amraemyia* produce many types of leaf galls; the nymphs and adults of the mango leaf scale-*Chionaspis vitis,* a whitish, hard scale, covers the young mango leaves, suck and feed on the sap, resulting in yellowing and fading up of leaves; the red tree ants-*Oecophylla smaragdina* bind many leaves together, make nests, live inside them and cause many problems, especially at the time of harvesting; the termites-*Odontotermes obesus* feed on the bark and roots and tunnel upwards, especially in ill-maintained orchards; the mango mites-*Aceria mangiferae* attack mango shoots and inflorescence causing malformation and presenting a bunchy, broom-like appearance to them.

CITRUS

The important citrus plants that are cultivated on a large scale include, lime *(Citrus aurantifolia),* lemon *(Citrus limon),* sweet orange *(Citrus sinensis),* mandarin orange *(Citrus reticulata),* kamala orange *(Citrus aurantium),* grape fruit *(Citrus paradisi),* malta *(Citrus medica)* and pomelo *(Citrus grandis)*

1. Citrus fruit-sucking moths

Othreis fullonica and *O. materna*

Order - Lepidoptera

Family - Noctuidae

Nature of damage. These insects are examples of adult Lepidopterous moths doing direct damage to plants. The moths puncture the ripening fruits, suck and feed on the juice from the fruits. Unlike most of the other moths, the feeding tube of these moths is well developed and provided with sharp spines to pierce the fruit surface. In a single fruit, the moth may puncture in 20-25 different places. The damaged fruits gradually begin to rot and drop down. In badly infested gardens, many fruits are found dropped under each tree. When a rotten fruit is squeezed, a fermented, frothing juice comes out through the fine holes made by the moth. Besides citrus, the pest attacks fruits of pomegranate, grapes, mango, guava, tomato etc. (Fig. 75).

Life cycle of the pest. The moths are found to pass their early stages on some wild plants growing in the vicinity of the orchards. The female moth lays ovoid, light greenish-colored, smooth eggs on the under surface of leaves of the host plants, weed plants or hedge plants. One female lays about 250 eggs singly during its lifetime. The young larvae, which hatch out of the eggs in 5-8 days, feed on the leaves of the weeds and hedge plants and become full-grown in 21-25 days. The grown-up larva is stout, cylindrical, a typical semi-looper, 5.0-6.0 cm. in length, with a velvety body and one or two pairs of ocelli. The larva has a dark bluish body, with lateral bright spots of yellow and red and a dorsal hump on the last abdominal segment. The caterpillar presents a threatening posture, with its hind portion raised up, the head region curved round and the eyespots made conspicuous, when provoked. The full-grown caterpillar pupates inside a cocoon made up of fine silken thread in-between leaf folds. The pupa is stout and reddish-brown in color. The adult moth emerges from the pupa in 7-14 days. The moth is large-sized, stoutly built, 6.0-9.0 cm. across the wing span, with prominent upwardly directed palpi. The moth of *Othreis fullonica* has a pale orange-brown body, with the forewings dark grayish and the hindwings orange-yellow, with 2 black, curved crescent-shaped markings. The moth of *Othreis materna* also has a orange-brown body and pale greenish-gray forewings, with reddish and pale white wavy markings. The hindwings are orange-yellowish in color, with a marginal dark band mixed with white spots and a circular dark spot at the center. The moths are active during the nighttime, visit and damage the fruits at night and are not generally seen during the daytime. The smell emanating from the ripened, as well as rotten fruits attracts the moths. Two generations appear in a year (Fig.75).

Control measures

Cultural methods (i) The orchards and the surrounding areas should be kept free from weeds and other wild plants that may harbor the caterpillars (ii) Bonfires may be set up to attract and destroy the moths (iii) Covering the fruits with paper or polythene covers affords protection against the attack by the moths.

Chemical control (i) Poison baits can be used to attract and destroy the moths. Lead arsenate as poison and sugar syrup as bait are mixed in the ratio of 1:160 (1.0 kg. Lead arsenate and 160 liters of dilute sugar syrup). To this mixture, a few drops of vinegar, an attractant is also added. This mixture is taken in small mud pots and hung from the branches of trees (ii) Another poison bait containing a mixture of malathion-20 ml. or diazinon-50 ml. and molasses or sugar syrup-200 gm. in 2.0 liters of water may also be used to attract and destroy the moths.

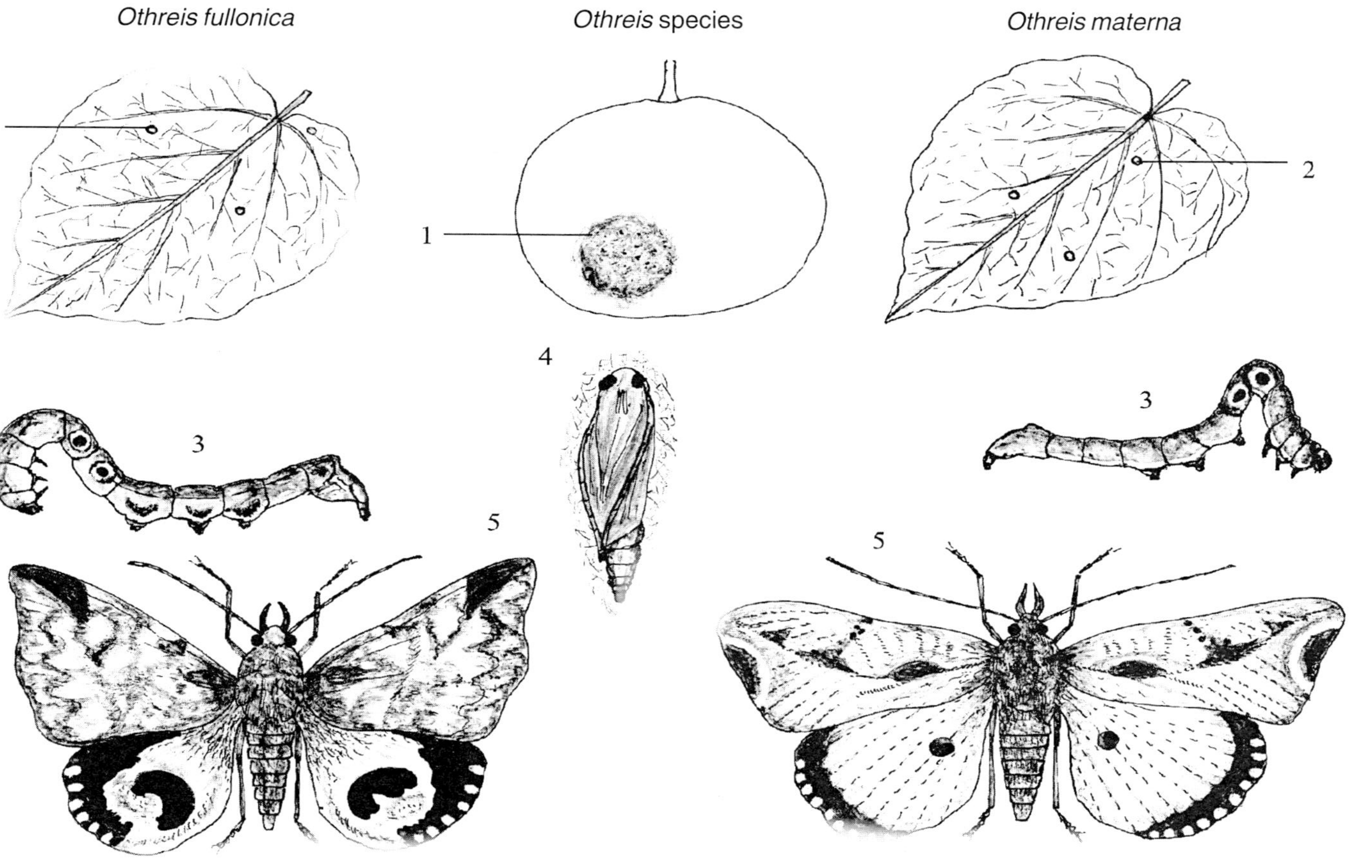

Fig 75. Citrus fruit sucking moths
1. Infested fruit 2. Egg 3. Larva 4. Pupa 5. Moth

2. Lemon butterfly

Papilio demoleus

Order - Lepidoptera

Family - Papilionidae

Nature of damage. The leaf-feeding caterpillar is found to infest all types of citrus plants. The pest is of common occurrence and has a wide distribution. The caterpillars bite and feed on the leaves of the host plants voraciously. Occasionally the pest appears in large numbers and defoliates the plants completely, leaving behind only the petioles and the twigs. The pest attacks curry leaf plants and a few other hosts (Fig. 76).

Life cycle of the pest. The mother butterfly lays spherical, seed-like, greenish-yellow, shining, smooth eggs singly on the under surface of young leaves and tender parts of the plant. The eggs hatch in 4-6 days and the dark caterpillars emerging from the eggs start feeding on the tender leaves in the beginning and then on the older leaves also. The young caterpillar is dark greenish-black in color, with whitish markings and looks like excreta of birds. It becomes full-grown in 14-28 days. The grown-up caterpillar is stout, cylindrical, 3.75-4.00 cm. in length, dark green in color, with brownish bands across the back. When the larva is provoked, a red, split tongue-like structure, known as the **'osmaterium'**, suddenly protrudes out from the back of the head in a threatening gesture. The grown-up caterpillar changes into a pupa as a naked chrysalis, which is attached to the plant surface by a fine, silken girdle. In 8-12 days, the butterfly emerges from the chrysalis. The adult insect is a large-sized, beautiful, swallow-tail butterfly. The wings are large, ornamented with yellow spots, bands and markings on a black background. The pest occurs throughout the year on citrus plants and other collateral hosts (Fig. 76).

Control measures

Physical methods. The caterpillars are large-sized and can be spotted easily. They can be hand picked and destroyed.

Chemical control. Foliar spraying with ensosulfan-0.035% or methyl parathion-0.05% fenitrothion-0.05% or monocrotophos-0.04% or phosphamidon-0.05% controls the pest.

Biological control. The egg parasitoids-*Trichogramma evanescens* and *Telenomus* spp. parasitize and destroy the eggs; the larval parasitoid-*Apanteles* spp. are capale of parasitizing and destroying the larvae.

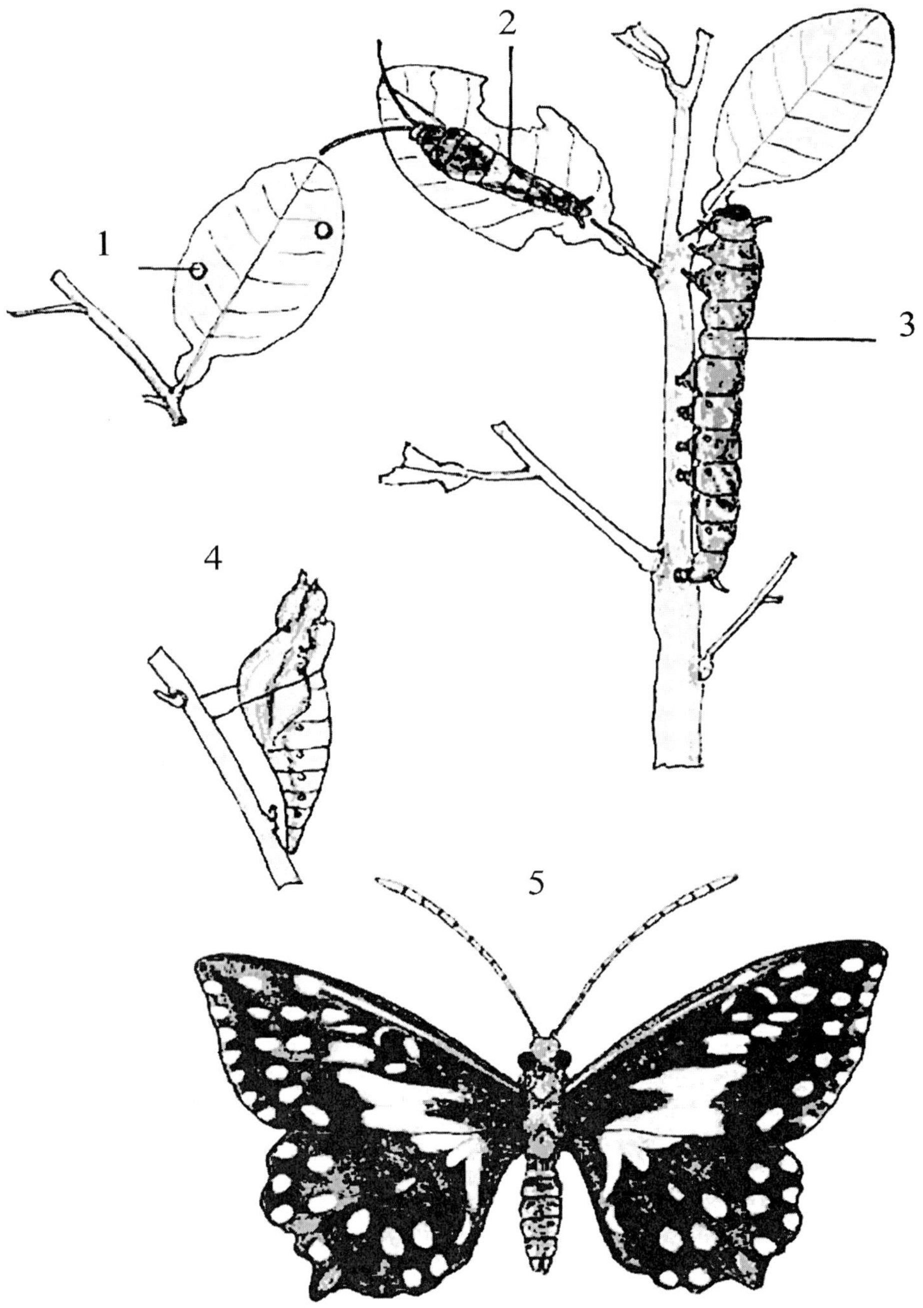

Fig. 76. Lemon butterfly-*Papilio demoleus*

1. Egg 2. Young Larva 3. Grown up Larva 4. Pupa 5. Butterfly

3. Citrus leaf miner

Phyllocnistis citrella

Order - Lepidoptera

Family - Phyllocnistidae

Nature of damage. This is one of the serious pests of all citrus plants and causes extensive damage to the plants, especially the young plants. The caterpillars mine into the young leaves and feed on the inner tissues, as a result the leaves curl, fade and often dry up. Attacked leaves show peculiar, characteristic, glistening, white, irregular, zigzag, streak-like mines and very often the tiny caterpillars are found inside the mines and the excreta of the caterpillars is also seen as minute black dots. All the growth stages of the pest are completed within the leaf mine itself (Fig. 77).

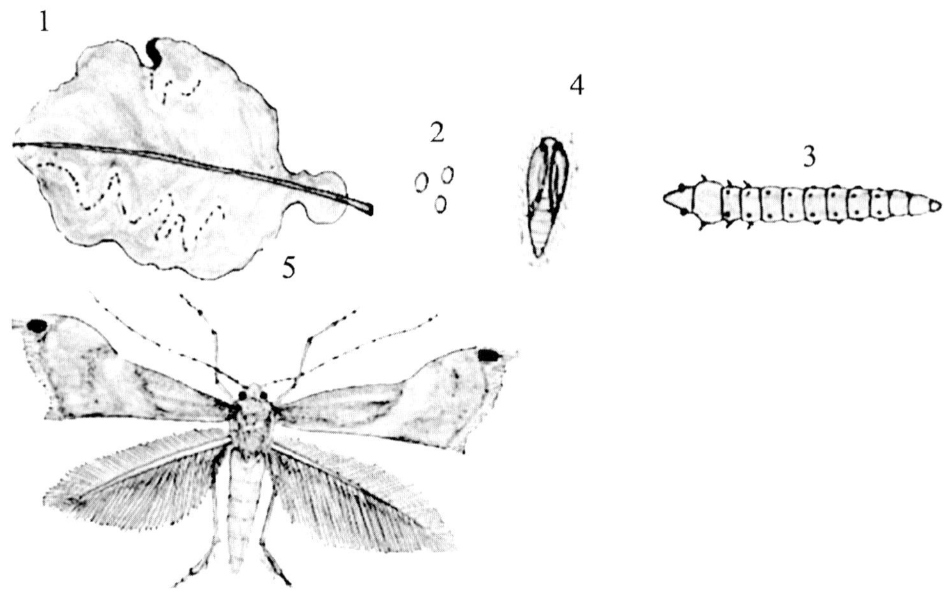

Fig. 77. Citrus leaf miner-*Phyllocnistis citrella*

1. Infested leaf 2. Eggs 3. Larva 4. Pupa 5. Adult moth

Life cycle of the pest. The female moth lays minute, round, flattish eggs singly on the tender leaves or on the tender shoot portions. The young larva, which emerges from the egg in 2-5 days, mines into the leaf, starts feeding on the inner tissues and grow. The larval stage lasts for 7-15 days. The grown-up caterpillar is very tiny and pale whitish in color. The full-grown caterpillar pupates inside the mine near the margin of the leaf within a cocoon made up of fine, silken thread. The adult moth emerges from the pupa in 10-20 days. The moth is tiny and straw-colored. The wings are fringed with fine, dense hairs around the margin and near the tip of each of the forewings, a black spot is present. Several generations may appear in a year. During the summer months, the infestation is generally low. The plants infested by this pest are found to be more susceptible to **'Tristeza'** virus disease (Fig. 77).

Control measures

Chemical control (i) Spraying the foliage with methyl demeton-0.025% or dimethoate-0.06% or phosphamidon-0.1% or formothion-0.05% or monocrotophos-0.04% affords effective control of the pest (ii) Spraying the crop with neem seed kernel extract-5.0% gives good control of the pest. To prepare neem seed kernel extract-5.0%, 10 kg. of neem seed kernel powder is taken and soaked in 50 liters of water for 12 hours. Then the extract is filtered and made up to 200 liters. To this extract liquid soap-200 ml. is added, stirred thoroughly and then used for spraying.

4. Citrus psyllid bug

Diaphorina citri

Order - Hemiptera
Suborder - Homoptera
Family - Psyllidae

Nature of damage. The pest is found in all parts of India and infects all types of citrus plants. Both the nymphs and adults harbor on the tender shoot portions or on young leaves in large numbers, suck and feed on the plant sap from tender shoots, young leaves, flower buds and flowers and cause severe damage. Severely infested shoots are blighted. The newly emerging shoots are more vulnerable to attack by this pest and in such shoots, the leaves curl, shrivel and ultimately the entire shoot dries completely. The infested plants become stunted and weak.

Life cycle of the pest. The female bug lays very small, long, cylindrical, yellowish eggs, singly on the young shoots and tender leaves. One female lays 150-200 eggs. The eggs hatch in 5-6 days. The just emerged young nymph, after crawling about for sometime fixes itself in one place, becomes sedentary and appears as a scale. The short legs and antennae present in the young nymphs disappear in about 30 days and for about 3 months, the insect remains in the same place and continues to feed on the sap. The full-grown nymph attains a pupal stage in the same place and emerges as an adult bug in about 14 days. The bugs are very small, about 1.0 mm. in length, light yellowish in color and with very small, delicate wings. The entire body and the wings of the bugs are covered with a white, powdery, waxy coating. They generally hop around on the plants and cannot fly long distances. The females are parthenogenetic and the males are rarely found (Fig. 78).

Control measures

Chemical control (i) Spraying the crop with tobacco decoction or Nicotine sulfate 40 S-0.08% is effective in controlling the pest (ii) Spraying with monocrotophos-0.04% or methyl demeton-0.025% also controls the pest.

5. Citrus white fly

Aleurocanthus spiniferus

Order - Hemiptera
Suborder - Homoptera
Family - Aleyrodidae

Nature of damage. The nymphs and adults of the pest harbor on the tender shoots and young leaves, suck and feed on the plant sap continuously and in case of severe infestation, the leaves become yellow and drop off prematurely. The fruit production and the quality of the fruits are also adversely affected. The thick, sugary, honeydew secreted by the insects spreads on the surface of plant parts, which invites many other insects that come to feed on the liquid. On these areas, **'sooty mould'**, a fungal disease appears, which interferes with the photosynthetic activities of the host plants.

Life cycle of the pest. The female white fly lays very small, yellow-colored eggs singly on the under surface of young leaves. The eggs are borne at the end of long, fine, hard, thread-like stalks. One female lays about 100 eggs.during its life time. The eggs hatch in 7-10 days. The young nymph after emerging out from the egg, crawls about for sometime, selects a particular place, fixes itself in that place permanently, sucks and feeds on the sap continuously. The nymph is ovoid in shape, flattish, scale-like, brown-colored and with spiny appendages on the body and all around the margin. It exudes a honeydew-like liquid continuously. The nymph becomes full-grown in 3-4 weeks, turns into a pupa in the same place and emerges as an adult fly within a few days. The adult white fly is very small, delicate, white or brownish in color and is capable of flying short distances. The body and forewings are covered with a white, powdery, waxy coating. The adults live only for a few days (Fig. 79).

Control measures. The measures suggested for the control of psyllid bug may be adopted.

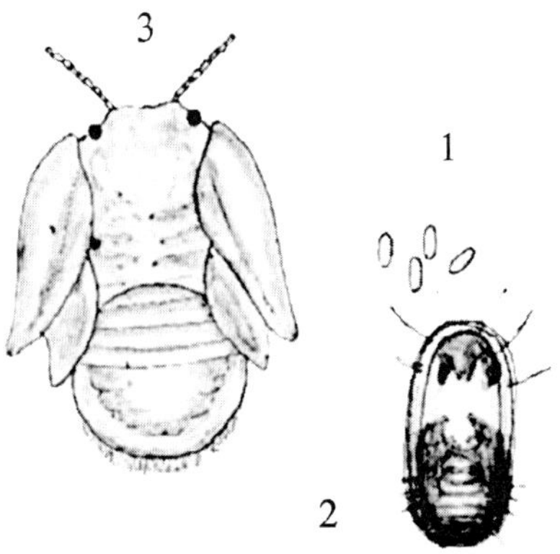

Fig. 78. Citrus psyllid bug-*Diaphorina citri* 1. Egg 2. Nymph 3. Adult bug

Fig. 79. Citrus white fly-*Aleurocanthus spiniferus* Adult white fly

Other pests. Citrus plants are subjected to attack by many other pests. However, they occur only occasionally and are not very serious. The citrus aphids-*Toxoptera citricidus* and *T. aurantii* and the citrus mealy bug-*Pseudococcus citri* infest the young shoots and tender leaves, suck and feed on the sap. The citrus aphid *Toxoptera citricidus* is also the vector that transmits the virus causing **'tristeza virus disease'**; the citrus shoot and bark borer-*Indarbela tetraonis* bores through the soft portion, where the branches fork, constructs long galleries with silken thread, wood waste and frass, remains hidden inside and feeds on the bark and stem tissues; the citrus stem borer beetle or orange borer-*Chloridolum alcamene* and *Chelidonium cinctum,* both longicorn beetles, the grubs of which bore into the stem of plants and cause death of branches or the whole plant; the cottony cushion scale-*Icerya purchasi,* an exotic sucking pest also causes damage; the white fly-*Dialeurodes citri* also infests citrus plants; the leaf roller-*Tonica zizypi* feeds on the leaves; the fruit flies-*Bactrocera (Dacus) dorsalis* also damage the fruits; the nymphs and adults of the hard scale insect-*Aonidiella aurantii* fix themselves on the tender shoot and young leaves, suck and feed on the plant sap; the grubs of the dark blue jewel beetle-*Belinota prasina* bore into the stem or branches, burrow inwards and feed on the inner tissues, as a result the branches or even the tree may die; the red tree ants-*Oecophylla smaragdina* bind leaves together to form cages and cause annoyance.

BANANA *(Musa* species*)*

1.Banana rhizome weevil

Cosmopolites sordidus

Order - Coleoptera

Family - Curculionidae

Nature of damage. The grubs of the weevil bore into the rhizomes of banana plants, make criss cross tunnels inside the rhizomes, feed on the inner tissues and cause damage. When the grubs burrow and damage the root primordia, the suckers die. The infested grown-up plants show retarded growth and the affected rhizomes are eventually infected by soil-inhabiting fungi and bacteria leading to diseases such as, 'wilt', which kill the plants.

Life cycle of the pest. The female weevil bites small holes in the basal portion of the stem, leaf sheaths next to the ground level, rhizome or cut portions of the stem and lays eggs inside the holes singly. The eggs are long, cylindrical and whitish in color. One weevil lays 10-15 eggs. The eggs hatch in 3-4 days and the footless, young grubs emerging out of the eggs burrow into the rhizomes, feed on the inner content and become full-grown grubs in 14-42 days. The grown-up grub pupates in the tunnel inside the rhizome or in the soil and emerges as an adult weevil in 7-15 days. The newly emerged weevil is reddish in color and the color changes to black within a few days. The full-grown weevil is 1.0-2.0 cm. in length, shiny black in color and with a downward curved, long snout. The weevils hide in the daytime in the leaf sheath or near the suckers and become active during the nights and feed on the plant parts (Fig. 80).

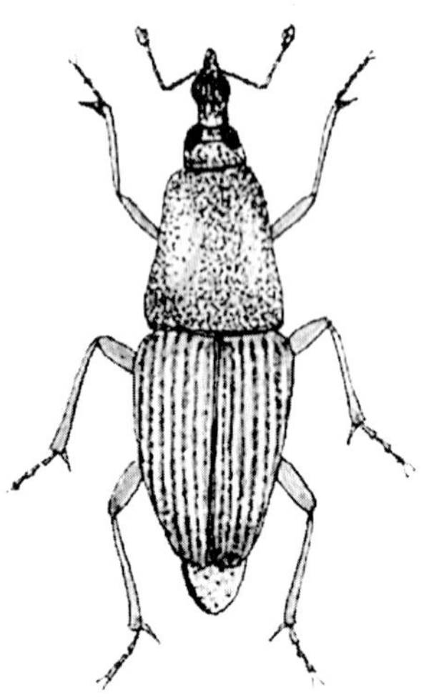

Fig. 80. Banana rhizome weevil-*Cosmopolites sordidus* Adult weevil

Control measures

Cultural methods. Care should be taken to select good suckers, which are free from pest infestation and preferably from uninfested fields.

Chemical control (i) Prior to planting, the suckers should be treated in dieldrin-0.1% (2.0 ml./liter of water) or aldrin-0.2% (4.0 ml./liter of water) solution by immersing the rhizomes of the suckers in the insecticidal solution for one minute (ii) Prior to planting, dieldrin 5 D or fensulfothion 10 G is spread inside the planting pit at the rate of 30 gm. per pit and incorporated into the soil (iii) Drenching the soil inside the pit with chlorpyrifos-0.1% (5.0 ml./ liter of water) to wet the soil to a depth of about 15 cm. is also effective in controlling the pest (iv) If the pest infestation is noticed in the planted crop, application of dieldrin 5 D or fensulfothion 10 G uniformly in the soil around the basal part of the stem at the rate of 30 gm. per plant and incorporating it into the soil controls the pest.

2. Banana aphid

Pentalonia nigronervosa

Order - Hemiptera

Suborder - Homoptera

Family - Aphididae

Nature of damage. Both the nymphs and adults congregate in-between the leaf sheath and stem on the basal portion of the plant and inside the whorls of unopened leaves of the shoot in colonies, puncture, suck and feed on the plant sap continuously. The constant drainage of sap from the plants, make them weak and unthrifty. The insects secrete a honeydew-like liquid through honey tubes, which invites many other insects and ants that come to feed on the sweet liquid. **'Sooty mould'**, a fungal disease also appears on the surface covered by the secretion. Besides causing direct damage to the plants, the pest is the vector of the virus causing **'bunchy top virus disease'**. The pest also infests cardamom, *Colocasia antiquorum* and a few other hosts.

Life cycle of the pest. The female aphids reproduce parthenogenetically and are viviparous. One female lays 32-50 nymphs during its life span. Both winged and wingless forms of females are found. Males are very rarely found. The full-grown aphids are very small, delicate, soft-bodied and dark brown in color. The nymphs attain adulthood within a short period of time and start reproduction and thus the pest multiplies very rapidly.

Control measures

Cultural method. Yellow sticky traps may be set up in the field at the rate of 10 traps per acre to attract and kill the winged aphids.

Chemical control. Spraying the plants from the shoot down to the basal part of the stem with methyl demeton-0.025% or dimethoate-0.06% affords satisfactory control of the pest.

Other pests. Several other pests are also noticed on banana plants. The nymphs and adults of the banana lace wing bug-*Stephanitis typica* pierce and suck the sap from under the surface of leaves causing white spots on the upper surface of the feeding spots (Fig. 81); *Aspidiotus destructor*-a sucking scale insect pest also infests banana leaves causing yellowing in patches and sometimes attacks the fruits also (Fig. 82); the leaf caterpillars-*Pericallia ricini* and *Spodoptera (Prodenia) litura* feed voraciously on the leaves; the adults of the leaf beetle-*Nodostoma subcostatum* remain hidden under unopened leaves and feed on the tender leaves and fruits; the cotton aphid-*Aphis gossypii* besides sucking sap from the leaves and causing direct damage, transmits the virus causing 'banana mosaic disease'; the maggots of the fruit fly-*Bactrocera (Dacus)* *dorsalis* infest banana fruits and cause damage; the nymphs and adults of the leaf thrips-*Helionothrips kaladiphilus* and *Panchaetothrips indicus* suck and feed on the sap from the leaves, while the flower thrips-*Thrips florum* suck and feed the sap from the floral parts.

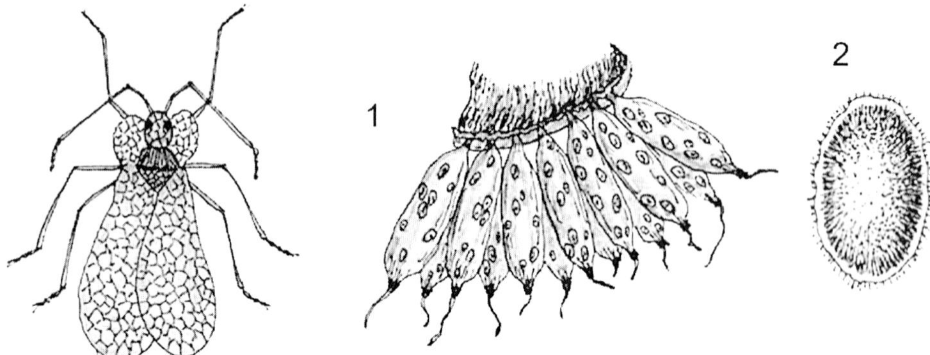

Fig. 81. Banana lace wing bug-*Stephanitis typica* Adult bug

Fig. 82. Banana scale insect-*Aspidiotus destructor* 1. Infected banana fruits 2. Adult female

CASHEW *(Anacardium occidentale)*

1. Cashew tree borer

Plocaederus ferrugineus

Order	-	Coleoptera
Family	-	Cerambycidae

Nature of damage. This is the most important borer pest of cashew. The grubs of the beetle bore into the stem or branches, make long tunnels up and down and feed on the inner tissues. When the vascular region of the branches are attacked and damaged, translocation of food and water to the upper parts above the affected areas is hindered, as a result the branches wilt and die suddenly. In case of severe infestation, many branches and even the whole tree may die. While making tunnels inside the branches or stem, the grub makes a few holes, which open outside and through these holes, the excreta of the grub and dry woody tissues are pushed out. Generally trees that are more than 15 years old are more vulnerable to attack by this pest.

Life cycle of the pest. The female beetle lays eggs singly under the loose barks or in-between the barks of branches. The young grubs hatching out of the eggs, scrape and feed on the bark initially. Later they burrow into the branches or stem and start feeding on the inner tissues and grow. The full-grown grub is whitish in color, large-sized, about 7.5 cm. in length, with a big, prominent, brown head. The body and the head are flattish and the hind portion of the body narrow. The full-grown grub tunnels down to the stem portion, makes a hard, woody cocoon and pupates within it inside the stem portion. The adult that emerges from the pupa is a longicorn beetle, about 5.0 cm. in length, dark brown in color, with long and conspicuous antennae and legs. The life cycle of the pest extends over a period of one month to several months. Many overlapping generations may appear in a year.

Control measures

Cultural methods. The infested and dried branches should be cut and destroyed

Chemical control (i) The bore-holes are cleaned and Petrol or Ether or Chloroform or Carbon disulfide is injected into the hole and the hole sealed with clay mixed with a fungicide such as Copper oxychloride (ii) Monocrotophos-10 ml. or dichlorvos-10 ml. or a mixture of monocrotophos-5.0 ml. + dichlorvos-5.0 ml. or pyrocone-1.0% solution-10 ml. is injected into the hole and the hole sealed with clay mixed with a fungicide (iii) A few crystals of Paradichlorobenzene may be put into the hole and the hole sealed with clay mixed with a fungicide.

2. Cashew stem and bark caterpillar

Indarbela quadrinotata

Order - Lepidoptera

Family - Metarbelidae

Nature of damage. The young caterpillar starts feeding on the soft bark for some distance. Then in that place, it constructs a long gallery with silken thread, frass and bark tissues. Later, it burrows a tunnel downward into the stem. The caterpillar stays hidden inside the tunnel during the daytime and during the nighttime, it comes out of the tunnel, remains inside the gallery and feeds on the bark. It goes on extending the gallery wherever it goes and thus the gallery becomes longer and longer gradually. Only one caterpillar is found in each gallery. Grown-up trees are more prone to attack by the pest than younger trees. In case of severe infestation, the tree becomes weak and may even die in course of time. The pest also infests mango, citrus plants, guava, casurina, curry leaf trees, moringa, jack, silk cotton trees, pomegranate and many other wild and roadside trees (Fig. 83).

Life cycle of the pest. During the months of April-June, the female moth lays eggs in clusters of 16-25 under loose barks or in cracks, mostly at the junction where the branches fork. One female lays 350-600 eggs. The young caterpillar that hatches out from the egg in 8-11 days, starts feeding on the bark, constructs a gallery and then burrows a tunnel in the stem, feeds on the bark by night and becomes a full-grown larva in 40-80 days. The grown-up larva is about 5.0 cm. in length and dark brown in color. It pupates in the tunnel inside the stem and emerges as an adult moth in 14-28 days. The entire life cycle is completed in 2-4 months. The moth is stout, light brown in color, shining and fringed with fine, dense hairs. The forewings are light brown in color, with deep brown, wavy markings across the wings. The hindwings are grayish-black in color (Fig. 83).

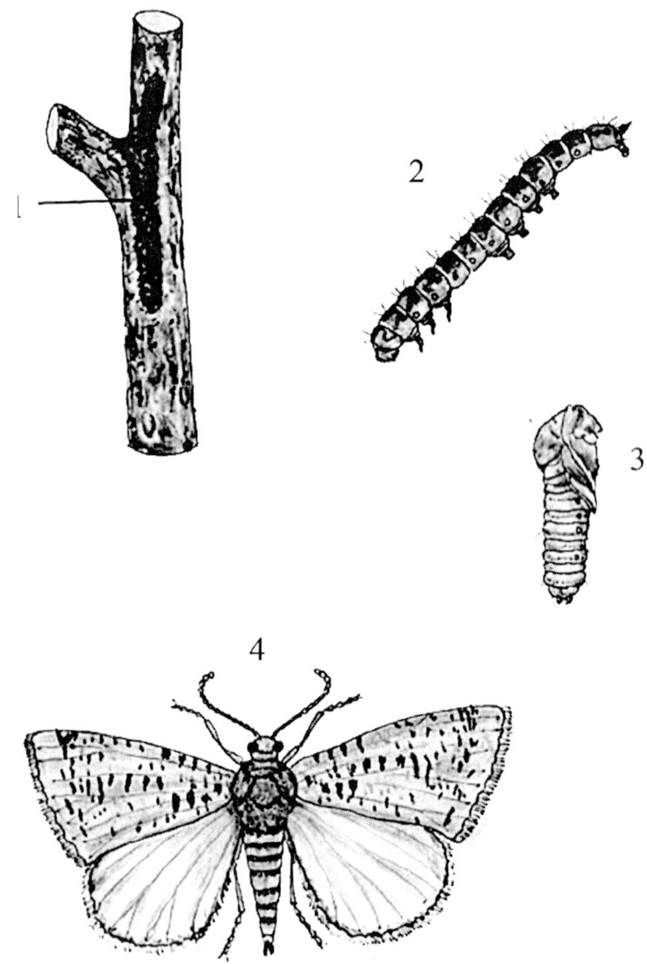

Fig. 83. Cashew stem and bark borer-*Indarbela quadrinotata*

1. Gallery made of silken thread, frass and bark tissue 2. Larva 3. Pupa 4. Adult moth

Control measures

Physical methods. The caterpillars stay under the protection of the galleries while feeding. So, the galleries that are found on the surface of the stem portions should be removed and burnt. Then the stem region is swabbed with lindane-0.04 % or carbaryl 50 WP-0.025 % solution.

Chemical control. The control measures suggested for the control of cashew tree borer may be adopted.

Pests of miner importance. A few other pests are also known to infest cashew trees. The caterpillars of the shoot and flower webber-*Lamida (Macalla) moncusalis* web the shoot and inflorescence together with silken thread, remain inside, bite and feed on the tender leaves and floral parts. In case of severe infestation, many webbed and dried up top shoots with the frass and dry plant matter sticking on to them are seen; the tea mosquito bug-*Helopeltis antonii* sucks and feeds on the sap from the young leaves; the leaf weevil-*Myllocerus viridanus* bites and feeds on the leaf tissues; the caterpillars of the leaf miner-*Acrocercops syngramma* mine into the tender leaves and cause white blotches; the slug caterpillar pest-*Parasa lepida* and the aphid pest-*Toxoptera adinae* also infest cashew; the nymphs and adults of the thrips-*Rhipiphorothrips cruentatus* suck the sap from the tender leaves

POMEGRANATE *(Punica granatum)*

1. Pomegranate fruit borer or Anar butterfly

Deudorix (Virachola) isocrates

Order - Lepidoptera

Family - Lycaenidae

Nature of damage. This is a very serious pest of pomegranate and causes severe damage to the fruits. The caterpillar bores into the ripening fruit, feeds on the seeds and makes the fruit to rot and fall down. The big, circular hole made by the caterpillar is clearly visible on the surface of the fruit and often characteristically plugged by the anal segment of the caterpillar or sometimes with its excreta. In a single fruit there may be 7-8 caterpillars. The pest also infests guava, orange, apple, tamarind, wood apple, soapnut etc., but is more severe on pomegranate (Fig. 84).

Life cycle of the pest. The female butterfly lays shiny white eggs, singly on the basal parts of flowers, immature tender fruits or on tender leaves near the fruits. The young caterpillar hatching out of the egg in 7-10 days, bores a hole into the fruit, burrows inside and feeds on the inner contents and immature seeds. The larval stage lasts for 18-47 days depending upon the weather

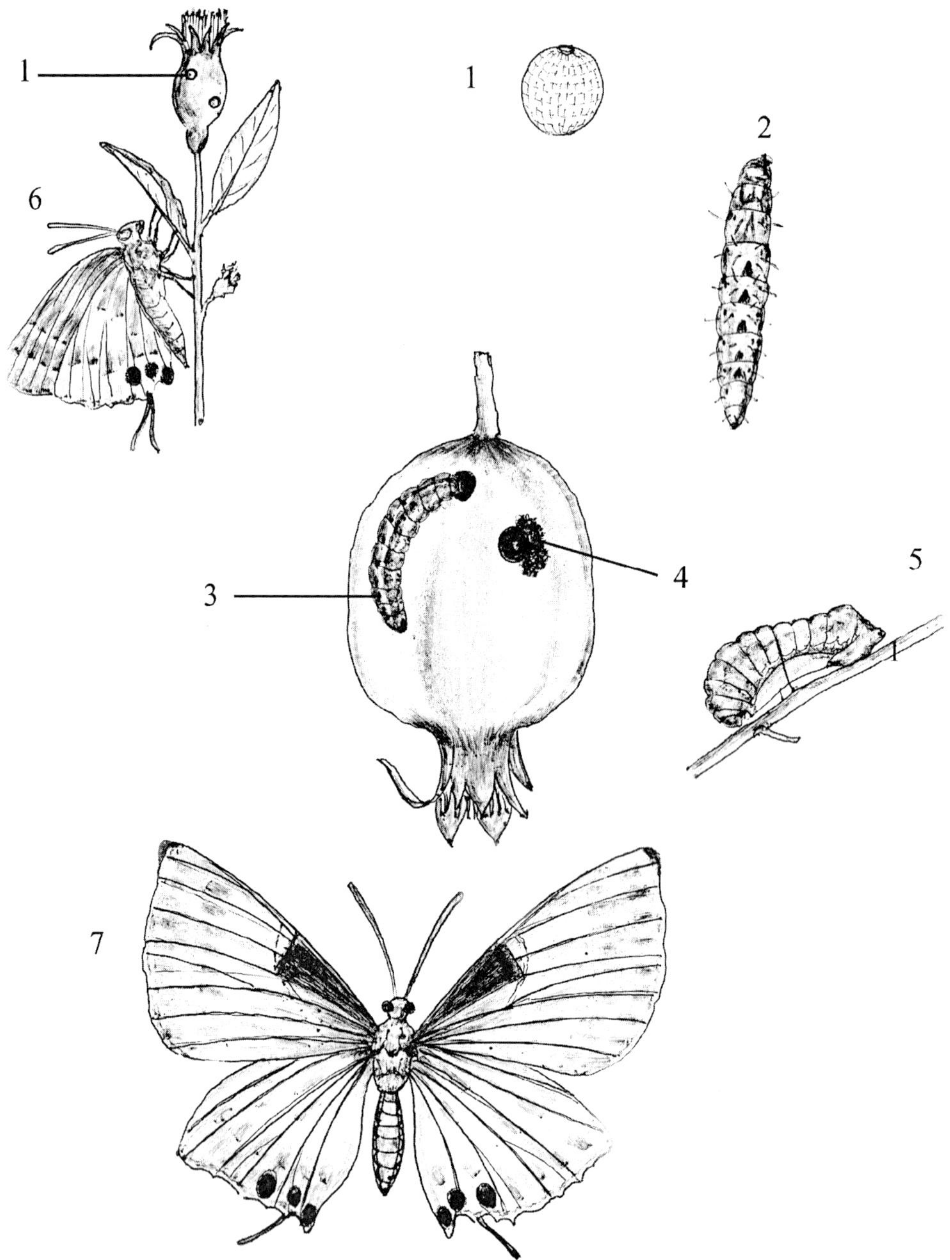

Fig. 84. Pomegranate fruit borer- *Virachola isocrates*

1. Egg 2. Larva 3. Larva boring the fruit 4. Excretory matter pushed out through the bore hole 5. Pupa 6. Butterfly about to lay eggs 7. Butterfly

conditions. The full-grown larva is 1.6-2.0 cm. in length, stoutly built, dirty dark brown in color, with a few short hairs on the body. It pupates either inside the damaged fruit or on the fruit stalk. The adult butterfly emerges from the pupa in 7-34 days. The butterfly measures 2.5 cm. across the wingspan. The male butterfly is bluish-violet in color and shining. The female butterfly is bluish-brown in color, with an orange-colored marking on each of the forewings and black spots and a tail-like extension on each of the hindwings. (Fig. 84).

Control measures

Cultural methods (i) The infested fruits should be removed and destroyed (ii) Covering the fruits with paper, cloth or polythene bags from the early stages of fruit setting prevents infestation of the fruits.

Chemical control. Spraying the crop with malathion-0.1% or endosulfan-0.035% or fenvalerate-0.01% or carbaryl 50 WP-0.2% at the time of fruit setting controls the pest effectively.

Miner pests. Besides the pest described above, the castor hairy caterpillar or tussock caterpillar-*Euproctis fraterna,* the castor semi-looper-*Achaea janata* and the slug caterpillar-*Parasa lepida* bite and feed on the leaves; The bark feeding caterpillar-*Indarbela quadrinotata* constructs galleries on the stem and branches with silken thread, wood particles and excreta, stays inside the gallery or inside tunnels burrowed in the stem and feeds on the bark; the castor capsule borer-*Conogethes (Dichocrosis) punctiferalis* and the maggots of the pomegranate fruit fly-*Bactrocera (Dacus) zonatus* attack the fruits and feed on the inner content; the nymphs and adults of the pomegranate white fly-*Siphoninus phyllireae* and the pale green aphid-*Aphis punicae* colonize the tender shoots, suck and feed on the sap; the mealy bugs-*Ferrisia virgata* and *Pseudococcus lilacinus* infest the fruits; the nymphs and adults of the thrips-*Rhipiphorothrips cruentatus* suck and feed on the sap from the under surface of leaves; the scale insects-*Aspidiotus rossi* stick on to the leaves and young shoots, suck and feed on the plant sap; the mites-*Oligonychus punicae* form colonies on the ventral surface of leaves, suck and feed on the sap from the leaves, as a result brown patches are formed and the leaves curl and dry up.

GUAVA *(Psidium guajava)*

Several pests, which infest other fruit trees are found to attack guava also. The nymphs and adults of the tea mosquito bug-*Helopeltis antonii* infest the tender shoots, young leaves, flower buds, flowers and young fruits, puncture, suck and feed on the sap from the plant parts. The punctured areas become black, hardy spots. The shoots, leaves and flowers are deformed and dry. On the attacked fruits, small, black spots are formed on the surface.

These spots later become dried up, turn hard and appear as cork-like, scabby, elevated spots and make the fruits unfit for marketing. The pest, while feeding injects saliva into the tissues and the toxic substances present in the saliva are responsible for producing such symptoms. The pest can be controlled by periodical spraying with malathion-0.1 % from the time of fruit setting; the nymphs and adults of the scale insect-*Chloropulvinaria (Pulvinaria) psidii* and *Coccus viridis,* the mealy bug-*Ferrisia virgata,* the aleyrodid bug-*Aleurotuberculatus psidii* and the cotton aphid-*Aphis gossypii* colonize the tender shoots, young leaves, flower buds, flowers and fruit stalks, suck and feed on the sap; the shoot and bark caterpillar-*Indarbela quadrinotata* and *Indarbela tetraonis* scrape and feed on the bark of the stem and branches; the castor capsule borer-*Conogethes (Dichocrosis) punctiferalis* and the pomegranate fruit borer-*Deudorix (Virachola) isocrates* bore into the fruits, feed on the inner content and cause damage; the maggots of the fruit fly-*Bactrocera (Dacus) dorsalis* infest the fruits; the leaf weevil-*Myllocerus destructor* cuts and feeds on the leaves.

Jack *(Artocarpus heterophyllus)*

1. Jack fruit borer

Glyphodes (Diaphania) caesalis

Order - Lepidoptera

Family - Pyraustidae

Nature of damage. The caterpillars bore into the tender shoots, feed on the inner tissues, resulting in drooping, fading and dying of the shoots. When fruits are formed, the caterpillars bore in to the fruits, feed on the inner content and damage the fruits. Saprophytic fungi affect the infested fruits and they rot. The young fruits when attacked usually drop down.

Life cycle of the pest. The young caterpillars, hatching out of the eggs bore into the tender shoots or newly formed fruits, feed on the inner contents and grow. The grown-up caterpillar is reddish-brown in color with a light brown head. On the surface of the body, slightly raised, black warts are present and on these warts, thin hairs are found. The full-grown larva pupates inside the fruit and emerges as an adult moth. The moth is brown-colored and the brown-colored wings have gray-colored, elliptical patterns (Fig. 85).

Control measures

Cultural methods. Young fruits, which had fallen down should be collected periodically and destroyed.

Chemical control. Spraying the crop with malathion-0.1% or endosulfan-0.035% at the time of fruit setting controls the pest.

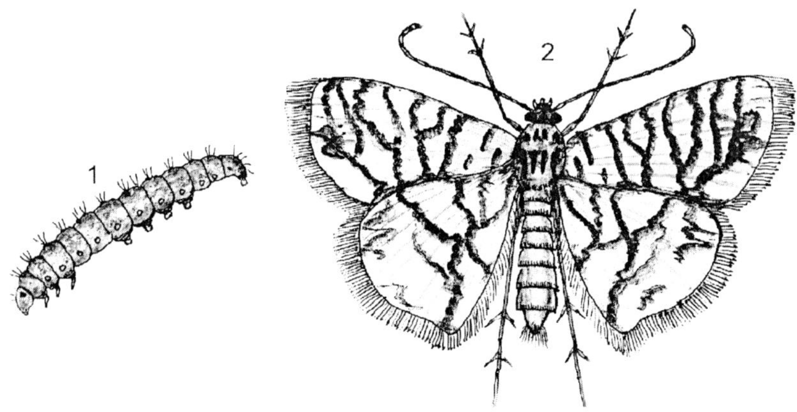

Fig. 85. Jack fruit borer-*Glyphodes caesalis*

1. Larva 2. Moth

Other pests. A few other pests attack Jack, but they are considered to be of miner importance. The small, whitish grubs of the small, gray-brown, jack bud weevil-*Ochyromera artocarpi* bore into the flower buds and young fruits and feed on the inner content, as a result the buds and fruits drop down (Fig.86); the shoot and bark caterpillar-*Indarbela quadrinotata* scrapes and feeds on the bark of the stem portion; the aphids-*Toxoptera aurantii* and *Greenidia artocarpi* and the mango mealy bug-*Drosicha mangiferae* suck and feed on the sap from the tender shoots and young leaves.

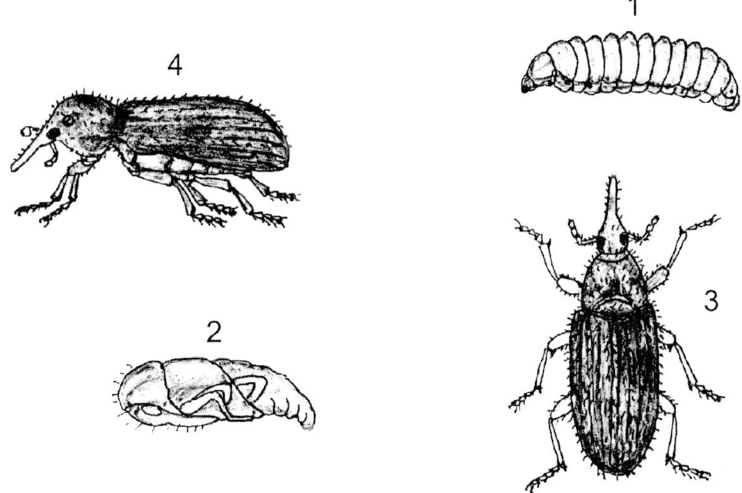

Fig. 86. Jack bud weevil-*Ochyromera artocarpi*

1. Grub 2. Pupa 3,4. Weevil

GRAPEVINE *(Vitis vinifera)*

1. Grape leaf beetle or Grape flea beetle

Scelodonta strigicollis

Order - Coleoptera

Family - Eumolpidae

Nature of damage. This is a specific pest of grapes. After pruning of the vines, when fresh shoots appear, the attack by the pest is more severe. The beetles bite and cut the newly emerging tender shoots. They bite long holes in the young leaves and feed on the leaf tissues. At the time of flowering the beetles attack and feed on the flower buds also. The grubs cut and feed on the roots from under the soil.

Life cycle of the pest. The mother beetle lays eggs singly or in small clusters under the bark of the stem portion or in the soil. The grubs that emerge from the eggs in about 4 days, start biting and feeding on the roots. The grubs can go into the soil up to a depth of 15 cm. under the soil and feed on the roots. The full-grown grub constructs an earthen cocoon under the soil and pupates inside it. The pupation period lasts for about 14 days and after that period, the adult beetle emerges from the pupa. The life cycle lasts for about 2 months. The beetle is very active, coppery-brown in color, about 8.0 mm. in length, with 6 dull black spots on the forewings.

Control measures

Physical methods. The beetles can be caught by using a hand net and destroyed

Chemical control (i) Soon after pruning, the crop should be sprayed with carbaryl 50 WP-0.1%. The sprayings should be continued at an interval of 15 days till fruit setting (ii) When the fruits are nearing maturity, if pest infestation is noticed, the crop should be sprayed with malathion-0.1%.

2. Grape stem girdler

Sthenias grisator

Order - Coleoptera

Family - Cerambycidae

Nature of damage. The beetles bite, cut and girdle the stem and branches and cause severe damage to the vines. The girdled or ringed branches or sometimes even the entire vine may wilt and die. The grubs also bore into the stem, feed on the inner tissues and cause damage. The pest also infests mulberry, casuarina, mango, jack, *Bougainvillea spectabilis* and crotons (Fig. 87).

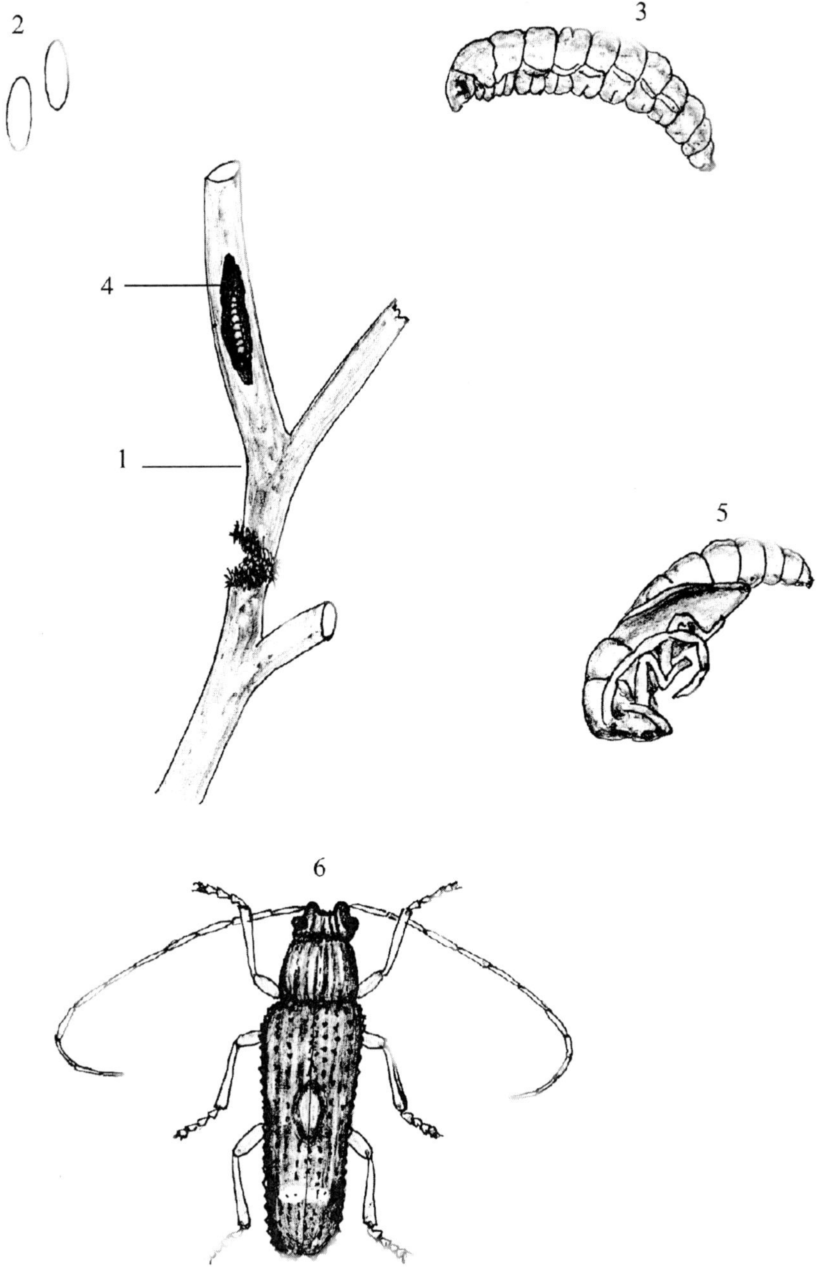

Fig. 87. Grape stem girdler-*Sthenias qrisator*

1. Stem girdled by the beetle 2. Eggs 3. Larva 4. Larva inside dried branch 5. Pupa 6. Beetle

Life cycle of the pest. The female beetle cuts or rings the outer skin of the stem or branch and lays eggs in the injured area. The young grub, which hatches out from the egg bores into the stem or branch through the dried portion, feeds on the inner tissues and becomes full-grown in 8-9 days. The grown-up grub pupates inside the stem or branch and emerges as an adult beetle in 10-12 days. The adult is a longicorn beetle, strongly built and medium-sized. The grayish beetle with long antennae and legs, has very strong, biting and chewing mouthparts (Fig. 87).

Control measures

Physical methods (i) The adult beetles may be caught by using hand nets and destroyed (ii) The girdled or ringed branches are cut off and destroyed.

Chemical control. The vines may be sprayed with carbaryl 50 WP-0.2 % solution, so as to cover the stem and branches thoroughly to prevent the beetles from girdling the stem or branches and laying eggs in the injured areas.

3. Grape leaf thrips

Rhipiphorothrips cruentatus

Order - Thysanoptera

Family - Thripidae

Nature of damage. Both the nymphs and adults infest the leaves, mostly the young leaves in large numbers, lacerate, suck and feed on the sap, as a result the leaves curl up, dry and drop down. The pest lacerates the fruits also and the fruits develop scabby patches. The infestation is more severe during the hot weather. The pest infests rose plants also.

Life cycle of the pest. The female insect inserts the eggs into the tissues of the tender shoots and leaves. The nymphs that hatch out of the eggs are very minute, pinkish in color and wingless. The adult insect is tiny, fragile and brownish in color with fringed wings like that of a comb (Fig. 94).

Control measures

Chemical control. The pest can be controlled by foliar spraying with methyl demeton-0.05% or dimethoate-0.03% or phosphamidon-0.1%.

Other pests. Grapevine is subjected to attack by a few other pests apart from the ones described above. Grape fruits are infested by the chilli thrips-*Scirtothrips dorsalis,* which lacerate the skin of the fruits, suck and feed on the juice, as a result the surface of the fruits become rough and coarse and mosaic-like patches are formed on the fruit surface. The size of the fruits is considerably reduced and the taste of the fruit is also spoiled; the mealy bugs-*Ferrisia virgata* and *Planococcus citri* colonize the young shoots and twigs, suck and feed on the sap; the fruit sucking moths-*Othreis materna, O. fullonica* and *Achaea janata* puncture the fruits suck and feed on the juice, as a result the fruits rot; the nymphs and adult females of the sedentary hard scale insect-*Aspidiotus cydoniae* suck and feed on the sap from the tender shoot portion; the leaf roller-*Sylepta lunalis* feeds on the leaves.

APPLE *(Pyrus malus)*

The apple wooly aphid-*Eriosoma lanigera* is a cosmopolitan and very serious pest of apple. Colonies of nymphs and adults, which look like white, cottony patches on the shoots and branches, suck and feed on the plant sap and cause severe damage. Young plants may die due to the attack by the pest. In the winter season, the migrating insects attack the roots below the ground, suck and feed on the sap, as a result small, gall-like swellings appear on the roots and impair the efficiency of the roots. The infested plants loose their vigor and become stunted. The purplish aphids are covered with a white, cottony mass. The severity of attack is maximum during July-August. Spraying the aerial parts with dimethoate-0.06 % or monocrotophos-0.04 % or phosphamidon-0.005 % during spring and again during summer controls aerial infestation. Soil application of paradichlorobenzene crystals-50 to 100 gm. or thiodemeton 5 G-300 gm./ tree controls root infestation.

The apple San Jose scale-*Quadraspidiotus perniciosus* is a cosmopolitan, polyphagous sucking pest. The nymphs and adults attach themselves to the leaves, twigs and fruits in colonies, suck and feed on the plant sap, resulting in loss of vigor of the host plants. Infestation of fruits causes purple discoloration, making them unfit for marketing. Severely infested young plants may die. The female scale is round and slightly convex, yellowish-orange in color, apoderous and apterous. The female scale insect is ovoviviparous and lays 300-400 nymphs during its life span. The males are slightly smaller than the females and linear. The adult males possess a pair of wings. There may be 3-4 generations in a year. Besides apple, the exotic pest infests cherry, plum, pear, peach etc. Spraying with dimethoate, monocrotophos or phosphamidon can control the pest.

Several other pests also infest apple. The hairy caterpillars-*Euproctis signata, E. fraterna* and *E. flava* feed voraciously on the foliage and defoliate the trees; the larvae of the Indian gypsy moth-*Lymantria obfuscata* feed gregariously on the leaves at night and defoliate the trees; the larvae of the apple leaf miner-*Gracillaria zachrysa* mine the leaves and later roll young leaves longitudinally into tubular or conical pouches, remain inside securely and feed on the leaf tissues; the leaf feeding beetles-*Holotrichia longiplennis, Macronota lineata, Melolontha furcicauda* and *Mylabris mecilenta* bite and feed on the leaves; the apple blossom thrips-*Taeniothrips rhopalantennalus* infest the flower buds, suck and feed on the sap from the flower buds.

PEACH *(Prunus persica)* and Pear *(Pyrus communis)*

A number of pests are known to infest peach and pear trees. The grubs of the peach borer beetle-*Sphenoptera lafertei* bore into the branches and stem, make irregular tunnels and feed on the inner tissues. Gum exudates are seen on the bark at the points of entry of the grubs as small globules. The leaves become pale and the growth of the tree is adversely affected; the caterpillars of the pear stem borer-*Sahyadrassus malabaricus* bore into the basal part of the stem, feed on the inner tissues and cause damage to the trees; the aphids-*Myzus persicae, Brachycaudus helichrysi, Hyalopterus pruni* and *Lachnus pyri* infest the leaves, petioles, blossoms and young fruits, suck and feed on the sap, as a result the leaves curl up and fall down. The infested flowers and fruits are also shed; the nymphs and adults of the San Jose scale-*Quadraspidiotus perniciosus* attach themselves to the leaves, twigs and fruits, suck and feed on the sap; the grubs of the fruit flies-*Bactrocera (Dacus) dorsalis* and *Bactrocera (Dacus) duplicatus* bore into the immature fruits and feed on the inner fleshy portions, as a result the fruits rot and fall down prematurely; the hairy caterpillar-*Euproctis fraterna* feeds gregariously on the foliage; the nymphs and adults of the anar white fly-*Siphoninus finitimus* aggregate in large numbers on the tender shoots, sucks and feeds on the plant sap, resulting in loss of vitality of the trees and reduction in fruiting capacity; the weevils-*Myllocerus viridanus* and *M. subfasciatus* bite and feed on the leaf tissues.

PLUM *(Prunus domestica)*

The nocturnal, dark green plum beetles, with yellowish-brown elytra-*Anomala aineatopennis* feed on the leaves and fruits, while the grubs bite and feed on the roots below the ground; the plum weevil-*Amblyrrhinus pricollis* and other polyphagous weevils-*Myllocerus lactivirens* and *M. undecimpustulatus* feed on the green matter of tender leaves; the nymphs and adults of the San Jose scale-*Quadraspidiotus perniciosus* suck and feed on the sap from the shoots; the hairy caterpillars-*Euproctis fraterna* and *E. signata* feed voraciously on the foliage; the grubs of the peach borer beetle-*Sphenoptera lafertei* bore into the stem and branches, feed on the inner tissues and cause damage to the trees.

ALMOND *(Amygdalus vulgaris)*

The almond weevil-*Myllocerus lactivirens* and other polyphagous gray weevils-*Myllocerus discolor, M. undecimpustulatus,* the aphids-*Myzus persicae* and *Brachycaudus helichrysi,* the peach borer-*Sphenoptera lafertei* and the San Jose scale-*Quadraspidiotus perniciosus* infest almond trees.

SAPOTA *(Achras sapota)*

1. Sapota leaf webber or Chikoo moth

Nephopteryx eugraphella

Order - Lepidoptera

Family - Phycitidae

Nature of damage. The caterpillars of this common pest web the young leaves together, remain inside the webs and feed on the leaf tissues. The caterpillars also feed on the flower buds and young fruits. Inside each fruit 3-4 caterpillars may be found.

Life cycle of the pest. The female moth lays pale yellow eggs singly or in groups of 2-3 on the tender leaves. The young caterpillars hatching out from the eggs in 3-5 days feed on the leaf tissues and become full-grown in 17-32 days. The grown-up larva is pinkish in color, about 2.5 cm. in length and with close-set, dark, longitudinal lines on the dorsum. It pupates inside the leaf web and emerges as an adult moth in 7-11 days.

Control measures

Physical method. The leaf webs with the larvae may be collected and destroyed.

Chemical control. Foliar spraying with monocrotophos-0.04% or cypermethrin-0.005% or phosalone-0.07% or endosulfan-0.07% is effective in controlling the pest.

Other pests. The caterpillars of the anar butterfly-*Deudorix (Virachola) isocrates* bore into the fruits, feed on the inner fleshy portion and damage the fruits; the sapota hairy caterpillar-*Metanastria hyrtaca* feeds voraciously on the foliage; the maggots of the fruit fly-*Bactrocera (Dacus) dorsalis* and *Bactrocera (Dacus) zonatus* bore into the fruits and feed on the inner fleshy pulp, as a result the fruits rot; the bark eating caterpillars-*Indarbela tetraonis* infest the stem and branches, feed on the bark tissues and cause damage to the trees; the scale insects-*Chloropulvinaria (Pulvinaria) psidii* and *Pseudococcus lilacinus* and the thrips-*Frankliniella dampfi* suck and feed on the sap from the leaves and shoot.

CUSTARD APPLE *(Anona squamosa)*

The black caterpillars of the moth pest-*Anonaepestis (Heterographis) bengalella* bore into the fruits, feed on the fleshy portion and damage the fruits; the mealy bugs-*Pseudococcus virgatis, P. lilacinus* and *Ferrisia virgata* colonize the young leaves, shoots, buds and fruits, suck and feed on the plant sap, as a result the leaves turn yellow and the quality of the fruits is also adversely affected; the maggots of the fruit fly-*Bactrocera (Dacus) zonatus* bore into the fruits, feed on the fleshy pulp and damage the fruits; the nymphs and adults of the scale insect-*Ceroplastes floridensis* attach themselves to the shoots, flower and fruit stalks and fruits, suck and feed on the plant sap.

PESTS OF PLANTATION CROPS

TEA *(Camellia sinensis)*

1. Tea mosquito bug

Helopeltis antonii

Order - Hemiptera

Suborder - Heteroptera

Family - Miridae

Nature of damage. The nymphs and adults harbor on the tender shoot region, buds and young leaves, make numerous punctures on the plant parts, suck and feed on the plant sap. The infested leaves turn black, curl, wilt and die. The insect, while feeding, injects saliva into the tissues, which contains certain toxic substances. Because of this, the area around the punctures turns black. When a number of punctures are thus made, the entire leaf area turns black, the leaves shrivel and fall off. A single insect can cause such injury in two shoots in a day. In case of severe infestation, most of the leaves die and fall down and the shoot appears blighted. Moderate temperature and high atmospheric humidity favor infestation on a large scale. The pest also attacks guava, cashew, neem and some other crops (Fig. 88).

Life cycle of the pest. The female bug inserts the eggs into the tissues of buds and branches from which leaves have been plucked. One female lays about 500 eggs. The eggs are long, cylindrical in shape and have two slender, hair-like processes of unequal length at the tip. The eggs hatch in 5-6 days during the summer months and in 25-27 days during the winter months. The young nymph hatching out of the egg is quite similar to the adult, but smaller in size, orange-red in color, wingless, long and slender, with long antennae and legs and is very active. The nymph becomes a full-fledged adult bug in about 2 weeks during the summer months and in about 8 weeks during the winter months, after moulting 5 times. The adult bug is long, slender, 6.0-8.0 mm. in length and of a mixed red, black and white colors. Two, pimple-like, raised pustules are seen on the prothorax. They are very active and are good fliers. During the daytime, they take shelter and hide under the lower surface of leaves and become active after dusk. They are easily blown and carried to distant places by wind and the infestation spreads in this manner (Fig. 88).

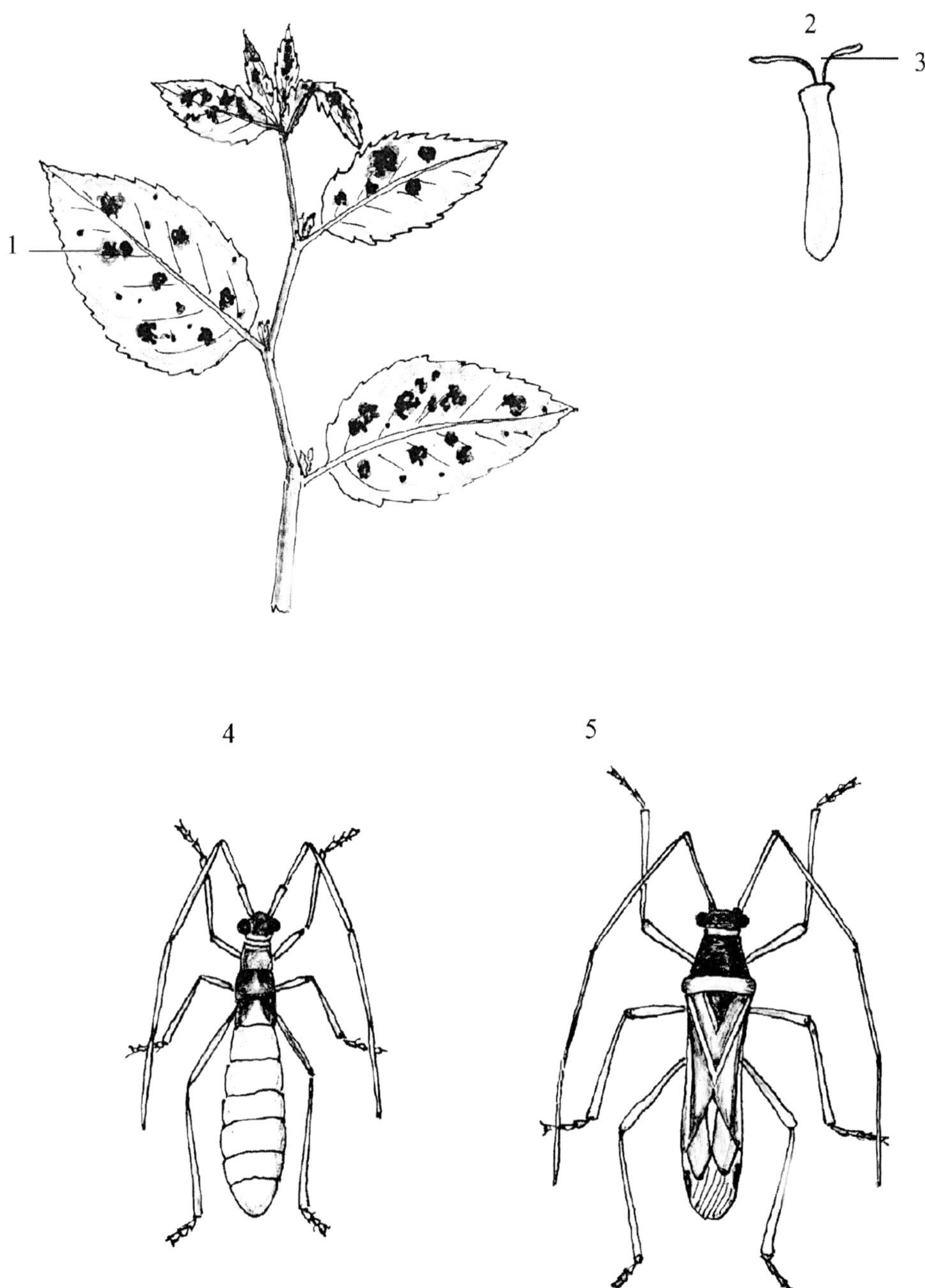

Fig. 88. Tea mosquito bug-*Helopeltis antonii*

1. Infested portions 2. Egg 3. Hair-like processes 4. Nymph 5. Adult bug

Control measures

Physical method. The nymphs and adults can be collected by using hand nets and destroyed.

Chemical control. Spraying the foliage with malathion-0.1% or endosulfan-0.035% or phosalone-0.07% or chlorpyrifos-0.07% is effective in controlling the pest.

2. Tea shot-hole borer

Xyleborus fornicatus

Order	-	Coleoptera
Family	-	Scolitidae

Nature of damage. The grubs of the beetle bore into the stem and branches, make criss cross tunnels inside, feed on the inner tissues and cause extensive damage. The adult beetles also make round holes similar to holes made by bullet shots in the stem and branches and feed on the inner tissues. Wood particles are pushed out through the holes as powder. In case of severe infestation by both grubs and adult beetles, the branches or stem become very weak and breaks easily by strong winds (Fig. 89).

Life cycle of the pest. The adult beetles mate inside the tunnels of live stem or branches. The mother beetle lays eggs on the stem portion. The young grubs, which hatch out of the eggs, bore into the stem or branches, start feeding on the inner tissues and become full-grown grubs. The grown-up grub pupates inside the tunnel in the branches or stem and emerges as an adult beetle. The entire life cycle is completed in 30-50 days (Fig. 89).

Control measures

Cultural method. Severely affected and dried up branches should be cut and destroyed by burning.

Chemical control. The stem portions and branches should be washed with carbaryl 50 WP-0.2 % solution to prevent the beetles from laying eggs on these parts and also to prevent the grubs and beetles from boring and entering into the stem and branches.

Other pests. Several other pests also infest tea plants. The nymphs and adults of the thrips-*Scirtothrips dorsalis* and *Taeniothrips setiventris* lacerate the plant tissues, suck and feed on the plant sap from tender shoots, buds and young leaves, as a result the leaves turn yellow, wilt and ultimately die; the grubs of the coffee stem borer-*Zeuzera coffeae* bore into the stem, feed on the inner tissues resulting in drying of branches or the whole plant in the case of

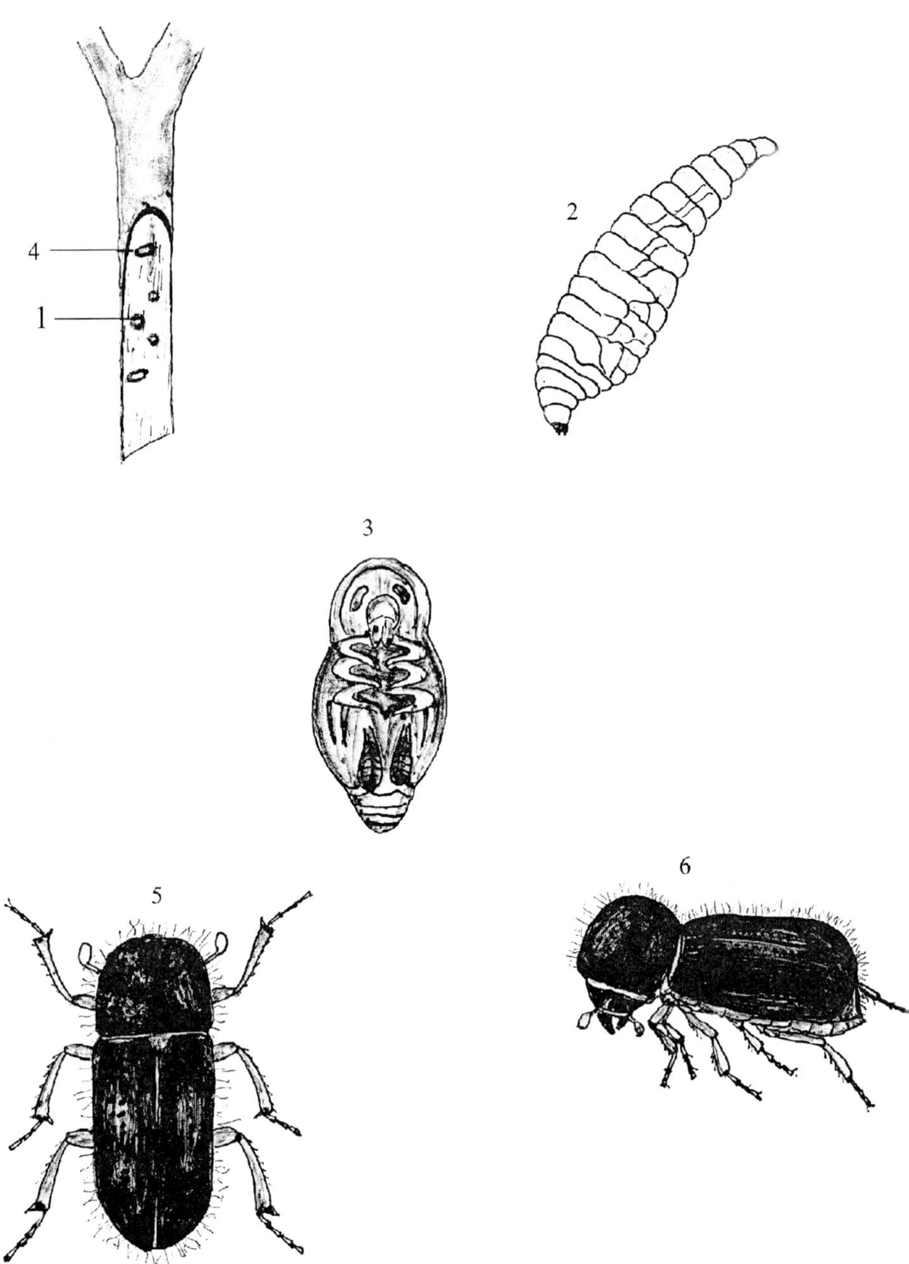

Fig. 89. Tea shot hole borer-*Xyleborus fornicatus*

1. Shot holes in the stem 2. Larva 3. Pupa 4. Pupa inside the bore hole 5. Beetle – Dorsal view 6. Beetle – Lateral view

severe attack; the caterpillars of the looper-*Biston (Buzura) suppressaria* feed on the leaves from the edges and defoliate the plants; the grubs of the beetle pest-*Holotrichia impressa* feed on the roots and damage young tea plants and seedlings in the nurseries; the slug caterpillar-*Contheyla rotunda, Eterusia cingala* and the bag worm-*Clania crameri* feed on the leaves; the aphid-*Toxoptera aurantii,* the scale insect-*Aspidiotus camellae* and the mites-*Tetranychus bioculatus* and *Acaphylla indica* attach themselves on the plant parts, suck and feed on the sap from the shoots and young leaves.

COFFEE *(Coffea arabica* and *Coffea robusta)*

1. Coffee white stem borer

Xylotrechus quadripes

Order - Coleoptera

Family - Cerambycidae

Nature of damage. This is one of the most serious pests of coffee plants. The grubs bore into the stem and branches, remain inside for 8-9 months, make criss cross, irregular tunnels, feed on the inner tissues and cause serious damage to the stem and branches. Severely infested branches fade, dry and die. Such infested branches become very weak and may break easily by strong winds. When many such branches die, the whole plant may die ultimately. Because of severe infestation by this pest, sometimes all the plants may have to be removed and replanted afresh.

Life cycle of the pest. The beetles usually appear in large numbers during April-May and October-November every year. The mother beetle lays eggs in the cracks of barks, under loose barks or on the branches. One female lays 50-100 eggs in 3-4 weeks. The young grubs on hatching out of the eggs in 8-10 days, initially feed on the soft bark tissues for sometime and later bore into the stem, feed on the inner tissues and grow. The larval stage lasts for 8-9 months. The grubs are whitish in color, fleshy, and flattish, with the head region broader and the hind portion narrower. The full-grown grub makes an exit hole in the stem and pupates near it inside the stem and emerges as an adult beetle in 25-30 days through the exit hole already made. The beetle is dark brown in color, about 1.25 cm. in length, with long, conspicuous antennae and long legs. The dark brown forewings (elytra) have yellowish bands across them. Generally there is only one generation in a year, however the beetles emerge in large numbers during April-May and again in October-November (Fig. 90).

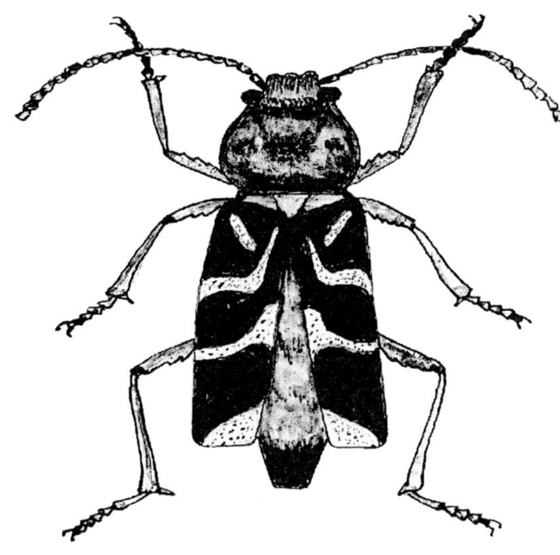

Beetle

Fig. 90. Coffee white stem borer-*Xylotrechus quadripes*

Control measures

Cultural methods (i) The severely infested and dried branches and plants should be removed and burnt (ii) During the time of egg laying, the beetles may be caught by using hand nets and destroyed.

Chemical control. Carbaryl 50 WP at the rate of 1.0 kg. or lindane 20 EC at the rate of 350 ml. is mixed with 50 liters of water and the stem portion and branches are washed with the insecticidal solution once in April and again twice in October-November at an interval of one month to prevent egg laying by the beetles and also to prevent the grubs from boring into the stem and branches.

2. Coffee red stem borer

Zeuzera coffeae

Order - Lepidoptera

Family - Zeuzeridae

Nature of damage. The caterpillars of this pest bore into the main shoot and branches and feed on the inner tissues. The affected branches fade and dry out. In case of severe infestation, the whole plant may wither and die.

Life cycle of the pest. The female moth lays eggs on the shoots and leaves. The young larvae hatching out of the eggs bore into the stem, make long tunnels, feed on the inner tissues and grow. The grown-up larva is orange-red in color, smooth, with blackish head and prothorax and an anal shield. The full-grown caterpillar pupates in the tunnel within the stem and emerges as an adult moth. The moth is dull whitish in color, with dark blue or blackish, small, fine markings on the whitish wings. The life cycle is completed in 2-4 months (Fig. 91).

Control measures. The control measures suggested for the control of coffee white stem borer may be adopted.

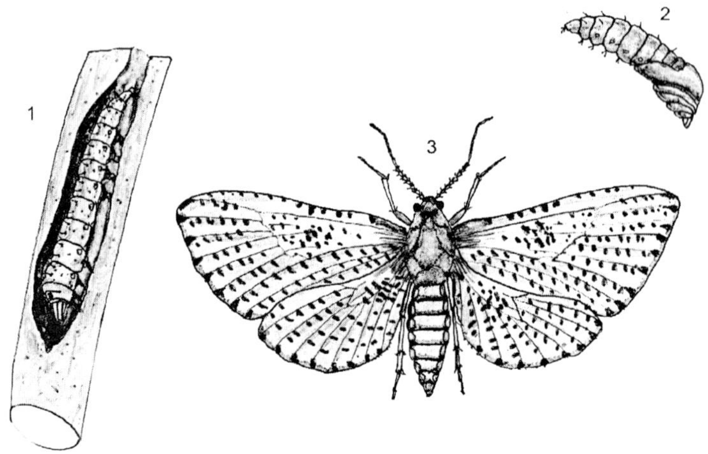

Fig. 91. Coffee red stem borer-*Zeuzera coffeae*
1. Larva inside the stem 2. Pupa 3. Moth

3. Coffee green scale or Green bug

Coccus viridis

Order	-	Hemiptera
Suborder	-	Homoptera
Family	-	Coccidae

Nature of damage. The pest also known as 'green bug' is found in almost all the coffee-growing parts of the world. The nymphs and adults stick on to the succulent shoot region and under surface of leaves, suck and feed on the plant sap. The infested plants become stunted and their vitality is also adversely affected. The sweet, honeydew secreted by the insects spread on the plant parts and on these surfaces, **'sooty mould'** grows and hamper the photosynthetic activity of the leaves, thereby causing additional damage. The pest has several other alternate hosts such as, lime, lemon, pomelo, coconut, teat, jasmine and cinchona.

Life cycle of the pest. The scale insect is ovoviviparous and the young ones hatch out of the eggs just after the eggs are laid. A single female may produce about 500 young bugs. For a short while, the young nymphs remain under the scale of the mother and then crawl out for feeding. The young ones, as well as the adults do not fix themselves in any place permanently. When the food supply becomes unsatisfactory at a particular site, they move elsewhere. The insects prefer the under surface of the leaf, but may occasionally colonize the tender shoot region and also the coffee berries. The adult female scale insect is about 3.0 mm. in length, ovoid in shape, with a hard, waxy scale-like covering, yellowish-green in color, slightly elevated and lacks antennae and legs. A winged stage is lacking. The life cycle is completed in about 2 months (Fig. 92).

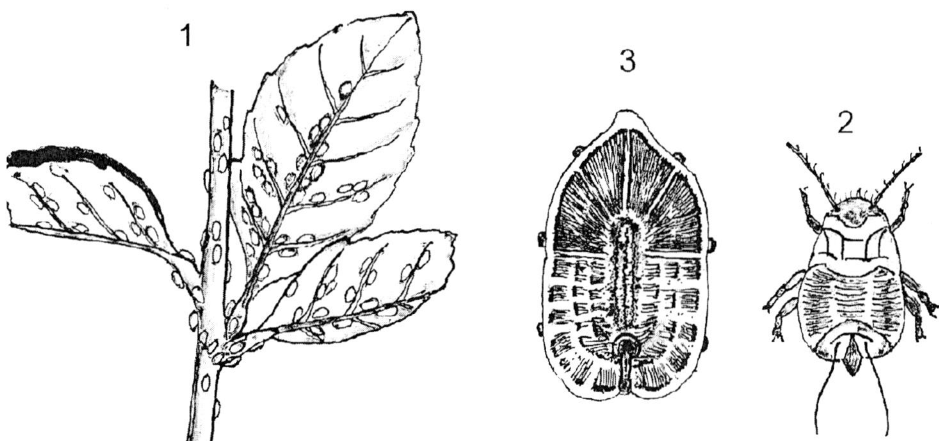

Fig. 92. Coffee green scale-*Coccus viridis*

1. Infested coffee plant 2. Nymph 3. Adult scale

Control measures

Chemical control. Spraying the plants with malathion-0.1% or fenitrothion-0.05% or monocrotophos-0.04% or profenofos-0.05% effectively controls the pest.

Biological control. The fungal parasite-*Cephalosporium lecani* infects and destroys the pest.

Other pests. Several other pests are also known to attack coffee plants. The hairy caterpillar-*Eupterote fabia* and the cut worm-*Agrotis segetum* appear in large numbers occasionally, feed voraciously on the leaves and defoliate the plants; the maggots of the fly-*Melanagromyza coffeae* mine the leaves, feed on the inner tissues and cause damage; the grubs and adults of the shot-hole borer beetle-*Xyleborus conidens* bore into the stem and branches, make tunnels, feed on the inner tissues and cause drying of branches and sometimes the whole plant; the grubs of the cockchafer beetle-*Holotrichia conferta* cut and feed on the roots, especially in the nurseries and cause death of seedlings and young plants; the sucking pests viz., the thrips-*Scirtothrips bispinosus* and *Heliothrips haemorrhoides* and the aphid-*Toxoptera aurantii* infest the shoots and young leaves; the coffee bean weevil-*Araecerus fasciculatus* infests the berries in the field and dry beans in storage; the mite-*Oligonychus coffeae* colonizes the under surface of leaves, suck and feed on the sap.

Cocoa *(Theobroma cacao)*

The tea mosquito bug-*Helopeltis antonii* punctures the cocoa pods, sucks and feeds on the sap, as a result the pods shrivel and the quality of the seeds is adversely affected; the cotton aphid-*Aphis gossypii* infests the tender succulent shoots and young leaves, sucks and feeds on the sap; the larvae of the castor shoot and capsule borer-*Conogethes (Dichocrosis) punctiferalis* bore into the cocoa pods, feed on the inner content and seeds and cause serious damage; the beetles of the grapevine stem girdler-*Sthenias grisator* cut and ring the stems and branches of cocoa plants; the caterpillars of the bark borer pest-*Indarbela quadrinotata* scrape and feed on the bark from the stem portion; the nymphs and adults of the citrus mealy bug-*Pseudococcus citri* remain on the shoots, branches and pods in large colonies, suck and feed on the sap from the infested parts (Fig. 93); the leaf weevil-*Myllocerus viridanus* cuts and feeds on the leaf tissues.

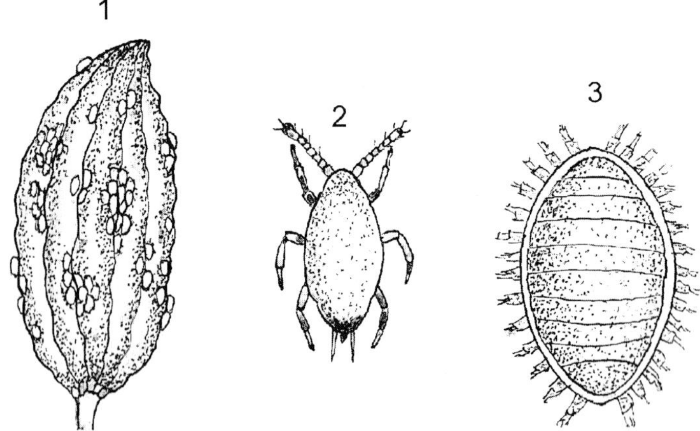

Fig. 93. Citrus mealy bug-*Pseudococcus citri*

1. Infested cocoa pod 2. Nymph 3. Adult bug

ARECANUT *(Areca catechu)*

Comparatively only a few pests are known to attack arecanut. The cosmopolitan, small, pale gray beetles-*Araecerus fasciculatus* bite and feed on the leaves and ripening fruits. This pest also attacks coffee berries and Taphrosia pods; the nymphs and adults of the arecanut spindle bug-*Carvalhoia arecae* live in colonies on the unopened spindles, suck and feed on the sap, as a result the spindles fail to open normally and often dry out; the coconut white grub or root grub-*Leucopholis coneophora* cuts and feeds on the roots of arecanut palms; both the nymphs and adults of the aphid-*Cerataphis tetaniae* infest the tender shoots, inflorescence and fruit bunches, suck and feed on the sap; the hard scale insects-*Aonidiella orientalis* and *Chinaspis dilatata,* the mealy bug-*Icerya aegyptiaca* and the thrips-*Oligonychus biharensis* and *Rhipiphorothrips cruentatus* infest arecanut plants, suck and feed on the sap from the tender shoot portion.

RUBBER *(Hevea brasiliensis)*

Only a few pests infest rubber trees and cause damage. The mango stem borer-*Batocera rufomaculata* bores into the stem region, feeds on the inner tissues and cause severe damage, sometimes even death of the tree; the minute, shot-hole bark borer beetles-*Xyleborus biporus* bore into the bark and interfere with the flow of latex; the scale insects-*Aspidiotus cyanophylli* colonize the twigs and foliage, suck and feed on the plant sap leading to general weakening of the trees; the termites-*Cryptotermes dilatatus* attack the woody parts and make tunnels in the stem region.

CARDAMOM *(Elettaria cardamomum)*

1. Cardamom thrips

Sciothrips cardamomi

Order - Thysanoptera

Family - Thripidae

Nature of damage. This is the most important pest of cardamom and is responsible for heavy loss in yield and quality of the produce.

Thousands of nymphs and adults harbor on the pedicels, sepals, petals, flower buds, peduncles and tender shoots, suck and feed on the plant sap. On the infested capsules, rough, cork-like patches develop and they become malformed and shrunken. The natural flavor of the capsules is lost and the seeds also do not develop normally. In case of severe infestation, the flowers and young capsules drop down. The loss due to this pest attack ranges from 10-90 per cent.

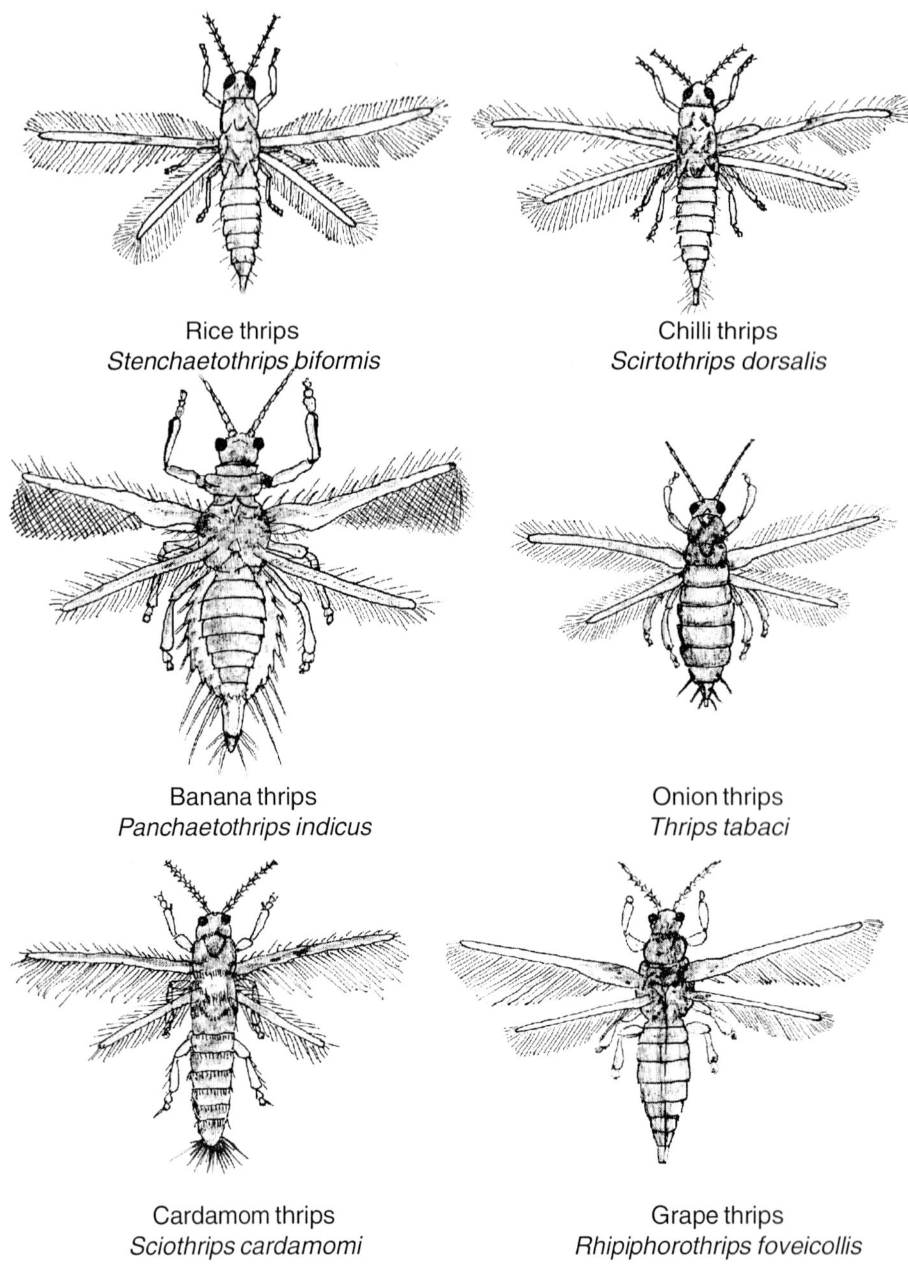

Fig. 94. Thrips

Life cycle of the pest. The female insect lays eggs after copulation or parthenogenetically. The mother inserts the eggs singly into the tissues of the inflorescence, unopened young leaf or inside the leaf sheaths. The egg laying extends over a period of about 30 days, at 2-3 eggs per day. The young nymphs come out of the eggs in about 10 days, feed on the plant sap and become full-grown in about 15 days. The grown-up nymph pupates on the plant parts and emerges as winged thrips in 5-6 days. The adult is very minute, delicate, brownish in color, with fringed wings like a comb (Fig. 94).

Control measures

Chemical control (i) Spraying the crop with malathion-0.1% or nicotine sulfate-0.02% or toxaphene-0.04% or quinalphos-0.05% or phosalone-0.05% is effective in controlling the pest (ii) In cardamom mostly cross pollination takes place by insects, especially honey bees. So, insecticides such as, toxaphene, which do not harm honeybees should be used and the spraying should be avoided during the time the honeybees visit the flowers in large numbers.

Miner pest. Besides the above mentioned pest, the caterpillars of the shoot and capsule borer-*Conogethes (Dichocrosis) punctiferalis* bore into the stem portions of cardamom plants and feed on the inner tissues; the banana lace wing bug-*Stephanitis typicus* sucks and feeds on the sap from the leaves; the banana aphid-*Pentalonia nigronervosa* infests the young leaves and tender shoots, sucks and feeds on the plant sap. Besides causing such direct damage, the aphid is the vector of the virus, which causes **'marble mosaic disease'**; the hairy caterpillars-*Eupterote fabia, E. cardamomi* and *E. testacea* feed voraciously on the leaves and defoliate the plants; the cardamom white fly-*Kanakarajiella cardamomi,* besides causing direct damage by sucking and feeding on the sap from the foliage, indirectly leads to the development of the fungal disease, 'sooty mould' on the honeydew secreted by the nymphs.

PEPPER *(Piper nigrum)*

1. Pepper pollu beetle or Pepper flea beetle

Longitarsus nigripennis

Order - Coleoptera
Family - Alticidae

Nature of damage. It is a specific and most important pest of pepper. The grubs bore into the matured fruits before they are ripe, feed on the inner contents of the seeds, leaving behind only the outer covering. One grub may damage 2-3 berries. The hole made by the grub is clearly visible on the infested berry. The yield loss due to the pest infestation may range from 30-40 per cent (Fig. 95).

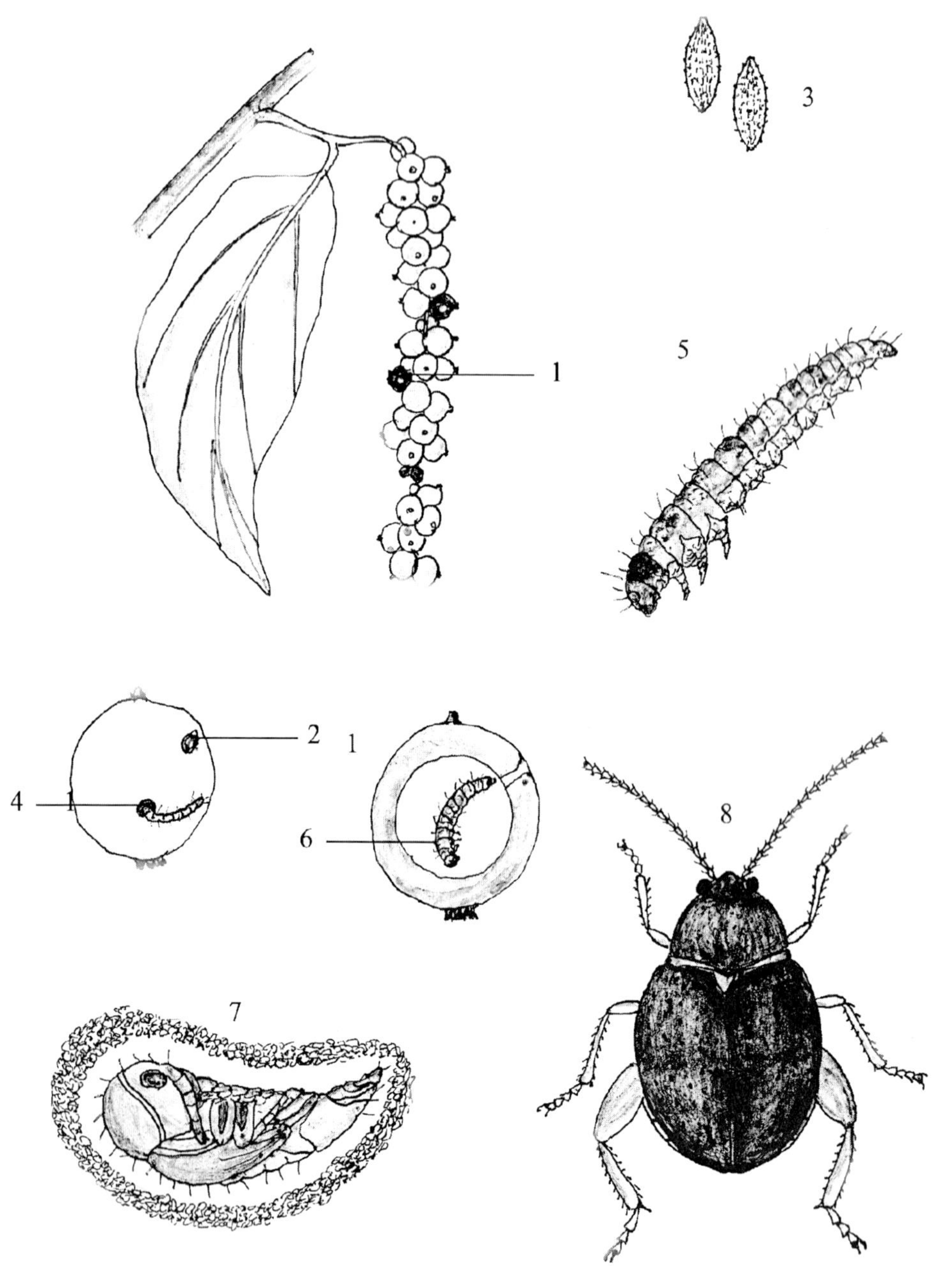

Fig. 95. Pepper pollu beetle-*Longitarsus nigripennis*

1. Infested fruit 2. Egg on the fruit surface 3. Eggs 4. Larva boring the fruit 5. Larva 6. Larva inside the seed 7. Pupa inside earthen cocoon 8. Adult beetle

Life cycle of the pest. The female beetle bites small holes on the skin of the young berries and lays eggs inside the holes singly. The eggs hatch in 7-10 days and the young grubs coming out of the eggs are very small, pale yellowish in color, with 3 pairs of legs on the thoracic segments. They grow by feeding on the inner contents of the seeds and become full-grown in about 30 days. The grown-up grub is about 5.0 mm. in length, yellowish in color, with a black head. The full-grown grub comes out of the berry, falls down to the ground, pupates under the soil in an earthen cocoon and emerges out as an adult beetle in about 7 days. The beetle is small, about 2.5 mm. in length, shining yellowish-blue in color and with strong hind legs. It is capable of jumping long distances like a grasshopper (Fig. 95).

Control measures

Cultural methods. The soil around the base of the stem of the vines should be stirred well periodically to bring out the pupae from under the soil and expose them to direct heat of the sun, so as to kill them

Chemical control (i) Incorporation of carbaryl 10 D or lindane 0.65 D at 20 kg./acre into the soil is effective in eliminating the pupae inside the soil (ii) Spraying the vines with lindane-0.1%, so as to cover the inflorescence and young berries thoroughly also affords control of the pest by destroying the young grubs hatching out from the eggs and also by preventing the adult beetles from laying eggs on the developing berries.

Other pests. A few other pests also attack pepper, but to a lesser extent and only occasionally. The caterpillars of the pepper shoot borer-*Laspeyresia hemidoxa* bore into the tender terminal shoots, as a result the shoots dry and die; the nymphs and adults on the pepper mealy bug-*Pseudococcus virgatis* colonize the shoots and leaves, suck and feed on the plant sap and devitalize the plants; the pepper scale insects-*Lepidosaphes piperis* and *Pinnaspis aspidistrae* and the aleyrodid-*Aleurocanthus valpariensis* infest pepper plants, suck and feed on the sap from the plant parts.

PESTS OF GREEN MANURE CROPS

Sunnhemp (*Crotalaria juncia*)

1. Sunnhemp hairy caterpillar

Utetheisa pulchella

Order - Lepidoptera

Family - Arctiidae

Nature of damage. It is one of the most serious and important leaf feeding caterpillar pests of sunnhemp. The caterpillars bite and feed on the foliage voraciously. After pod formation, the larva bores the pods, inserts its head inside the pod and feeds on the immature seeds. The larva never enters the pod completely. One larva can damage many pods (Fig. 96).

Life cycle of the pest. The female moth lays small, spherical, whitish eggs on the young leaves and tender shoots. The young caterpillars hatching out of the eggs start feeding on the leaves and later feed on the immature seeds and grow. The grown-up caterpillar is large, 3.5-4.0 cm. in length and has conspicuous color pattern of red and brown tubercles carrying short, tufts of hairs on them and a mid dorsal white stripe, broken by red between successive segments and a brown head. The full-grown caterpillar pupates in a silken cocoon in the folds of leaves or in the soil and the adult moth emerges from the pupa in a few days. The life cycle is completed in about 5 weeks. The adult moth is medium-sized and very beautiful, with a bright color pattern of red, black and white. The forewings have red and black spots against a white background. The hindwings are whitish-gray in color, with marginal black patches (Fig. 96).

Control measures

Physical and cultural methods (i) The moths fly about during the daytime and may be caught by using hand nets and destroyed (ii) As the moths are attracted by light, light traps may be set up to attract and destroy the moths (iii) The caterpillars, which can be easily spotted, may be hand picked and destroyed.

Chemical control (i) The young caterpillars can be controlled by dusting with carbaryl 10 D or endosulfan 4 D at 10 kg./acre (ii) Slightly grown-up larvae can be controlled by spraying with carbaryl 50 WP-0.1% or methyl parathion-0.05% or endosulfan-0.07%.

2. Sunnhemp caterpillar

Argina syringa

Order - Lepidoptera

Family - Arctiidae

Nature of damage. This is another caterpillar pest, which attacks sunnhemp crop. The caterpillars besides feeding on the leaves, bore into the pods, feed on the immature seeds and cause considerable damage.

Life cycle of the pest. The female moth lays eggs in clusters on the under surface of leaves. The young larvae, which hatch out of the eggs feed on the leaves and then the developing seeds inside the pods and grow. The grown-up larva is yellow in color, with black markings across and white spots on the

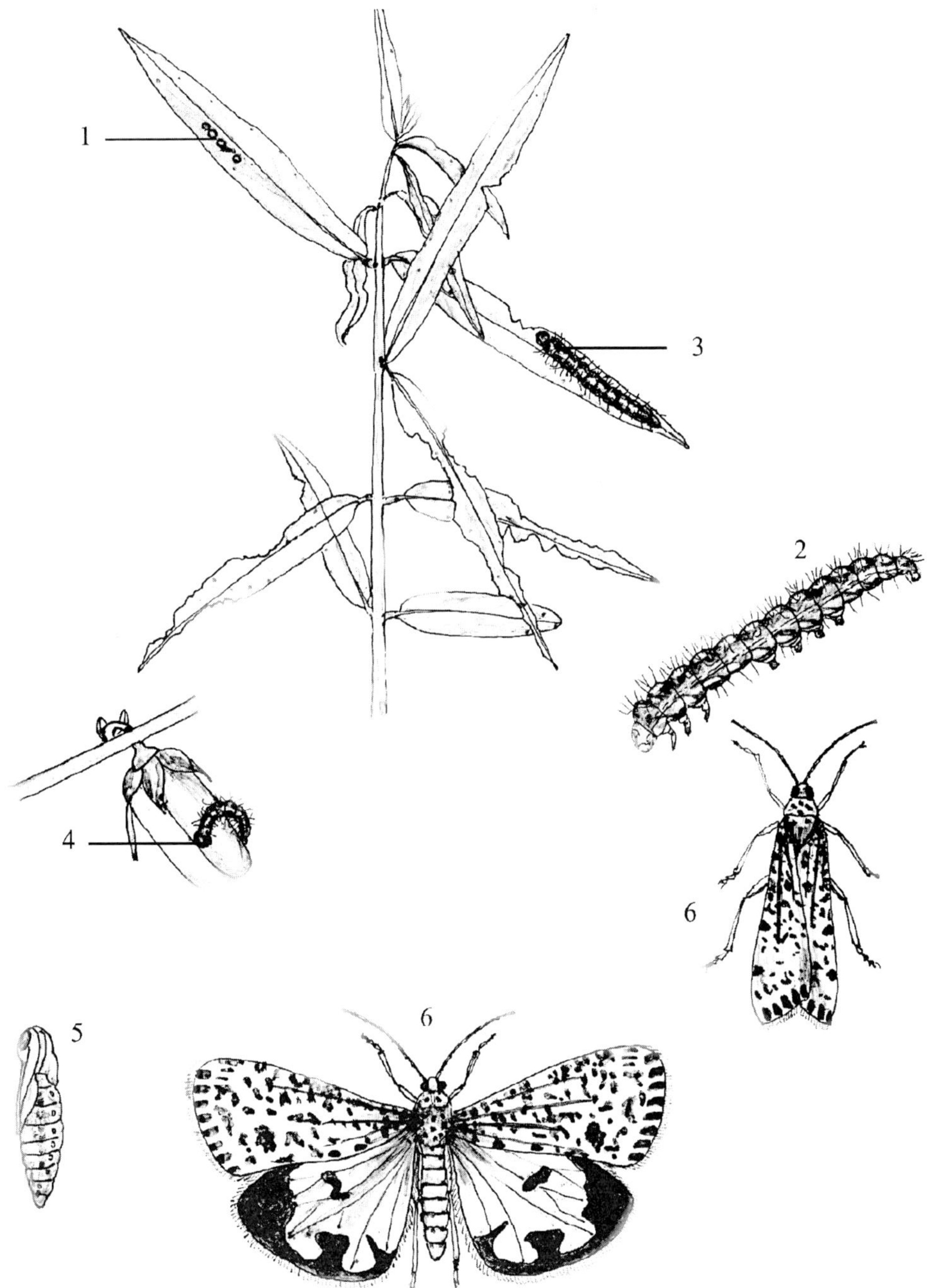

Fig. 96. Sunnhemp hairy caterpillar-*Utetheisa pulchella*

1. Eggs 2. Larva 3. Larva feeding on the leaf 4. Larva boring the pod 5. Pupa 6. Moth

body and tufts of long hairs. The full-grown caterpillar pupates under the soil and the moth emerges from the pupa after a few days. The life cycle is completed in 22-30 days. The moth is crimson in color and the wings are of the color of burnt bricks with black markings and spots.

Control measures. The control measures suggested for the control of sunnhemp hairy caterpillar-*Utetheisa pulchella* may be adopted.

3. Sunnhemp flea beetle

Longitarsus belgaumensis

Order - Coleoptera

Family - Alticidae

Nature of damage. The beetles bite holes in the leaves, feed on the leaf tissues and damages the foliage. The grubs bite, cut and feed on the young roots.

Life cycle of the pest. The mother beetle lays eggs on the soil under the canopy of the host plants. The young grubs hatching from the eggs enter into the soil, cut and feed on the young roots and grow. The grown-up grub pupates inside an earthen cocoon under the soil and the adult beetle emerges from the pupa after a few days. The entire life cycle is completed in 23-28 days. The beetle is small-sized, yellowish-brown in color with stout and strong hind legs. It is capable of jumping like a grasshopper.

Control measures

Chemical control (i) Incorporation of carbaryl 10 D or lindane 0.65 D at 20 kg./acre into the soil is effective in eliminating the grubs and pupae under the soil (ii) Dusting the crop with carbaryl 10 D or endosulfan 4 D is effective in controlling the beetles.

Other pests. Besides the above-mentioned pests, a few more pests infest sunnhemp. The caterpillars of the pulses pod borer-*Etiella zinckenella* bore into the pods and feed on the immature seeds; the thick-set, greenish caterpillars of the blue butterfly-*Lampides boeticus* also bore into the pods, feed on the tender pods and immature seeds; the larvae of the moth-*Laspeyresia tricentra* bore into the stem and cause small, gall-like swelling at the place of attack.

DAINCHA *(Sesbania bispinosa)*

1. Daincha green semilooper

Pericyma glaucinans

Order - Lepidoptera

Family - Noctuidae

Nature of damage. The caterpillars of the pest remain on the leaf petioles, bite and feed on the leaves voraciously and cause extensive damage to the plants.

Life cycle of the pest. The young caterpillars hatching out of the eggs, start feeding on the leaves and grow. The grown-up caterpillar is a typical semilooper, light green in color with yellow, broad stripes on the lateral sides of the body. Of the 5 pairs of thoracic legs, the first two pairs are vestigial and so when the larva crawls, a loop is formed. Full-grown larva pupates in-between folds of leaves or dry leaves fallen to the ground and emerges as an adult moth within a few days. The moth is dark brown in color with black wavy lines and markings on the brown wings.

Control measures

Physical methods. The large-sized caterpillars, which are quite visible can be hand picked and destroyed.

Chemical control. Spraying with carbaryl 50 WP-0.1% or endosulfan-0.04% or fenitrothion-0.05% controls the pest.

Other pests. A few other pests also infest Daincha. The borer caterpillar pest-*Azygophleps scalaris* bores into the stem, feeds on the inner tissues and sometimes causes death of the plant; the tobacco caterpillar-*Spodoptera (Prodenia) litura* feeds on the leaves and tender shoots and defoliates the plants; the caterpillars of the butterfly pests-*Terias hecabe* and *Catopsilia pyranthe* feed on the leaves.

SESBANIA *(Sesbania speciosa)*

The daincha green semilooper-*Pericyma glaucinans* feeds on the leaves of sesbania plants; The tobacco caterpillar-*Spodoptera (Prodenia) litura* feeds on the leaves and tender shoots and may defoliate the plants; the borer caterpillar pest-*Azygophleps scalaris* bores into the stem, feeds on the inner tissues and may kill the plants; the caterpillars of the butterfly pests-*Terias hecabe* and *Catopsilia pyranthe* feed on the foliage.

GLIRICIDIA *(Gliricidia maculata)*

The lab lab aphid-*Aphis craccivora* infests the young leaves, tender shoots, inflorescence and tender pods, sucks and feeds on the plant sap; the leaf weevil-*Myllocerus viridanus* bites the leaves from the margins, feeds on the tissues and causes damage.

PESTS OF FLOWERING AND ORNAMENTAL PLANTS

ROSE *(Rosa* species*)*

1. Rose hairy caterpillar

Euproctis fraterna

Order - Lepidoptera

Family - Lymantriidae

Nature of damage. The tussock caterpillars bite and feed on the leaves and cause extensive damage to the plants. Besides rose, the pest attacks castor, pomegranate, cotton, mango, red gram etc. (Fig. 97).

Life cycle of the pest. The female moth lays flattish, round, yellow-colored eggs in clusters on the under surface of leaves and covers them with yellow-colored anal hairs. The young caterpillars on hatching out of the eggs in 5-9 days are small, long and slender with fine hairs on the body. They start feeding gregariously on the foliage and become full-grown in 27-35 days. The grown-up larva is 3.75-4.00 cm. in length, reddish-brown in color, with a reddish head and short, brown hairs all over the body. A pair of pencils of long, brownish-yellow hairs on either side of the head and a tuft of brownish-yellow hairs, directed backwards at the anal end are also present. The full-grown larva pupates on the plant surface in a thin cocoon of silken thread and hairs. The moth emerges from the pupa in 10-12 days. The tussock moth is medium-sized and has orange-yellow wings with light wavy markings (Fig. 97).

Control measures

Physical and cultural methods (i) The egg masses, which are quite conspicuous may be collected and destroyed (ii) The young caterpillars, which are found in large numbers in groups after hatching may be collected and destroyed (iii) The grown-up caterpillars, which can easily be seen, may be hand picked and destroyed.

Chemical control (i) The young caterpillars can be controlled by dusting with carbaryl 10 D or endosulfan 4 D (ii) The gregariously feeding, somewhat grown-up caterpillars can be controlled by foliar spraying with malathion-0.1% or methyl parathion-0.05 % or endosulfan-0.07% or fenitrothion-0.05% or chlorpyrifos-0.05% solution.

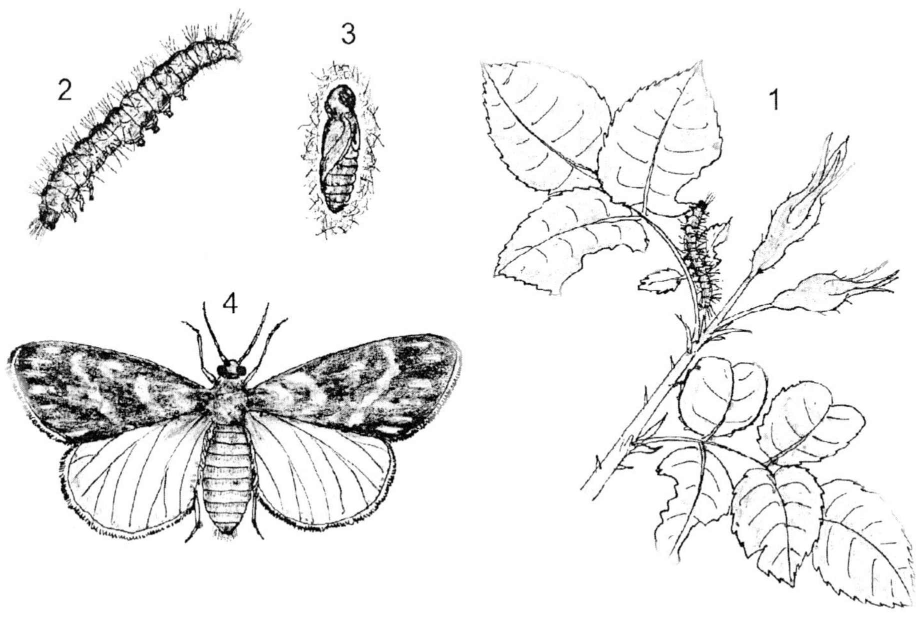

Fig. 97. Rose hairy caterpillar-*Euproctis fraterna*

1. Larva feeding on the leaves 2. Larva 3. Pupa 4. Moth

2. Rose hard scale insect

Aonidiella aurantii

Order	-	Hemiptera
Suborder	-	Homoptera
Family	-	Diaspididae

Nature of damage. The pest causes very serious damage to rose plants. The nymphs and adults fix themselves to the stem, shoots, petioles, leaves and flower stalks in large numbers, suck and feed on the plant sap continuously, as a result the plant becomes very weak, unhealthy and stunted. Severely affected plants may die ultimately. The pest also attacks citrus plants.

Life cycle of the pest. The young nymphs, which hatch out of the eggs crawl about for sometime on the plant parts, select a suitable place and start feeding on the sap from that place. Once a nymph settles in a particular place, it stays there in a sedentary state throughout its life span and never leaves that place. The female looses its legs and antennae and is covered by a reddish-brown, hard, waxy scale. The females have no wings. The male insect pupates inside the scale itself and emerges as a winged adult (Fig. 98).

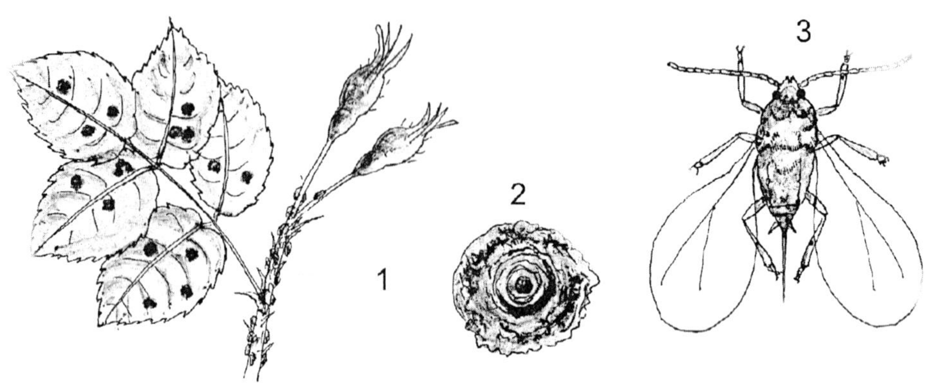

Fig. 98. Rose hard scale insect-*Aonidiella aurantii*

1. Scale infested plant 2. Female insect 3. Male insect

Control measures

Cultural methods. Severely affected branches may be cut and destroyed.

Chemical control. Spraying the plants with phosphamidon-0.1% or dimethoate-0.06% or monocrotophos-0.04% controls the pest.

3. Rose white fly

Aleurocanthus rosae

Order	-	Hemiptera
Suborder	-	Homoptera
Family	-	Aleyrodidae

Nature of damage. The nymphs, as well as the adults harbor on the under surface of leaves in large numbers, suck and feed on the sap from the leaves and shoots. Due to continued drainage of sap, the leaves turn yellow, fade and drop down. The growth of the plant is also stunted. The insects secrete a honeydew-like liquid, which spreads on the plant surface and attracts many other insects, especially ants and **'sooty mould'**, a fungal disease appears on the surface of the plant parts.

Life cycle of the pest. The female white fly lays yellowish and stalked eggs on the leaves and tender shoots. One insect lays about 100 eggs during its life span. The young nymphs, which hatch out of the eggs in 7-10 days fix themselves up at a particular place and continue to feed on the sap. The nymphs are scale-like, sub-oval, dark-colored and have minute spines on the body. The full-grown nymph pupates in the same place and the adult, winged white fly emerges out from the pupa. The life cycle is completed in 3-4 weeks. The adult white fly is very small, whitish in color, delicate and the body and forewings are covered with a white, waxy, powdery coating. The flies are quite active, can fly only short distances and are short-lived.

Control measures

Physical methods. Yellow sticky traps may be set up in rose gardens to attract and destroy the adult white flies.

Chemical control. Spraying the plants with methyl demeton-0.05% or dimethoate-0.06% or phosphamidon-0.05% or acephate-0.05% or phosalone-0.07% or ethion-0.05% or monocrotophos-0.04% or neem oil-2.0% or neem seed kernel extract-5.0% is effective in controlling the pest.

4. Rose aphid

Macrosiphum rosaeformis

Order	-	Hemiptera
Suborder	-	Homoptera
Family	-	Aphididae

Nature of damage. Both the nymphs and adults infest the tender shoots, young leaves, flower buds and flower stalks in colonies of thousands, suck and feed on the plant sap. The leaves become curled up, the shoots fade and the infested plants become unhealthy and stunted. The insects secrete honeydew, which spreads on the infested portions and 'sooty mould' infects these areas.

Life cycle of the pest. The insects are very small, greenish in color and delicate. Both winged and apterous forms are present. The females are ovoviviparous or parthenogenetic. Males are very few in numbers and are smaller than the females. The life cycle is completed in about 10 days and the insects multiply very rapidly.

Control measures

Physical methods. Severely infested branches may be cut and destroyed

Chemical control. The control measures suggested for the control of rose white fly may be adopted.

Other pests. Rose plants are subjected to attack by several other pests besides the ones described above. The caterpillars of the reddish-brown tussock moth pest-*Porthesia scintillans,* the castor slug caterpillar-*Latoia lepida and* the castor semilooper-*Achaea janata* bite and feed on the foliage voraciously and defoliate the plants; the nymphs and adults of the grapevine leaf thrips-*Rhipiphorothrips cruentatus* suck and feed on the sap from young leaves and shoots; the leaf cutter bee-*Megachile anthracina* cuts the leaves from the margin into semicircular bits and uses them to build nest; the chafer beetle-*Oxycetonia versicolor* bites and feeds on the leaves; the scale insects-*Lindingaspis rossi* completely cover the stem, suck and feed on the plant sap and may even cause death of the plants; the aleyrodid-*Aleurocanthus spiniferus* also sucks and feeds on the sap from the leaves and shoots; the flower chafer beetle-*Oxycetonia versicolor* cuts and feeds on the flowers; the stems of rose plants are sometimes girdled by the grapevine stem girdler-*Sthenias grisator*; young rose cuttings may sometimes be attacked by white ants.

JASMINE *(Jasminum sambac)*

1. Jasmine bud worm

Hendecasis duplifascialis

Order - Lepidoptera

Family - Phycitidae

Nature of damage. The caterpillars of this pest bore into the flower buds, feed on the inner content, as a result the flowers fail to open and are spoiled. The caterpillar after boring and destroying one bud moves on to the next bud. Then the larva webs together 2-3 buds with fine silken thread to form a gallery or nest, remains inside the gallery, bores and feeds on the inner contents of the developing buds. On plants infested with this pest, several webbed and dried up shoots with the frass are seen sticking to the galleries.

Life cycle of the pest. The female moth lays eggs on the flower buds singly or in small clusters of a few eggs in each cluster. The young larva, which hatches out from the egg starts feeding on the floral parts by boring into the flower buds. The grown-up larva is green in color with a black head. The full-grown larva pupates inside the soil and emerges as a small-sized adult moth.

Control measures

Chemical control. Spraying the plants with endosulfan-0.07 % or monocrotophos-0.08 % or cypermethrin-0.001 % controls the pest effectively.

2. Jasmine gallery worm

Elasmopalpus jasminophagus

Order - Lepidoptera

Family - Pyraustidae

Nature of damage. The young caterpillars bore into the flower buds, feed on the inner content and destroy them. Later the caterpillar webs the young leaves, shoots and flower buds together with fine silken thread, makes a gallery, stays inside and feeds on all the parts. The excreta of the larva is seen sticking to the webbed nest. The infested plants have several such webbed and dried up shoots. The pest attacks all varieties of jasmine.

Life cycle of the pest. The caterpillars are small, green in color with red head and prothorax and brown streaks on the lateral sides of the body. The full-grown caterpillar pupates inside the gallery and the moth emerges from the pupa after a few days. The moth is small-sized and dark grayish in color.

Control measures

Physical method. The webbed shoots may be pruned and destroyed

Chemical control. Spraying the plants with endosulfan-0.07% or monocrotophos-0.08% or fenitrothion-0.05% affords effective control.

3. Jasmine leaf web worm

Nausinoe geometralis

Order - Lepidoptera

Family - Pyraustidae

Nature of damage. The leaf webbing caterpillar webs the leaves together with fine silken thread, remains within the web, feeds on the green matter of the leaves and skeletonizes them completely. Severely affected shoots dry and die.

Life cycle of the pest. The female moth lays eggs singly or in small groups of 2-10 on the shoot portion. One moth may lay about 100 eggs. The eggs hatch in 3-4 days and the young caterpillars start feeding on the young leaves and then web the leaves together, stay inside the webs and feed on the leaves. The grown-up caterpillar is green in color with black streaks on the sides of the thorax and dark warts on the body. On the warts thin hairs are seen. Pupation takes place inside the web and the adult moth emerges from the pupa within a few days. The moth is medium-sized and the light brownish wings have white spots. The entire life cycle is completed in 21-30 days,

Control measures. The measures suggested for the control of the gallery worm may be adopted.

Other pests. Jasmine is attacked by a few more pests. The caterpillars of the leaf roller pests-*Glyphodes unionalis* and *G. celsalis* roll the leaves, remain inside the leaf rolls and feed on the leaves; the horned caterpillar pest-*Acherontia*

styx feed on the leaves and defoliate the plants; nymphs and adults of the white fly-*Dialeurodes kirkaldyi* harbor on the under surface of leaves in large numbers, suck and feed on the sap; the maggots of the blossom midge-*Contarinia maculipennis* feed on the floral parts; the flower thrips-*Thrips orientalis* infest the flowers, suck and feed on the sap from the floral parts; the eriophyid mite-*Aceria jasmini* and the red spider mite-*Tetranychus neocaledonicus* suck and feed on the sap from the leaves and shoots; the nymphs and adults of the red scale of rose-*Aonidiella aurantii* suck up sap from the tender shoots and buds and make the plants to fade.

CHRYSANTHEMUM (*Chrysanthemum* species)

The leaf folder caterpillar, a slender, green caterpillar of a medium-sized, yellowish moth-*Lamprosema indica* feeds on the tender foliage and flower heads; the black aphid-*Macrosiphoniella sanborni* infests the tender shoots, sucks and feeds on the sap; the larvae of the gram caterpillar-*Helicoverpa (Heliothis) armigera* feed on the leaves and flower heads; the thrips-*Microcephalothrips abdominalis* colonize the tender shoots and under surface of leaves, suck and feed on the sap.

CROTONS

The black scale-*Saissettia coffeae*, the soft scale-*Lecanium depressum* and the mealy bugs-*Icerya aegyptiaca* and *Ferrisia virgata* infest all kinds of crotons, suck and feed on the sap and cause serious damage; the grape stem girdler-*Sthenias grisator* sometimes attack and girdle the stem and branches of croton plants.

PESTS OF TREES

TEAK (*Tectona grandis*)

1. Teak defoliator

Hyblaea puera

Order - Lepidoptera

Family - Hybleidae

Nature of damage. It is a pest, which causes very serious damage to the foliage of teak trees and affects the growth of the trees badly. The pest infestation is severe during the months of April-June and during August-September. The caterpillars feed on the young leaves completely. In the older leaves, the caterpillars scrape and feed on the green leaf tissues, leaving behind only the ribs and thus skeletonize the leaves.

Life cycle of the pest. The female moth lays oval, white-colored eggs on the young leaves in clusters. One female lays 500-600 eggs. The young caterpillars, which hatch out of the eggs in 2-4 days start feeding on the young leaves and then on the older leaves and become full-grown in 8-17 days. The full-grown caterpillar is about 3.5 cm. in length, fairly stout, cylindrical, brownish-green in color, with a dark head and faded lines on the lateral sides. It pupates in the leaf fold and in 5-13 days the adult moth emerges from the pupa. The moth has a reddish-brown head and thorax and dark brown abdomen with orange-colored bands at the end of each segment. The forewings are grayish-red with light wavy streaks. The hindwings are dark brown, with deep orange and red patches. There may be 12-14 generations in a year (Fig. 99).

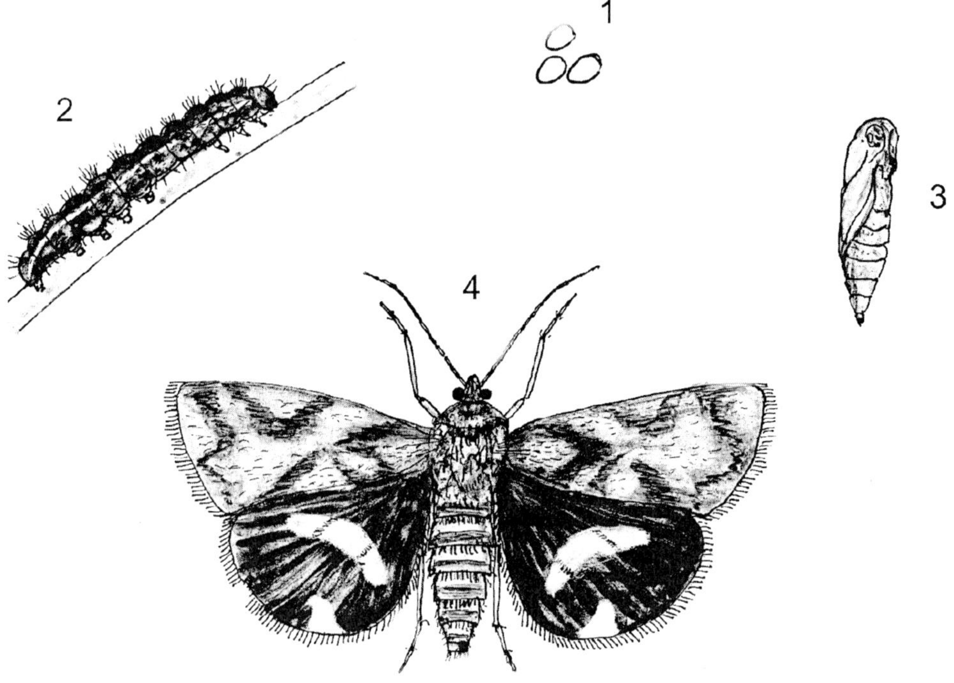

Fig. 99. Teak defoliator-*Hyblaea puera*

1. Eggs 2. Larva 3. Pupa 4. Moth

Control measures

Chemical control. Wherever possible, the foliage may be sprayed with quinalphos-0.05% or methyl parathion-0.05% or fenitrothion-0.05% or endosulfan-0.07% with a high tree sprayer.

2. Teak leaf skeletonizer

Eutectona (Hapalia) machaeralis

Order - Lepidoptera

Family - Pyraustidae

Nature of damage. This is another very serious leaf feeding caterpillar pest, which often occurs along with the teak defoliator. However, the pest attack is more severe during the months of April-May and during November. The caterpillars scrape and feed on the green leaf tissues, leaving behind only the ribs and the leaves of the infested trees appear as skeletons.

Life cycle of the pest. The female moth lays 250-400 eggs, singly on the leaves. The eggs hatch in 2-3 days. The caterpillars feed on the green matter of the leaves and become full-grown in 18-21 days. The grown-up larva is slender and elongated, about 2,5 cm. in length, light green in color, with a series of purplish spots on either side of the median dorsal yellowish, longitudinal streak. It pupates in the leaf fold or in-between fallen leaves on the ground. The adult moth emerges from the pupa in 4-11 days. The moth is small and bright yellow in color. The bright yellow wings have a few pinkish spots and zigzag, transverse, wavy markings and a marginal dark streak on the wings (Fig. 100).

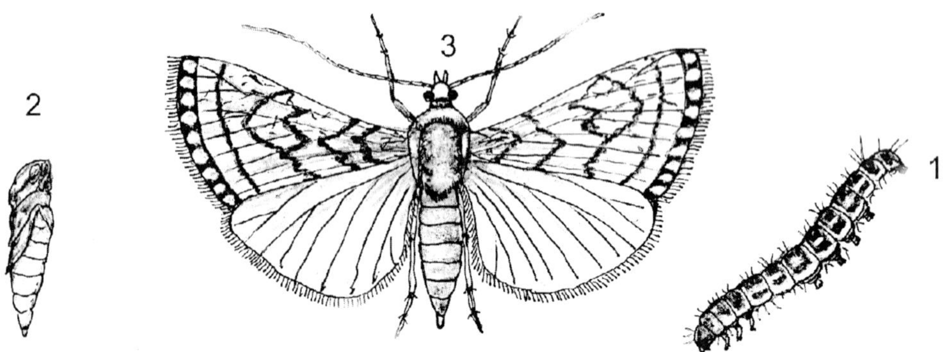

Fig. 100. Teak leaf skeletonizer-*Hapalia machaeralis*

1. Larva 2. Pupa 3. Moth

Control measures

Chemical control. The control measures suggested for the control of teak defoliator may be adopted.

Other pests. A number of other pests are also noticed on teak trees. The caterpillars of the fairly large, nocturnal moth-*Aleterogystia (Cossus) cadambae* bore holes on the bark and later bore and tunnel into the heart wood of trees of about 15 years old, feed on the bark and inner tissues of the wood, thus adversely affecting the quality of the timber; the leaf feeding caterpillar pests-*Acherontia lachesis, Hyposidra successaria, Dasychira grotei, Spodoptera (Prodenia) litura* and the hairy caterpillar-*Spilosoma (Diacrisia) obliqua* feed on the leaves and defoliate teak trees; the larvae of the leaf roller-*Sylepta straminea* roll the leaves, remain inside the roll and feed on the leaf tissues; the larvae of the bark and stem borers-*Indarbela quadrinotata* and *I. tetraonis* infest the stems of teak trees; the pods and seeds are eaten and destroyed by the pod borer-*Conogethes (Dichocrosis) punctiferalis*; the shot-hole borer beetles-*Xyleborus andrewsi, X. butamali* and *X. noxius* infest the woody stems; the cotton aphid-*Aphis gossypii,* the bug-*Tettigoniella ferruginea* and the mealy bug-*Phenacoccus hirsutus* suck the sap from the tender shoots and young leaves and cause damage.

NEEM *(Azadirachta indica)*

The pest, which attacks and causes severe damage to neem trees, is the tea mosquito bug-*Helopeltis antonii.* The nymphs and adults infest the young shoots and tender leaves in large numbers, suck and feed on the sap, leading to drying up of the distal shoots. Severely infested trees appear to be blighted. The pest can be controlled by spraying the foliage with malathion-0.1% or endosulfan-0.035%; the small, pale gray anthribid beetle pest-*Araecerus fasciculatus* bores the fruits, burrow into the seeds, feeds on the inner content and damages the seeds; the nymphs and adults of the mealy scale-*Pulvinaria maxima* attach themselves to the leaves, petioles, twigs and shoots in large numbers, suck and feed on the plant sap continuously. The leathery, brown female scale insects, white male puparia and white ovisacs of the females are found sticking on to the infested trees in profusion. The honeydew secretion of the insects drips down on to the ground below the infested tree and wets the soil; the thrips-*Dolichothrips indicus* infest the leaves, while the thrips-*Taeniothrips chaetogastra* infest the flowers and leaves, suck and feed on the sap.

PORTIA *(Thespesia populnea)*

The nymphs and adults of the specific portia mealy scale insect-*Pulvinaria thespesiae* attach themselves to the shoots, twigs and leaves in large numbers, suck and feed on the plant sap, as a result the leaves turn yellow and fade; the black scale-*Saissetia nigra,* another sucking pest, which is also found commonly on crotons and cotton, sucks and feeds on the sap from the tender shoot region; the moringa hairy caterpillar-*Eupterote mollifera* often infests portia trees, feeds on the leaves voraciously and defoliates the trees within a short period of time. During the daytime, the caterpillars congregate in large groups in hollows or depressions on the stem or in-between forks of branches and remain almost motionless. After dusk they move out and feed on the foliage. The excreta of the caterpillars are found in large quantities as black pellets under the infested trees.

Casuarina *(Casuarina equisetifolia)*

The caterpillar pest-*Eumenodera tetrachorda* mines into the needles, feeds on the inner tissues, leaving behind only the epidermis; the caterpillars of the coffee red stem borer-*Zeuzera coffeae* and the grubs of the longicorn beetles-*Coelosterna scabrator* and *Stromatium barbatum* bore into the stem region, feed on the inner tissues and cause damage to the stem; the bark and stem borers-*Indarbela tetraonis* and *I. quadrinotata* infest the stem region; the bag worm-*Clania crameri* feeds on the needles; the mealy bugs-*Icerya purchasi* and *Icerya aegyptica* suck and feed on the sap from the young shoots and tender leaves; the termites-*Odontotermes obesus* and *Microtermes obesi* infest the stem, scrape and feed on the bark and woody portion and cause extensive damage.

EUCALYPTUS *(Eucalyptus citricdora)*

The termites-*Odontotermes obesus, Microtermes anandi* and *M. obesi* cause severe damage to eucalyptus trees. These termites attack, feed and destroy the bark of the tap root, which may cause death of the infested trees; the adults of the beetle-*Celosterna scabrator* bore and feed on the tissues of tender shoots and bark up to the sapwood, thus killing the shoots, while the grubs bore into the stem and roots, as a result the leaves turn yellow and fade; the mango stem borer beetle-*Batocera rufomaculata* girdles the stem leading to drying and death of the trees.

SUBABUL *(Lucaena leucocephala)*

The nymphs and adults of the exotic psyllid bug-*Heteropsylla cubana* suck and feed on the sap from young shoots and leaves causing severe damage to the trees.

Tamarind *(Tamarindus indicus)*

Only a few pests are known to attack tamarind. The caterpillars of the fruit borer pest-*Deudorix isocrates* bore and feed on the inner contents of fruits from the early stages of fruit formation; the inflorescence is attacked by the webber pest-*Eublemma angulifera*, which webs the flowers in the inflorescence together, remains inside the web and feeds on the floral parts; the tamarind scale-*Aspidiotus tamarindi*, the hard scales-*Aonidiella orientalis, Saissetia oleae* and the mealy bugs-*Planococcus (Pseudococcus) lilacinus* and *Pseudococcus viridis* literally cover the tender shoots and fruits, suck and feed on the sap from the plant parts; the stout-legged, dirty brown, tamarind bruchid beetle-*Pachymerus gonagra* burrows and feeds on the ripening fruits and seeds in the trees, but more often in stored tamarind.

SANDALWOOD *(Santalum album)*

The larvae of coffee red stem borer-*Zeuzera coffeae* bore the stem region of young sandalwood saplings and cause their death; the long-horned grasshopper-*Letana inflata* damage the foliage. This pest is the vector of **'spike disease of sandalwood'**; both the nymphs and adults of the thrips-*Mesothrips manii* suck the sap from tender leaves resulting in the formation of galls.

GENERAL PESTS

LOCUSTS

Patanga succincta, Locusta migratoria and Schistocerca gregaria

Order - Orthoptera

Family - Acrididae

Nature of damage. Locusts belong to the same family as that of grasshoppers and are closely allied to grasshoppers, but are polyphagous and voracious feeders. Unlike grasshoppers, which are more specific in their food habits, the locusts feed on almost any green vegetation including agricultural crops, forest trees, fodder crops etc. that come across their path. Further, they differ from grasshoppers in their capacity to aggregate under suitable conditions, forming bands in their immature stages and swarms in their adult stage and migrate in millions from country to country or from one place to another place within the country.

The three species of locusts found in India are the Bombay locust-*Patanga succincta,* the migratory locust-*Locusta migratoria* and the desert locust-*Schistocerca gregaria,* of which the desert locust is most important. The adult locusts, which fly in swarms migrate over long distances, settle on any green vegetation on their way, feed and destroy the standing crops in vast areas extending over several square kilometers within a short span of time and move on to other places.

Phases of locusts. Locusts are found in two phases. The first one is the solitary phase in which the individuals are inactive and they live scattered. The second phase is the gregarious phase in which the individuals are very active and they tend to remain together, breed rapidly and form swarms, which leave the breeding grounds and move to distant tracts and may even cross many countries. There may be some individuals, which are intermediate between the two phases in morphology and habitats and they are considered to be in a transient phase.

The desert locust - *Schistocerca gregaria* lives in dry grasslands or deserts and is found in many countries of the world. In India, the solitary phase of the locust is restricted to the desert breeding areas of Rajasthan, Sourashtra, Baroda, Cutch, Hissar, Punjab etc. These permanent dwelling homes of the desert contain certain outbreak areas, where under some favorable climatic conditions, the locust enters the gregarious phase, multiplies rapidly, forms bands and swarms and then migrates to other distant places. The desert locust's belt extends from Rajasthan and other breeding areas in India through Pakistan, Arabia and other Middle-East countries up to many parts of Africa. In the gregarious phase, the swarms spread more extensively and cover parts of Europe, Africa, Persia, Palestine, Russia, Afghanistan etc. In India the swarms invade whole of Northwest region, including the hilly tracts in the North and sometimes covering Assam in the East and Tamil Nadu in the South. Wherever conditions are found suitable, breeding takes place and swarms are multiplied

The Bombay locust - *Patanga succincta* is found in India, Sri Lanka and Malaysia. The migratory locust-*Locusta migratoria* breeds during the spring in Baluchistan and the adults migrate to the desert areas of India and breed there again during the summer.

Life cycle of the pest. Soon after copulation, the female locust lays eggs in clusters of 50-100 in each cluster or egg pod in holes 10-15 cm. deep, drilled by it in sandy soils, which have sufficient amount of moisture for proper incubation and hatching of the eggs. After laying the eggs, it closes the mouth of the hole with a frothing liquid, which hardens subsequently into a waterproof plug. A female may lay eggs thrice at weekly intervals. Locusts in the gregarious phase fly or rest together in swarms and the egg masses are generally laid very close to one another. Thus, about 1,000 egg pods may be found in a square meter area. The eggs hatch in 3-4 weeks in the spring and in about 2 weeks in the summer. The young nymphs emerging out of the eggs are very small, wingless and black in color in the gregarious phase. They hop around and feed on all kinds of vegetation. A locust hopper moults its skin 5 times in succession, once in 3-5 days in the early stages of its growth and in about a week at the later stages. The hopper stage lasts for about 4 weeks in the summer and 6 weeks in the spring. After the last and final moult, the nymph acquires wings and becomes a full-fledged adult locust. As the hoppers moult, they usually undergo a gradual change in their coloration from black to patterns of black and yellow, then to bright pink in the young adult stage and finally to a bright yellow color in the sexually matured, full adult stage.

The young hoppers unite in large numbers into bands and march from place to place, voraciously eating up all vegetation along their path. Hopper bands and locust swarms generally rest between sunset and sunrise in bushes, crops, trees etc. They also take shelter and rest during the hot mid-day in the summer. The adult locusts swarm at a speed of 12-16 km./h. and swarms have been seen 2,000 km. at sea.

The hoppers of the solitary phase vary in color depending on the color of the surrounding vegetation and may be green, gray or brown. The adults of the solitary phase are gray in color throughout their life, but their color may change to yellow if they enter into the gregarious phase. They have longer and crested pronotum and longer hind femur and the antennae are 27-30 segmented. The individuals are more dispersed and lead a solitary, sedentary life and are more juvenile. They breed in desert areas during the recession period until transformation to the gregarious phase. On the contrary, in the gregarious phase the individuals are more active and are crowded together to form bands. The adults have shorter, saddle-shaped pronotum and have much shorter hind femur and the antennae are 26 segmented. They are sexually more matured.

Locust breeding depends very much on rainfall and soil condition. There are usually two breeding seasons in a year, the spring and the monsoon. In countries such as, India, Egypt, Sudan etc., where rainfall is mostly received in the summer and the monsoon season, locusts generally breed at that time of the year. If there is rainfall again in the spring season and if the soil conditions are suitable for breeding as in Punjab and parts of Uttar Pradesh, breeding takes place during that season also. The swarms produced during the spring and the summer in the Middle East and Baluchistan usually fly eastward into Sind, Punjab, Rajasthan, Uttar Pradesh etc. during May-July and breed there. From these breeding grounds new swarms that are produced fly westward into Baluchistan, Persia, Arabia etc. and breed there. Thus the cycle goes on season to season and year to year. When conditions become unfavorable for rapid breeding throughout the year, the locusts get scattered and revert to the solitary phase till favorable conditions return for the gregarious phase (Fig.101).

Control measures

Locust control measures include the involvement of 'Anti-locust organization' and direct control of locusts at all stages of development from the egg stage to the flying adult stage.

Forecasting and forewarning. 'The Anti-locust organization' is more or less a locust intelligence agency that collects and circulates information about the time of locust migration, sites of resting, breeding and development and the probable paths of their migration, so that appropriate control measures can be taken up to destroy them at various stages of their growth. 'The locust warning organization' of the Government of India collects and monitors regularly the information about locust population, swarms and swarm migration in Pakistan, Iran and other countries. Within India also such information is collected and circulated to the states likely to be affected by locusts. The states in turn supply similar information to the 'Central locust warning organization' about locust occurrence in their own territories. In this way information about locusts and their occurrence is monitored continuously within the states of India and between India and other countries interested in locust control.

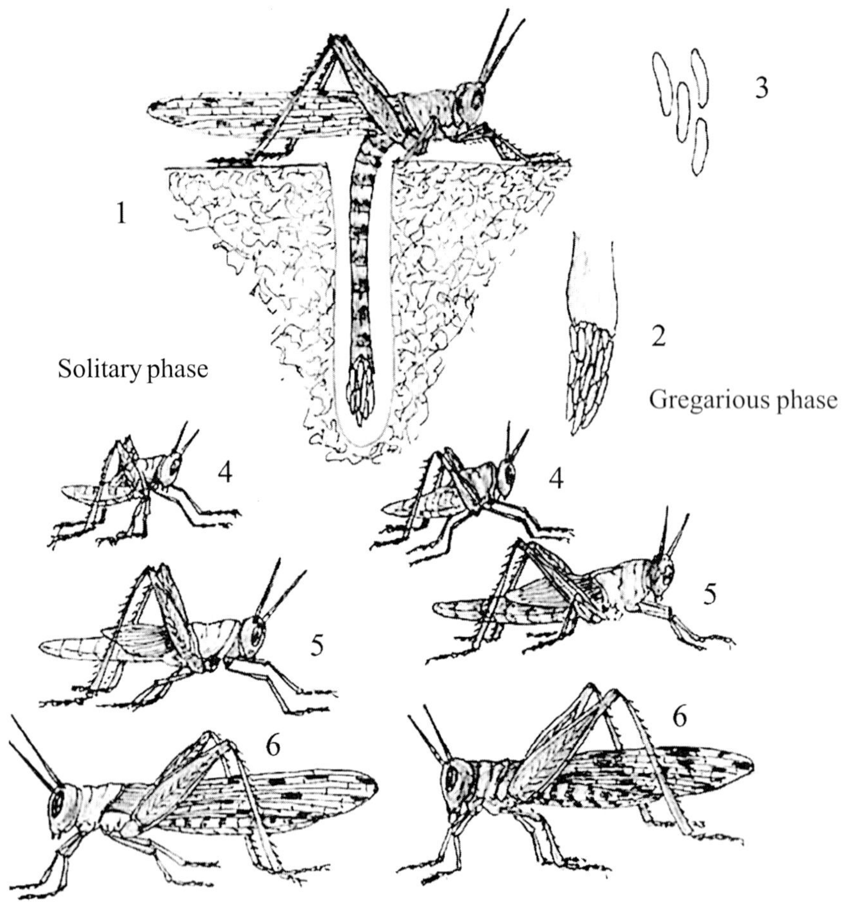

Fig. 101. Desert locust-*Schistocerca gregaria*

1. Adult locust laying eggs 2. Egg case 3. Eggs 4. Early stage nymph 5. Final stage nymph 6. Adult locust

Direct control. Direct control involves destruction of eggs at the breeding sites, destruction of hopper bands at various growth stages, as well as full-fledged adult locusts by cultural, chemical and other possible methods.

Cultural methods (i) Destruction of eggs at the breeding sites, if possible is the best and easiest method of controlling locusts. This can be done by deep ploughing, harrowing or hand digging and exposing the eggs to the vagaries of nature. However, this method is very laborious, costly and in many cases rather impossible because the eggs are frequently laid in inaccessible, desert areas (ii) The young hoppers can be destroyed by the trap method. Trenches 30 cm. wide and 45 cm. deep are dug across and ahead of the path of the marching hopper bands. The hoppers are driven into the trenches with the help of brooms or by waving pieces of white cloth. The hoppers, which fall into the trenches, cannot come out of the trenches easily. Then the hoppers can be collected and destroyed or they may be buried under the soil (iii) The hopper bands and adults, which rest in bushes, trees etc. can be killed by using flame throwers (iv) The swarms can be prevented from alighting on crops, especially on valuable crops by waving pieces of white cloth and by producing smoke screens by burning partially dried grass, dried cowdung etc.

Chemical control (i) Young hoppers can be controlled by dusting with carbaryl 10 D or chlordane 5 D or heptachlor 5 D or aldrin 5 D at 10 kg./ acre, using power operated dusters, so that large areas can be covered within a short time. In case of large scale occurrence over vast areas, aerial dusting may be resorted to (ii) Aerial spraying with diazinon controls the pest effectively (iii) Spraying with neem seed kernel extract 5 % at 200 liters per acre is also effective in controlling the pest (iv) Poison baits have also been used successfully to control the pest, especially the hopper bands. Poison bait is prepared by mixing a poison, Sodium fluosilicate-1.0 kg. or chlordane 50 WP-1.0 kg. or diazinon 40 WP-1.0 kg. or dieldrin 50 WP-1.0 kg. + a food material, wheat bran or rice bran-30 kg. + an attractant, gur or molasses-3.0 kg. in sufficient quantity of water to make a moist dough. The bait is distributed in the infested areas at the rate of 20 kg./ acre. The young and adult locusts, which are attracted by the smell of the attractant, feed on the poison bait and are killed.

Biological control. Several carnivorous birds such as kites, crows, mynahs, sterlings, storks etc. are natural enemies of locusts and feed on them.

Termites

Odontotermes obesus and **Microtermes obesi**

Order - Isoptera

Family - Termitidae

'Termites' commonly called 'white ants' are social insects resembling ants in that they both have a somewhat similar life, living in big colonies in which there are different forms or castes. Termites are mostly tropical or subtropical, but occur all over the world, including the temperate countries. There are two main groups of termites viz., subterranean termites, which live mostly in underground nests and the dry wood termites, which are confined entirely to wood. The subterranean termites are responsible for causing much more damage. Some of the subterranean species live in nests just under the ground, while some others construct mounds known as **'termitaria'** built with earth excavated, while making underground chambers. These termitaria vary in size and some of them may be as high as 3.0 meters.

Nature of damage. Several species of termites are responsible for causing extensive damage to crops and other properties, of which *Odontotermes obesus* and *Microtermes obesi* are the most important ones. Termites are destructive to both living and dead vegetation. They feed on sound or decaying wood or other plant materials such as, grasses, humus, fungi etc. The termites living in soil damage roots of various cultivated crops, grasses and other vegetation. Wheat, sugarcane, sunflower, groundnut, barley, oats, vegetables, roses, fruit seedlings, coconut trees etc. are some of the crops subjected to attack by termites on a large scale. Mainly the workers feed on these and cause damage. Termites also eat and destroy household articles, paper, cloth and almost anything that contain cellulose and cause serious damage.

Termite colony. A termite colony has four types of individuals or castes, which can be grouped into reproductive and sterile forms. The reproductive forms include fully winged ones whose function is the formation of colonies. They are the king and the queen, which have developed from the winged forms and have subsequently lost their wings before settling down to form a new colony. The sterile castes include soldiers and workers. They are wingless males and females and they perform all other functions except reproduction. The soldiers protect the colony from enemies and intruders, while the workers do the entire work in the nest such as, excavating and building up the termitarium, cleaning the nest, procuring and storing food, feeding and caring the young ones and the queen etc. The queen is the mother of the entire colony and lives in a specially prepared chamber situated almost at the center of the nest at a depth of 0.3-1.8 m. below the ground level. She is very big, creamy white in color and measures 5.0-7.5 cm. in length and her whitish abdomen is marked with transverse dark brown bands. She always remains in her chamber and her only function is to lay eggs. She lays 40,000-80,000 eggs in a day. She is taken care of constantly and fed by the workers. There is usually a single queen in each colony. The workers in a termite colony are responsible for causing damage to crops, wooden structures etc. and collect food for all the other castes. Some subterranean termites cultivate a crop of fungus on various plant materials inside the chamber and use the fungus as a supplementary food and is considered to be a source vitamins and organic nitrogen for the termites. The fungus garden is composed of spongy, dark reddish-brown, coral-like combs constructed by the workers Besides collecting food, they are also responsible for feeding the soldiers, young nymphs in the colony and the queen, which live exclusively on stomodaeal food (a mixture of salivary secretion and regurgitated intestinal content) and proctodaeal food (contents of the rectal pouch) supplied by the workers (Fig.102).

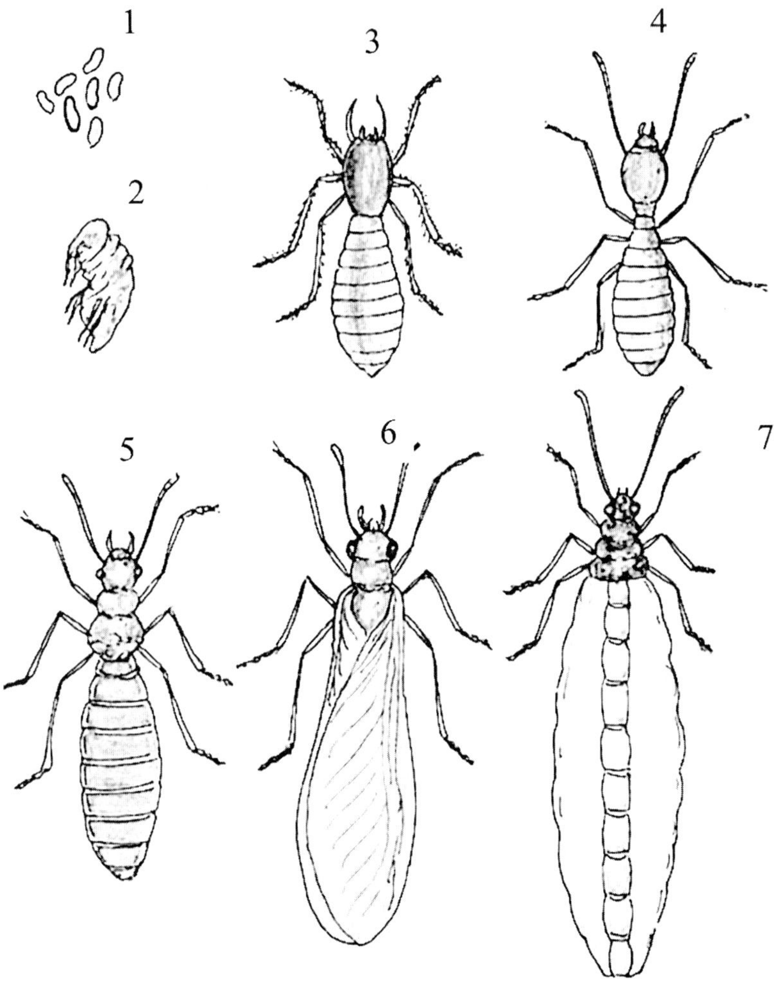

Fig. 102. Termites-*Odontotermes obesus*

1. Eggs 2. Nymph 3. Soldier 4. Worker 5. King 6. Winged adult 7. Queen

Castes of termites. There are two types of reproductives viz., primary reproductives and supplementary reproductives. The primary reproductives are the male and female macropterous adults with two pairs of large, well developed, membranous wings of almost equal size. Their genital organs are well-developed. They lead a short aerial life after leaving the old colony and their main function is to locate and develop new colonies. Supplementary reproductives are either brachypterous or apterous and they have no aerial life. They have only wing pads instead of normal wings and their genital organs are reduced and are smaller. They generally appear in a colony after the death of the primary reproductive. The queens developed from supplementary reproductives do not attain such a large size as that of the primary reproductives and they are capable of laying much lesser number of eggs, but help to keep the colony going. The other castes in a termite colony are the workers and soldiers.

Formation of a new colony. Swarming generally takes place after the first showers of monsoon rains, when the weather is hot and moist. Under such favorable conditions the primary reproductives leave the old nests in thousands through holes in the parent colony to form new colonies. The winged reproductives are generally attracted to light. Before leaving the nest the reproductives attain sexual maturity. After completion of the first and only flight (nuptial flight), they cast off their wings and search out for a similar individual of the opposite sex. Then the male and the female pair seeks a suitable place in the ground or wood, excavates a small chamber in which mating takes place. Copulation takes place frequently several times. Soon after copulation, the female starts laying eggs to form a new colony. Thus, the colonizing forms become the queen and king of the future colony. The number of eggs laid increases with the age of the queen, reaching several thousands per day and a colony comprising of over a million individuals develop. The earlier broods of nymphs are generally sterile castes of workers and soldiers, which attend on the queen and the king. The life of termites is very long ranging from 15-50 years. The death of a queen is followed by the development of supplementary reproductives. The life of a termite colony may be 100 years or more. The capacity of egg laying by the supplementaries is not so large as that of the primaries and more than one supplementary queens may be found in one colony. Rarely both kinds of queens may be found in a colony. The nymphs that hatch out of the eggs in about 7 days develop into adult workers and soldiers in about 6 months. In the summer, the eggs may give rise to reproductive forms, which reach maturity in 1 or 2 years. These then emerge and fly out to form new colonies.

Control measures

Cultural methods (i) The fields and surrounding areas should be kept clean and dry leaves, dry sticks, trash and other dry materials, which may serve as food for termites should be removed and destroyed by burning (ii) Use of partially decomposed green or farm yard manure to crops should be avoided (iii) To eliminate and prevent termite in buildings, breeding of termites in the vicinity should be prevented by removing all sources of infestation and by destroying all existing termite nests.

Chemical control (i) Termite nests should be destroyed and the termite mounts broken. Then carbaryl 10 D or chlordane 5 D or heptachlor 5 D or aldrin 5 D is blown into them liberally (ii) After breaking open the termite mounds carbaryl 50 WP-0.01% or lindane 20 EC-0.04% solution is sprayed inside the nest thoroughly (iii) To prevent termite attack in sugarcane fields, carbaryl 50 WP-0.1% or lindane 20 EC-0.04% solution is poured uniformly in the planting furrows before planting. A quantity of 1000 liters of insecticidal solution may be required to cover one acre (iv) Carbaryl 10 D or chlordane 5 D or heptachlor 5 D or aldrin 5 D should be applied uniformly over the sugarcane setts after planting at the rate of 20 kg./acre and then covered with soil (v) To prevent termite attack in wheat, barley, oats and other crops and in fruit seedling beds, carbaryl 10 D or chlordane 5 D or heptachlor 5 D or aldrin 5 D is dusted on the surface of the soil at 20 kg./acre and incorporated into the soil (vi) In irrigated crops and in nurseries, crude oil emulsion is mixed with irrigation water slowly at the rate of 5.0 liters/acre (vii) All wood works in buildings, which are in contact with the ground should be periodically treated by spraying with carbaryl 50 WP-0.01% or lindane 20 EC-0.04% after scraping off the soil tunnels constructed by the termites.

PESTS OF STORED GRAINS AND STORED PRODUCTS

Besides infesting and causing direct damage to the standing crops in the fields, several insect pests cause severe damage to the stored grains and other stored products. It has been estimated that out of the total quantity of cereals, oilseeds and pulses, insect pests damage about 5.0 per cent in storage and this amounts to about 2.5 million tons every year. These pests are found in large numbers in granaries, storehouses, godowns, storage bins, provision and grocery shops, flour mills etc. Generally these pests do not attack the crops in the fields.

Most of the storage pests belong to the Orders-Coleoptera and Lepidoptera. However, the pests belonging to the Order-Coleoptera cause more extensive damage, as the grubs and adults of these pests are capable of causing damage.

1. Rice weevil

Sitophilus oryzae

Order - Coleoptera

Family - Curculionidae

Nature of damage. The grubs as well as the adult weevils burrow into the seeds, feed on the inner content and cause severe damage. Generally one grub is found inside each grain. In case of bigger grains like maize, more than one grub may be found in a single grain. Severely affected grains become powdery and emit a bad odor. When the moisture in the granaries is high, the pest infestation is also high. Generally, weather conditions such as, high temperature and high humidity are favorable for the pest attack and under such conditions the pest intensity is very high. Besides rice, the pest attacks wheat, sorghum, maize and several other grains in storage (Fig. 103).

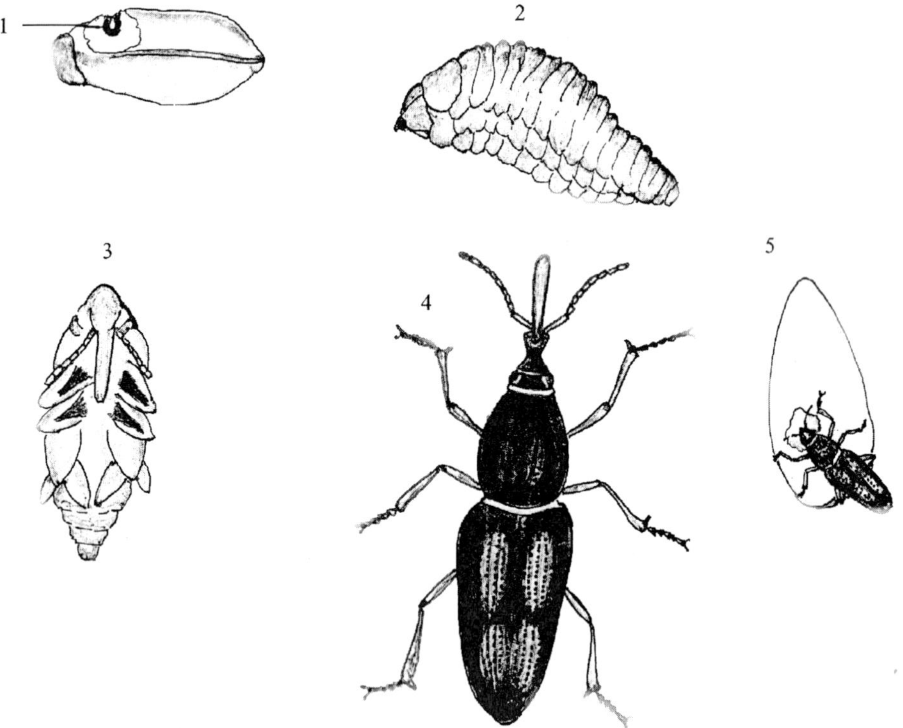

Fig. 103. Rice weevil-*Sitophilus oryzae*

1. Egg laid inside wheat grain 2. Larva 3. Pupa 4. Adult weevil 5. Weevil feeding on the grain

Life cycle of the pest. The female weevil scoops a small groove on the grain surface, lays an egg inside the groove and covers it with a gummy substance. Likewise, a female lays 100-400 eggs in 3-4 months. The eggs are whitish and short-cylindrical in shape. The eggs hatch in 4-9 days and the young grubs, which emerge from the eggs are pale white in color, footless (apodous) and have strong biting mouthparts. The grub knaws into the grain, feeds on the inner content and becomes full-grown in 2-3 weeks. The grown-up grub is whitish in color, stout with wrinkled body segments. It pupates inside the grain itself and emerges as an adult weevil in 10-20 days. The weevil is 4.0-5.0 mm. in length, reddish-brown in color with a conspicuous, curved snout and 4 dull yellow or reddish-brown spots on the forewings. The weevil lives for about 5 months. Several generations may appear in a year. The pest multiplies much faster during the summer months than in the winter months. In the cold season, the weevils remain in a dormant state in the cracks and crevices in the storehouses (Fig.103).

2. Pulse beetle

Bruchus (Pachymerus) chinensis

Order - Coleoptera

Family - Bruchidae

Nature of damage. The pest infests gram, beans, red gram, soybean, black gram, green gram, peas and other allied grains. The grubs bore into the grains by cutting a neat round hole and feed on the inner content, leaving behind only the outer seed coat. The beetles also damage and feed on the grains in a similar manner. More than one grub may be found in each grain (Fig.104).

Life cycle of the pest. The female beetle lays oval, scale-like, pale yellow or whitish, shining eggs on the surface of the grains and pastes them with a gummy substance. Sometimes the eggs are laid in the field itself. One insect lays about 100 eggs in batches. The eggs hatch in 3-4 days and the thickset grubs bore into the seeds by cutting a round hole. The grubs become full-grown in 2-3 weeks during the summer months. In the winter season, it takes several weeks for the grubs to grow up. The full-grown grub is cylindrical, stout, fleshy, whitish with brown mouthparts, wrinkled body segments and are footless (apodous). The grub pupates inside the hollowed out grain and emerges as an adult beetle in 5-6 days in summer and in about 4 weeks in winter. They are small, active, roundish beetles, about 6.0 mm. in size, the body abruptly rounded behind at the anal end, with stout hindlegs, dark brown in color and with 2 whitish spots on the prothorax. They live for about 2 weeks. There may be 7-8 generations in a year (Fig. 104).

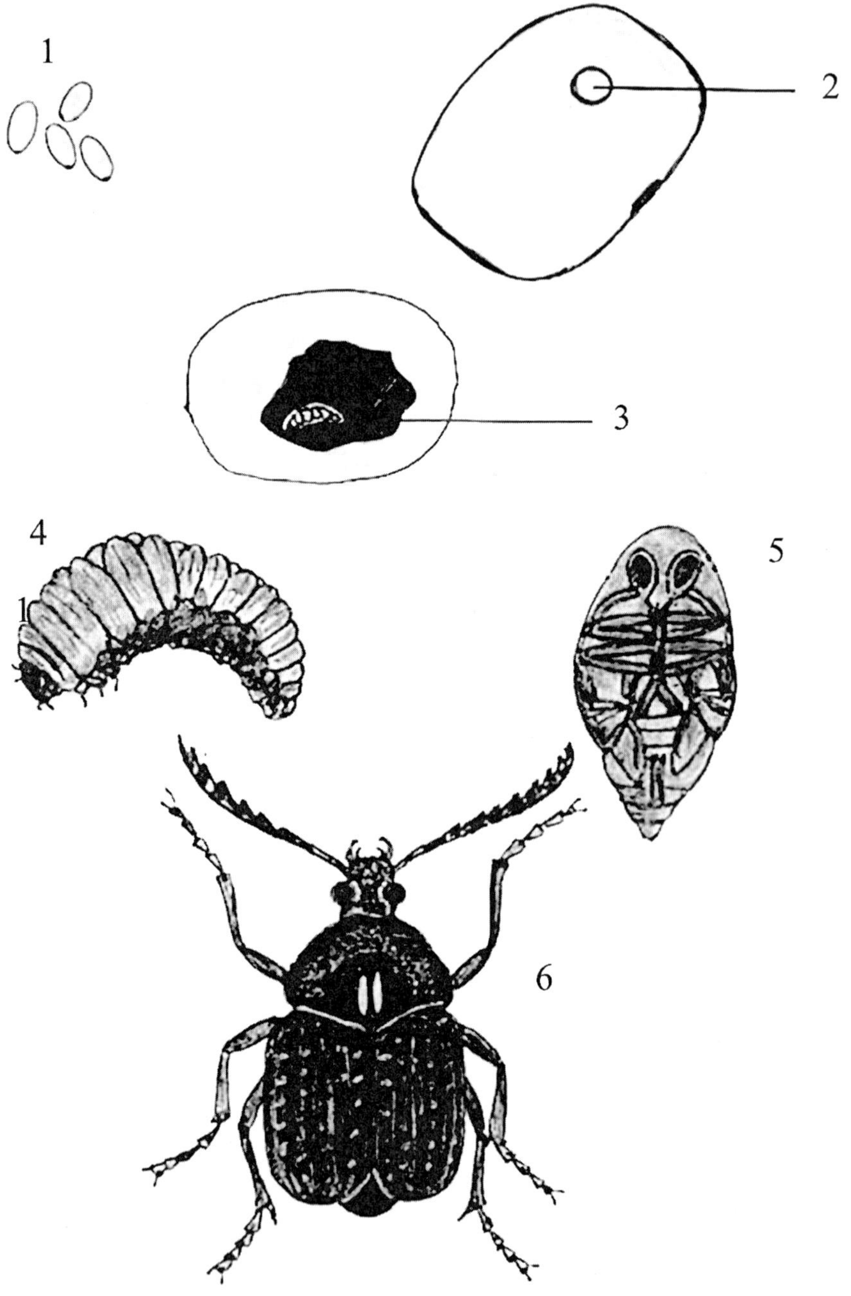

Fig. 104. Pulse beetle-*Bruchus chinensis*

1. Eggs 2. Beetle infested grain 3. Larva inside the grain 4. Grub 5. Pupa 6. Adult beetle

3. Rice meal moth

Corcyra cephalonica

Order - Lepidoptera

Family - Galleriidae

Nature of damage. The caterpillars of the pest infest paddy, cholam, pearl millet, groundnut, wheat, cashewnut and many such grains, as well as shredded and powdered grains. The larva remains inside tubular gallery made up of the powdered matter, shredded grains, loose grains etc., webbed together in a connected mass with silken thread and feeds on the grains and stored products. The infested grains become a compact mass and emit a bad odor.

Life cycle of the pest. The female moth lays eggs on the grains or flour. One insect lays 200-300 eggs. The larva, which comes out of the egg, constructs a tubular gallery with powdery matter, grains etc., remains inside the gallery, feeds on the grains or flour and becomes full-grown in 10-12 days. It pupates inside the gallery and emerges as an adult moth in 7-8 days. The moth is small, grayish-brown in color and very active (Fig. 105).

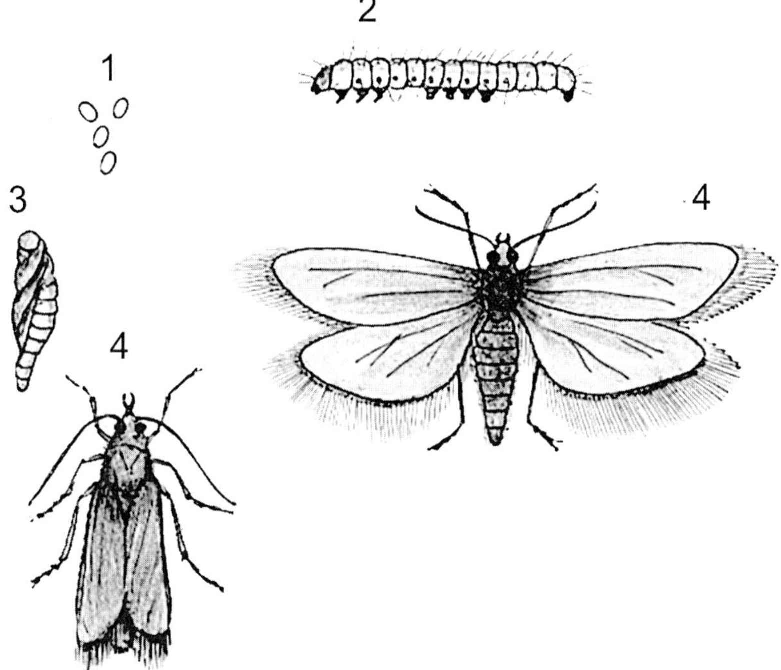

Fig. 105. Rice meal moth-*Corcyra cephalonica*

1. Eggs 2. Larva 3. Pupa 4. Moth

4. Rice grain moth or Angoumois grain moth

Sitotroga cerealella

Order - Lepidoptera

Family - Gelechidae

Nature of damage. The caterpillars of the pest infest rice, cholam, pearl millet, finger millet, maize etc. in the fields, as well as in the granaries. The larva burrows into the unhusked paddy grain, feeds on the kernel and turns it into chaff or bores into grains of other crops and feeds on the inner content. The moths can be seen flying in large numbers in the granaries and storehouses.

Life cycle of the pest. The female moth lays eggs on the paddy grains in the field at the milk stage. After the harvest, when the grains are stored in the granaries, the pest infestation is manifested. The damage is not noticed in the initial stages. The young caterpillars, which emerge from the eggs, burrow into the grains, cover the bored holes with silken mesh, remain inside and feed on the inner content and become full-grown in 2-3 weeks. The grown-up larva pupates inside the grain in a silken cocoon and emerges as a moth in 7-10 days. The pest infestation is noticed only after the emergence of the moths, which are seen flying in large numbers. The attacked grains become chaffy and only the excreta and the remains of the silken cocoon are seen inside the grains. The moth is small, pale yellow or straw-colored, shining with narrow, pointed forewings and the hindwings are fringed with long, fine hairs. Several generations may appear in a year (Fig.106).

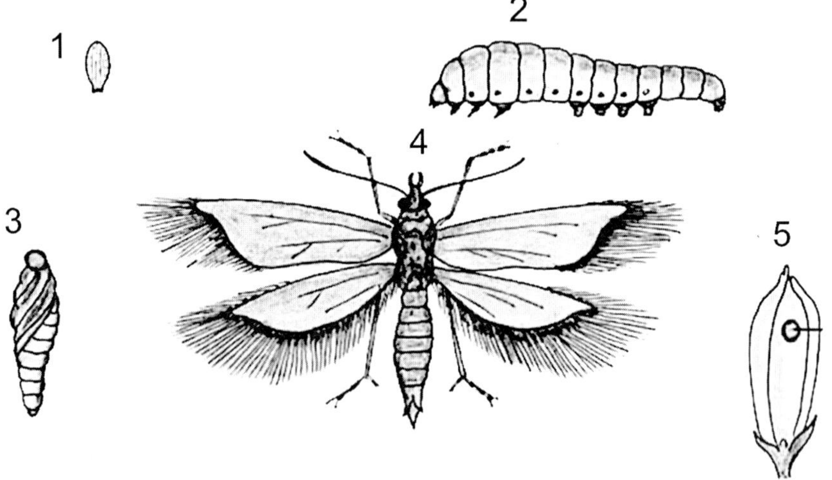

Fig. 106. Rice grain moth-*Sitotroga cerealella*

1. Egg 2. Larva 3. Pupa 4. Moth 5. The hole through which the moth has emerged

5. Red flour beetle

Tribolium castaneum

Order - Coleoptera

Family - Tenebrionidae

Nature of damage. It is an extremely common storage pest and the grubs and adult beetles are found to infest almost all stored products like food grains, oilseeds, vegetable powders, oil cakes, dry fruits, nuts and museum specimens of dry insects, stuffed birds, mammals etc. It is a special pest of materials such as, rice, wheat flour, pulses, powders, biscuits, bran etc. Severely infested flour turns grayish-yellow in color and emits a foul odor.

Life cycle of the pest. The female beetle lays white, cylindrical eggs, singly on the surface of grains or other stored products. Because of the sticky nature of the eggs, dust or flour settles on them and makes them invisible. One female lays about 450 eggs. The eggs hatch in 4-5 days during the hot season and in 10-12 days during the cold season. The grubs are slender, elongate, cylindrical, pale yellowish in color with fine hairs on the surface of the body. The larval stage lasts for 27-90 days depending upon the weather conditions, the duration being longer in the colder months. The grown-up grub pupates on the grain or flour. Thin, bristle-like hairs are seen on the surface of the pupa also. The beetle emerges from the pupa in 6-9 days. The beetle is small, elongate, about 3.0 mm. long, flattish and reddish-brown in color. The head and thorax are distinct and the antennae are quite prominent (Fig. 107).

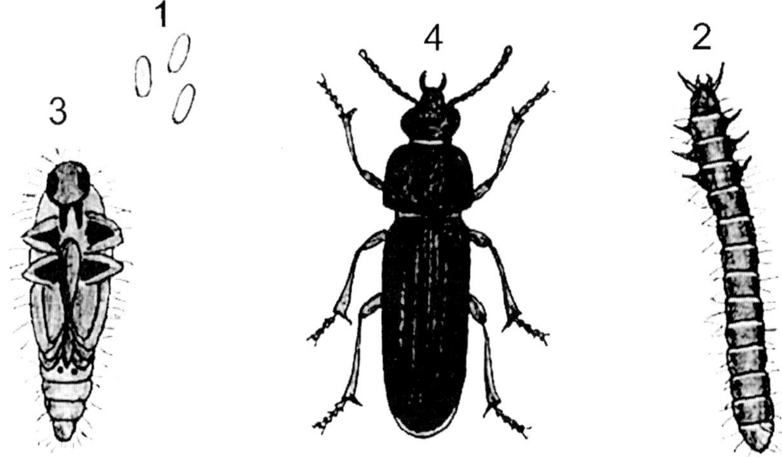

Fig. 107. Red flour beetle-*Tribolium castaneum*

1. Eggs 2. Larva 3. Pupa 4. Adult beetle

6. Khapra beetle or Wheat beetle

Trogoderma granarium

Order - Coleoptera

Family - Dermestidae

Nature of damage. Though it is a serious pest of wheat, occasionally it infests sorghum, barley and pulses. Only the grubs are responsible for causing damage, while the adult beetles are harmless. The grub generally bores through the germ pore of the seed and gradually burrows into the seed and feeds on the inner content. The attacked grains break and become powdery. The grubs do not go into the deeper layers, but generally attack the grains at the top portion.

Life cycle of the pest. The mother beetle lays cylindrical eggs singly or in clusters of 2-5 on the grains. One female lays about 125 eggs. The eggs hatch in 6-16 days depending upon the weather conditions. The grub is pale brownish in color, with yellowish-brown bands across the body segments and reddish-brown, long hairs on the body surface. The larval period extends for 3-7 weeks. During unfavorable weather conditions, the larva hibernates for long periods without feeding. The full-grown grub pupates between the grains. Spiny hairs are found on the surface of the pupa also. The adult beetle emerges from the pupa in 4-6 days. The female beetle is reddish-brown in color, cylindrical, 4.0-8.0 mm. in length with prominent club-shaped antennae. The male beetle is smaller than the female, about 2.5 mm. in length and dark brown in color (Fig. 108).

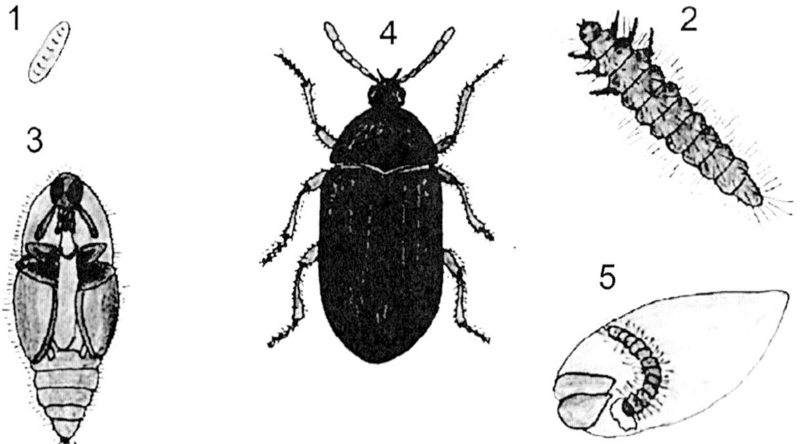

Fig. 108. Khapra beetle-*Trogoderma granarium*

1. Egg 2. Larva 3. Pupa 4. Adult beetle 5. Larva feeding on the grain

7. Lesser grain beetle

Rhizopertha dominica

Order - Coleoptera

Family - Bostrichidae

Nature of damage. The pest infests rice, sorghum, maize, barley, gram, wheat, dry fruits etc. Both the grubs and adult beetles cause damage, however the beetles are capable of doing more extensive damage. When whole grains are attacked, they are broken and become powdery. The beetles burrow and damage much more than they can consume. The grubs also burrow into the grains, feed on the inner content, leaving behind only the outer seed coat. The young grubs, which cannot bore into the whole grains, feed on the broken and powdered remnants of grains damaged by the beetles.

Life cycle of the pest. The grubs and beetles, which hibernate in cracks, crevices and damaged parts of the granaries and godowns during the unfavorable cold season, become active during the hotter months. The female beetle lays pear-shaped eggs, generally near the germ pore of the seeds, singly or in small clusters and pastes them on to the seed surface with a gummy substance. One female lays 300-500 eggs. The eggs hatch in 5-6 days during the summer season, but may take several days during the winter months. The young grub, which hatches out from the egg is very active and bores into the grain through the soft portion near the germ pore. Slightly grown-up grubs do not bore into the grains. The larval stage extends for about 40 days. The grown-up grub is dirty white in color, with a light brown head, slightly curved abdomen and thin hairs on the body. The full-grown grub pupates inside the grain or in-between the grains and emerges as an adult beetle in 7-8 days. The beetle is dark brown in color, about 4.0 mm. in length with a stout, globular head. The beetle lives for many months. In one year, about 5 generations may appear (Fig.109).

8. Saw-toothed grain beetle

Oryzaephilus surinamensis

Order - Coleoptera

Family - Cucujidae

Nature of damage. Both the grubs and adult beetles, like the red flour beetle pest infest several grains, flour, dry fruits etc. They scrape and burrow the seeds, feed on the inner content and may cause extensive damage.

Life cycle of the pest. The female beetle lays cylindrical eggs on the grains or stored products. One female lays about 300 eggs in about 70 days. The

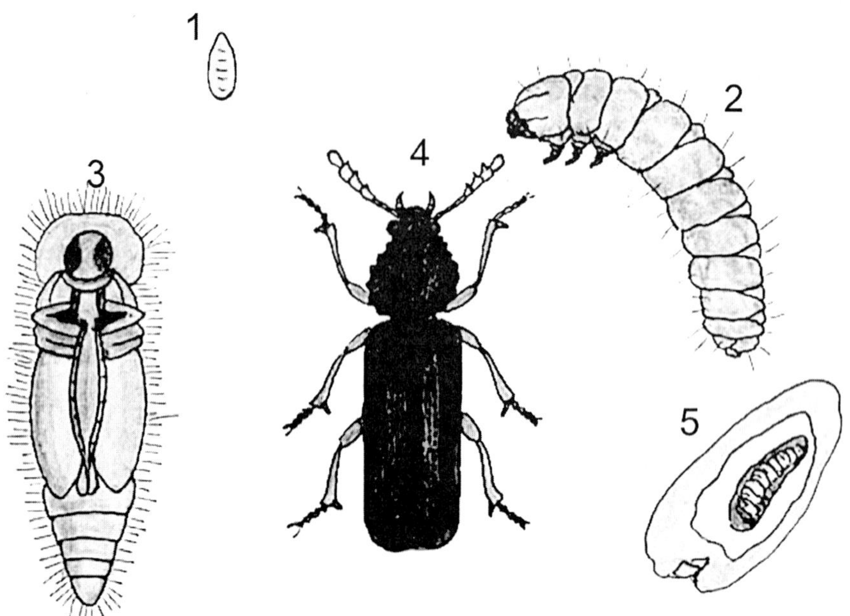

Fig. 109. Lesser grain beetle-*Rhizopertha dominica*

1. Egg 2. Larva 3. Pupa 4. Adult beetle 5. Larva inside the grain

beetle is small-sized, dark brown in color, slightly smaller than the red flour beetle, slender and flattish. The beetle has 6 serrated, tooth-like edges on both the sides of the prothorax, which gives the beetle its name and 3 irregular ridges on the dorsal surface of the prothorax. The life cycle is completed in 28-35 days. The beetle lives for many months (Fig.110)

Other pests. Besides the pests discussed above in detail, a number of other pests also attack and ravage stored grains and stored products, including horticultural products. The grubs of the sweet potato weevil-*Cylas formicarius* bore into sweet potato tubers, both in the field and in storage and make them unfit for consumption; the grubs of the cigarette or tobacco beetle-*Lasioderma serricorne* feed on stored tobacco and various tobacco products, besides ginger, turmeric and chillies; the grubs of the drug store beetle-*Stegobium paniceum* tunnels into stored products like turmeric, ginger, coriander and dried vegetables, dried fruits and animal matter; the grubs of the tamarind beetle-*Pachymerus gonagra* bore into tamarind fruits both in the tree, as well as in storage and feed on the seeds; the caterpillars of the potato tuber moth-*Phthorimaea operculella* bore into potato tubers, which are exposed in the field and potato tubers in store houses, feed on the inner fleshy portion and make the tubers unfit for consumption; the caterpillars of the Indian meal moth-*Plodia interpunctella* feed on grains and dried fruits; the larvae of the fig moth-

Cadra cautella attack stored grains, cashewnut, groundnut, wheat etc.; the caterpillars of the cabbage borer-*Hellula undalis* bore into the heads of cabbage and cauliflower in the field and continue to remain in the heads even after harvest, feed on the tissues and cause severe damage; the common house rats-*Rattus rattus rufescens* cause extensive damage to stored grains and stored products in granaries, godowns, store houses, grocery shops etc. They feed on all types of vegetables, animal feeds, poultry feeds and also non-vegetarian foods.

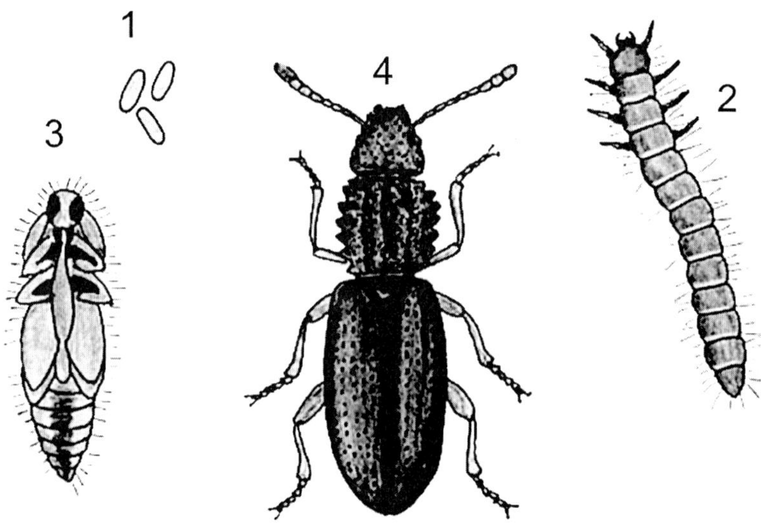

Fig. 110. Saw toothed grain beetle-*Oryzaephilus surinamensis*
1. Eggs 2. Larva 3. Pupa 4. Adult beetle

Control of storage pests

Almost all the storage pests are more or less similar in their life habits. They breed and multiply very rapidly in the midst of their food and all the different stages are found on the same food materials and once infested, the materials are very badly damaged in a very short time. The pests get dispersed very easily from place to place with consignments of infested materials, through gunny bags and other containers and are easily conveyed by road, rail or waterways. Except in a few specific cases, the measures to control these pests are quite similar, since their general habits are somewhat similar. The really practical and effective measures for controlling these pests and for preventing damage and loss by these pests are cleanliness and proper treatment of the materials before storage.

Rats, mites, insect pests and other microorganisms such as, fungi and bacteria ravage stored products. The damage due to insect pests and microorganisms depends to a large extent on the moisture content of the stored grains, oxygen content in the granaries or godowns, increase in the temperature of the stored products and such factors. Insect pests such as, rice weevil, rice meal moth and lesser grain moth, which are capable of causing enormous damage, breed and multiply very rapidly within a short period of time, when the moisture content of the grains is more than 9.0 per cent. So, before storing the grains, they should be dried thoroughly, so that the moisture content is less than 9.0 per cent. The grains should be stored in moisture-proof granaries, godowns or bins, otherwise the grains will absorb moisture from the surrounding atmosphere.

Insects need a certain amount of oxygen to sustain their lives. When the grains are stored in airtight granaries or storehouses, the density of oxygen is reduced, which prevents multiplication of the pests. The eggs of khapra beetles require 16-18 per cent oxygen to maintain their viability.

Temperature is another important factor, which favors multiplication of storage pests. In the stored grains or stored products, in some places pest infestation may be present. When the insects respire, the temperature and moisture in that particular place may increase, which may create a conducive atmosphere for the rapid multiplication of the pest. Opening the granaries on sunny days, increasing the aeration inside and stirring the stored grains may help to reduce the temperature in the stored grains, thereby prevent multiplication of the pests to some extent.

Stored grain pests gain access to the grains in many ways and cause serious damage. Pests such as, pulse beetle, rice weevil and rice meal moth may reach the granary along with the harvested produce from the fields. Many of the pests may reach the granary or godown through gunny bags in which old infested grains have been stored. Infestation may occur at the time of transit of grains and products from one place to another. The hibernating larvae and adult insects, which may hide in cracks and crevices in the walls, floors and roofs of granaries, godowns, warehouses etc. may become active and infest freshly stored grains and products. Further, the flying adult insects may find access to the granaries and infest the stored grains or products.

The various methods adopted to control storage pests are of two types viz., control measures taken prior to pest infestation or measures adopted prior to storage of fresh consignments of grains or products and measures adopted after pest infestation is noticed.

1. **Control measures taken prior to pest infestation** (i) The granary, godown, warehouse, storage bin or container should be thoroughly

cleaned and any traces of loose grains, broken grains, chaff, bran, wastes, dust etc. should be removed and burnt (ii) Cracks, crevices and damaged areas on the floor, walls and roof should be patched up with cement and white washed with lime. The inside walls may be painted with coal tar up to a height of 2.0 meters (iii) The floor, walls and roof of the empty granary or godown may be dusted with carbaryl 10 D or malathion 5 D at 2.0 kg. per 38 cu. m. volume or sprayed with malathion-0.05% fluid at 3.0 liters per 10 sq. m. area or fumigated with hydrogen cyanide gas (cyanogas) at 42 gm. per 38 cu. m. or ethylene dichloride-carbon tetrachloride gas, commonly known as EDCT (Killoptera) at 8-12 kg/38 cu. m or Aluminium phosphide (celphos), 3.0 gm. tablets at 7 tablets per 38 cu. m. volume. If fumigants are used, the store houses should be closed and kept air-tight for 24-48 hours (iv) As far as possible, new gunny bags should be used to store fresh produce. If old gunny bags are to be used, they should be kept immersed in boiling water for 30 minutes or kept immersed in malathion-0.1 % solution for 10 minutes or treated with a fumigant such as Killoptera under airtight conditions at 1.0 kg. per 10 gunny bags for 24 hours. The gunny bags are then dried and used for packing the produce (v) The carts or vehicles used for transporting the produce should be thoroughly cleaned, preferably washed with some antiseptic solution such as phenol or lysol (vi) The grains to be stored should be thoroughly cleaned and made free from chaff, husk or any extraneous matter, which may carry the pest into the store house. The grains should be dried well, so that the moisture content comes down to 8-10 per cent and then stored. If the moisture content is more, the grains will be vulnerable to attack by storage pests and microorganisms (vii) The grain-filled gunny bags should be arranged on wooden boards or bamboo poles and not on the floor, to prevent moisture from the floor being absorbed by the grains (viii) The gunny bags should be arranged in layers one above the other, leaving sufficient space in-between the stacks for surveillance. In the case of wheat grains, 18 bags may be arranged in one stack, for rice 15 bags may be arranged in one stack and for flour 12 bags may be arranged in a stack (ix) There should be a clearance of 50 cm. between the stacks and side walls of the store house and 60 cm. between the roof and the top layer of the bags (x) When the bags are being arranged, over each layer of bags, malathion 5 D is dusted at 500 gm. per 100 sq. m. or malathion-0.05% spray fluid is sprayed (xi) Freshly harvested grains should not be stored along with old, pest infested grains (xii) As far as possible, grains of only a single crop should be stored in one store house (xiii) During rainy days, the store house should not be kept opened (xiv) Only on bright sunny days the godowns should be opened for surveillance. If pest infestation is noticed, the grains should be taken out, dried and then stored again (xv) Neem seed kernel powder may be mixed with the grains at 1.0 kg. per 100 kg. of grains or activated clay may be mixed with the grains at 1.0 kg. per 100 kg. of grains. Activated clay has the capacity to absorb atmospheric moisture, but once moisture is absorbed to its maximum capacity, it looses its efficacy to afford further protection against the pests. Hence activated clay-treated grains should be stored in airtight granaries or godowns (xvi) The grains stored for seed purposes may be mixed with malathion 5 D or carbaryl 10 D at 1.0 kg. per ton of seeds. This treatment affords protection to the seeds from pest infestation for a period of about one year.

2. **Control measures to be taken after pest infestation is noticed** (i) Once pest infestation is noticed, it is better to treat the grains with a fumigant. For this purpose, the grains may be fumigated with Killoptera at 24-32 kg. per 100 cu. m. volume of storage space or with Celphos 3 gm. tablet at one tablet per ton of grains or with Methyl bromide at 2-3 kg. per 100 cu. m. volume. The fumigation should be done under airtight conditions for 24-48 hours, after which the store house is opened for some time to aerate it (ii) The infested grains may be taken out from the store house, dried, cleaned and then stored again.

Recent studies have established that the red flour beetle and the rice weevil pests have developed resistance to malathion, lindane and phosphene. Further, the lesser grain beetle has also developed resistance to malathion and lindane. So, care should be taken in using such chemicals for the control of these pests.

NON-INSECT INVERTEBRATE PESTS

PHYTOPHAGOUS NEMATODES

'Nematodes', which are otherwise known as **'eelworms'**, **'threadworms'**, **'roundworms'** or **'nemas'** are ubiquitous and can survive and live under any conditions and environments wherever life can exist. They are found associated with all living organisms, including man, animals, birds, plants etc. Nematodes are grouped under the **Phylum-Nemahelminthes** and **Class-Nematoda**. The plant parasitic nematodes are mostly microscopic, vermiform and measure from 0.5-1.0 mm. in length, while those parasitic to man and animals are macroscopic. The nematode parasitizing whale fish is about 0.8 m. in length. Nematodes are mostly cylindrical, the body tapering towards the head and tail ends. However, in many species the matured females assume swollen pouch-like, lemon, pear, kidney or other irregular shapes, while the males remain slender. Though plant parasitic nematodes or phytonematodes often cause severe damage to cultivated crops, they do not seem to be a limiting factor in crop production. Unlike insect pests and other disease causing microorganisms like fungi, bacteria, viruses etc., nematode population increases rather slowly and the diseases caused by them do not attain epidemic proportions within a short period of time. Nematode infestation is cumulative in nature and the gradual increase in population results in slow decline, sometimes spreading over many years. Plant nematodes mostly inhabit the soil in association with the roots of plants. However, a few species feed on the above ground parts of plants, such as stems, leaves, earheads, grains etc. The symptoms produced by nematode infestation are not so distinct and clear-cut as the symptoms produced by most of the pathogenic microorganisms like fungi, bacteria, viruses etc. However, the morphological, physiological and histological changes produced on the hosts as a result of nematode infestation is also referred to as **'diseases'**.

Plant parasitic nematodes are obligate parasites and are stylet feeders. They exist as ectoparasites, semi-endoparasites or exposed endoparasites and endoparasites. There are both migratory and sedentary nematodes. At the mouth end, they possess a characteristic, piercing stylet or spear, which is a sharp, pointed structure like a hypodermic syringe.

While feeding, the lips of the nematode are held oppressed to the feeding site and the stylet is worked back and forth into the plant tissue with the help of three sets of muscles attached to the base of the stylet After penetrating the protective wall of the host, the nematode sucks partially digested food along with the cell sap into the alimentary canal through the stylet tube by the vacuum created by the pulsating action of the median valve situated posterior to the stylet. Simultaneously, the nematode also injects a small quantity of saliva into the host tissue, which may contain enzymes necessary for dissolution of cell walls of the host and digestion of plant materials for ingestion. The enzymes injected into the host tissue contain certain toxic substances, which are responsible for developing visible disease symptoms in the host plants. Nematodes feeding on the contents of plant cells deprive the host of nutrients and water necessary for its growth. Further the conductive tissues responsible for the translocation of nutrients and water to the top portions of the host is also often blocked.

All phytonematodes have a soil phase at some growth stage or growth stages and so they are influenced by climate, soil and host conditions. Since moisture is essential for the nematodes, soil moisture becomes a limiting factor for their existence in soil. Plant nematodes are usually confined to the top 20-25 cm. layer of the soil, mostly in the rhizosphere of host roots. However, some nematodes are found much deeper in the soil. The citrus nematode is found at depths of 2.0-2.5 m. and the burrowing nematodes are found even at depths of 3.0 m. below the soil surface.

The nature of injury caused by nematodes differs with the species of nematode and the host. *Radopholus, Pratylenchus, Helicotylenchus* and *Tylenchulus* cause marked necrosis at the site of penetration. *Trichodorus* inhibits growth of the root meristems, which results in the formation of stubby roots that inhibits absorption of water and mineral nutrients from the soil. *Anguina* induces gall formation in various parts of the roots. The saliva injected into the root tissues by *Heterodera, Globodera, Meloidogyne, Tylenchulus, Anguina* etc. induces special enlarged cell formation from which the nematodes feed on the sap. In the case of *Meloidogyne,* hypertrophy of cells around the site of penetration also occurs, which results in the formation of root knots or root galls. In *Ditylenchus,* the enzyme pectinase injected into the host tissues along with the saliva dissolves the pectin cementing material in the middle lamella as a result the tissues collapse and become necrotic.

Besides causing such direct damage to the host plants, some of the nematode species act as vectors and transmit viruses that cause certain virus diseases. *Xiphinema index* is the vector that transmits 'grapevine fan leaf virus disease'. Species of *Longidorus* transmit the *polyhedral 'Nepo' viruses (Nematode polyhedral viruses),* while *Trichodorus* species transmit the *tubular 'Netu' viruses (Nematode tubular viruses).* A few fungal and bacterial diseases are also transmitted mechanically by some nematodes. The root knot nematode-*Meloidogyne hapla* predisposes linseed plants to attack by '*Fusarium* wilt' caused by *Fusarium oxysporum* f.sp. *lini.* The reniform nematode-*Rotylenchulus reniformis* and the root knot nematode-*Meloidogyne incognita* predispose cotton plants to '*Verticillium* wilt' caused by *Verticillium dahliae.* The rice root nematode-*Hirschmaniella oryzae* causes 'mentek' disease of rice. The wheat seed gall nematode-*Anguina tritici* causes 'ear cockle' disease of wheat. Further, this nematode in association with the bacterium-*Corynebacterium tritici* causes 'yellow ear rot' or 'tundu' disease in wheat.

Nematodes are mostly dispersed through contaminated soil, infested planting materials, irrigation and drainage water, floods etc.

A few of the very important nematode diseases are described below:

1. Red gram cyst nematode

Heterodera cajani

Order - Tylenchida

Family - Heteroderidae

Nature of damage. The nematodes attack the roots of young plants and cause serious damage. As a result of infestation, the entire root system is adversely affected. The xylem vessels in the infested plants are damaged to a large extent and their functioning is disrupted The symptoms produced by the pest include shallow root system, general stunting, yellowing of leaves and production of number of small, secondary roots. But, no root knots or galls are produced on the roots as in the case of *Meloidogyne.* The affected plants wilt during the hot part of the day, even when there is sufficient moisture in the soil. On the surface of the roots, white, pearl-like females, which turn brown later on, are found adhering to the roots. These are generally known as **'cysts'**. The cysts dislodge easily and fall down and remain in the soil in a dormant state for prolonged periods even in the absence of a host. Besides red gram, the pest attacks many other pulse crops, sesame etc.

Life cycle of the pest. The adult female lays about 500 eggs, which are retained in the body of the female. The eggs remain dormant until favorable conditions are present or when the eggs are stimulated by the root exudates of the host plant. The first moulting takes place inside the egg. The second stage larva, which emerges from the egg is found free in the soil until it finds a suitable host root. It enters the root just behind the root tip region and feeds on the cell contents. The larva that is to develop into a male, after its second and third moult becomes much elongated and coils. After its fourth moult, it comes out of the host as an adult male. The female nematodes are sedentary, endophytic and feed from a specific site. The host tissue around the feeding site undergoes certain histological changes. Large giant cells are formed around the head of the larva with many large nuclei in each of the cells. This group of cells is technically known as **'syncitium'** and these cells may provide more food to the feeding larva. The female larva grows up and enlarges in size considerably and attains almost a spherical shape. As it develops, it breaks open the cortical layer and remains attached to the root by the slender anterior end, while the remaining part of the body protrudes out of the root surface. Fully matured female nematode forms a tough-walled, leathery, almost spherical sac around it. The encysted female is called a cyst and this cyst encloses the eggs. The cysts, which protrude outside the host epidermis dislodge easily and fall down to the soil. The cysts are capable of withstanding adverse environmental conditions and chemical reaction and the eggs remain in a viable condition inside the cysts for a number of years. The entire life cycle is completed in about 3 weeks and many generations can occur during a single crop season. The cysts are carried from one field to other fields through drain and flood water, wind and through soil adhering to farm implements, cattle or humans (Fig.111).

Control measures

Cultural methods (i) Movement of soil from fields where the disease has occurred to other fields should be avoided (ii) A two to three year crop rotation without inclusion of a host helps to reduce the population of the cysts to a considerable extent.

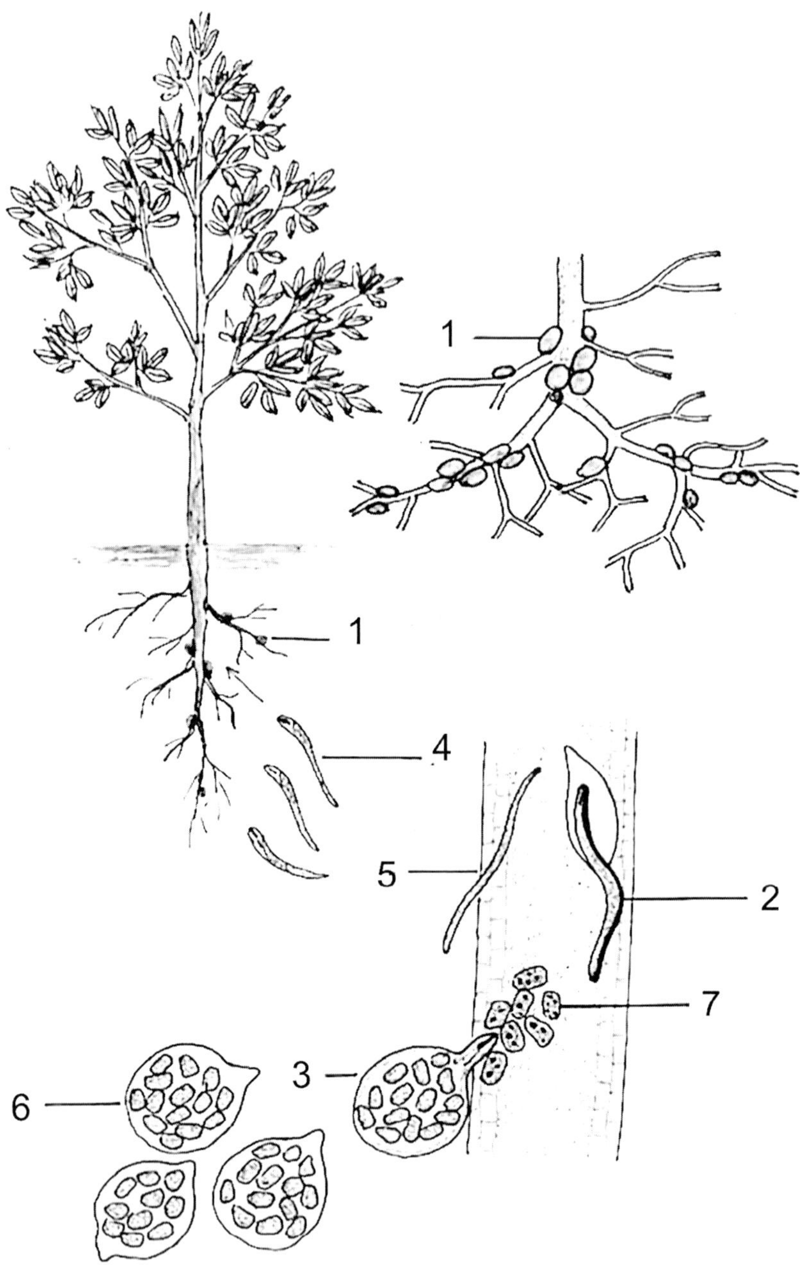

Fig. 111 . Red gram cyst nematode-*Heterodera cajani*

1. Cysts attached to the roots 2. Male nematode 3. Mature female nematode 4. Second stage larva 5. Young larva 6. Female lying in the soil 7. Giant feeding cells

Chemical control (i) The nematodes can be controlled by soil application of Dichloropropane-dichloropropene (DD mixture) at 80-100 lit. / acre by using a soil injector at 15.0-22.5 cm. depth and at points 30.0 cm. apart, one month prior to sowing (ii) Soil application of nemagon at 12 lit. / acre with a soil injector is also effective in controlling the pest.

2. Golden nematode of potato

Globodera rostochiensis and *G. pallida*

Order - Tylenchida

Family - Heteroderidae

Nature of damage. 'Golden nematode of potato'-*Globodera rostochiensis* is a very important cyst nematode and one of the most dangerous pests of potato and is responsible for causing considerable yield loss. An assessment made in England and Wales some time ago indicated that the annual loss of yield of potatoes caused by golden nematode of potatoes alone was of the order of 2,00,000 metric tons. Due to infestation by the pest, the growth of the infested plants become stunted, and the plants show wilting symptoms during the hotter part of the day even when there is sufficient moisture in the soil. Heavy infestation leads to stunted plant growth, unhealthy foliage, premature yellowing of leaves, poor development of root system and reduction in the size and number of tubers and as a result the yield is reduced drastically. Numerous small, secondary roots are produced around the collar region of infested plants, but no root knots are produced on the roots. The symptoms produced by the pest are very much similar to the symptoms produced by species of *Heterodera.* The pest mostly attacks potato, but attacks tomato and eggplant also. Golden nematode of potato, an exotic pest was introduced to India from Europe in 1961 along with seed potatoes.

Life cycle of the pest. The male nematodes are vermiform and motile, while the females are globose, sessile and sedentary in nature. The adult female swells greatly, rupture the root cortex and protrude out from the root, while the head and the neck region remains embedded in the root.

The adult male comes out of the root and fertilizes the female. The fully matured female nematode becomes a tough-walled cyst. The wall of the cyst, which is white in color in the beginning, turns to cream, light yellow, golden yellow and finally to brown, hence it is called 'golden nematode of potato'. The cyst can remain viable in the soil for about 30 years even in the absence of a host. *Globodera pallida* another cyst nematode of potato is also a serious pest of potato and is often found along with the golden cyst nematode.

The cysts usually spread through soil adhering to farm implements, cattle, man etc. The cysts can also spread through wind and water running down the slopes from infested to nematode free fields. Long distance dissemination of the cysts is through infested seed potatoes carried from one place to another or from one country to another

Control measures

Cultural methods. Once the soil gets infested with the cyst nematodes, it is very difficult to eradicate the pest completely. So every care should be taken to avoid introduction of the pest to nematode free areas (i) The nematode is highly host specific and so long term rotation extending over a period of 2-3 years with crops other than potato, tomato and eggplant, such as wheat, maize, beans etc. reduces the build-up of the pest population to a large extent (ii) Volunteer potato plants growing in infested field should be destroyed (iii) Potatoes raised in infested fields should not be used for seed purposes (iv) Vegetable nurseries and other plants meant for transplanting should not be raised in infested fields (v) Hot water treatment of seed potatoes at 44°-45°C for 10-65 minutes has been found to be effective in eliminating the seed-borne cysts.

Chemical control (i) Tuber disinfection by fumigation with methyl bromide for 5-6 hours or formalin vapor for 12 hours eliminates the seed-borne cysts (ii) Soaking the seed tubers in formalin-10 % solution for 1 hour is also effective in eliminating the seed-borne cysts (iii) Soil application of Dichloropropane-dichloropropene (DD mixture) at 80-100 lit./acre by using a soil injector at 15.0-22.5 cm. depth and at points 30.0 cm. apart, one month prior to sowing or soil application of nemagon at 12 lit./acre with a soil injector is also effective in controlling the pest (iv) Soil application of carbofuran 3 G at 13 kg./acre along the planting furrows at the time of planting affords good control of the pest **Quarantine**. Strict quarantine measures have to be taken to prevent movement of seed material from infested area to other areas.

3. Root knot nematode of tomato

Meloidogyne incognita **and** *M. javanica*

Order - Tylenchida

Family - Meloidogynidae

Nature of damage. These nematodes are distributed worldwide, but are mostly prevalent in the tropical and subtropical countries. The larvae penetrate into the cortex region of the roots, suck and feed on the cell contents. As a result of infestation by these nematodes, the plants become stunted, the leaves turn yellow and the plants show symptoms of wilting even when there is sufficient moisture in the soil. The general symptoms resemble those caused by nutrient deficiency. Root knots or galls of various sizes and shapes are produced at the tips and other areas of the roots. Formation of such galls in the roots is the most characteristic symptom. Inside the galls larvae at different stages of development are found. The infestation by these nematodes results in reduction in the rate of photosynthesis and water absorbing capacity of the plants. The infested plants are predisposed to attack by several foliar diseases and these nematodes serve as vectors for transmitting some diseases caused by certain pathogenic microorganisms. Besides tomato, these nematodes attack and cause severe damage to a number of crops such as, sugarcane, potato, cotton, tobacco, betelvine, grapevine, cucumbers, lady's finger, eggplant and many other vegetable crops. In tomato, the loss in yield due to infestation by these nematodes may go up to 46 per cent.

Life cycle of the pest. These nematodes are sedentary, root endoparasites. The adult female nematode lays 200-500 eggs in a gelatinous matrix extruded through the vulva. The matrix or egg sac containing the eggs can be seen protruding out of the galls on the roots. Differentiation of larvae takes place within the eggs. The second stage larvae hatching out of the eggs, which are long and cylindrical move about in the soil and are attracted by the root exudates of the host plant. The larvae move closer to the roots and are held in a mass around the roots and infest the roots. Usually the larvae penetrate the roots just behind the root tips. After penetration, the larvae migrate intercellularly in the cortex region. About two weeks after penetration of the roots, sex differentiation takes place in the larvae. When only a few larvae occupy a large area of tissues, most of them develop into females and in such cases reproduction is mostly parthenogenetic. The larva, which develops into a female penetrate the meristematic tissues and migrates inside until its head becomes established in an intercellular space near the endodermis. After establishing the feeding zone, it becomes sedentary and feeds on the cell contents. The site of the host tissue where the female feeds undergoes considerable histological changes. Around the mouth of the feeding female, large, giant parenchymatous cells are formed with many large nuclei in each cell due to host-parasite interaction. This group of cells is known as the **'syncitium'**. The tissues around the feeding zone undergo hypertrophy resulting in the formation of galls or root knots. The female larva swells up with each moult and the adult larva becomes obese, spherical or flask-shaped with a slender anterior region. While the head and neck region remains unaltered, the body may remain embedded within the host tissues or its posterior part protrudes outside the host tissues as a gelatinous matrix. The larva, which develops into adult male, is initially parasitic. After moulting three times it becomes a typical worm-like adult and leaves the host or enters the sac-like matrix of the female for the purpose of copulation. The males die rather soon. The female larva develops into an egg-laying adult in about 80 days (Fig. 112).

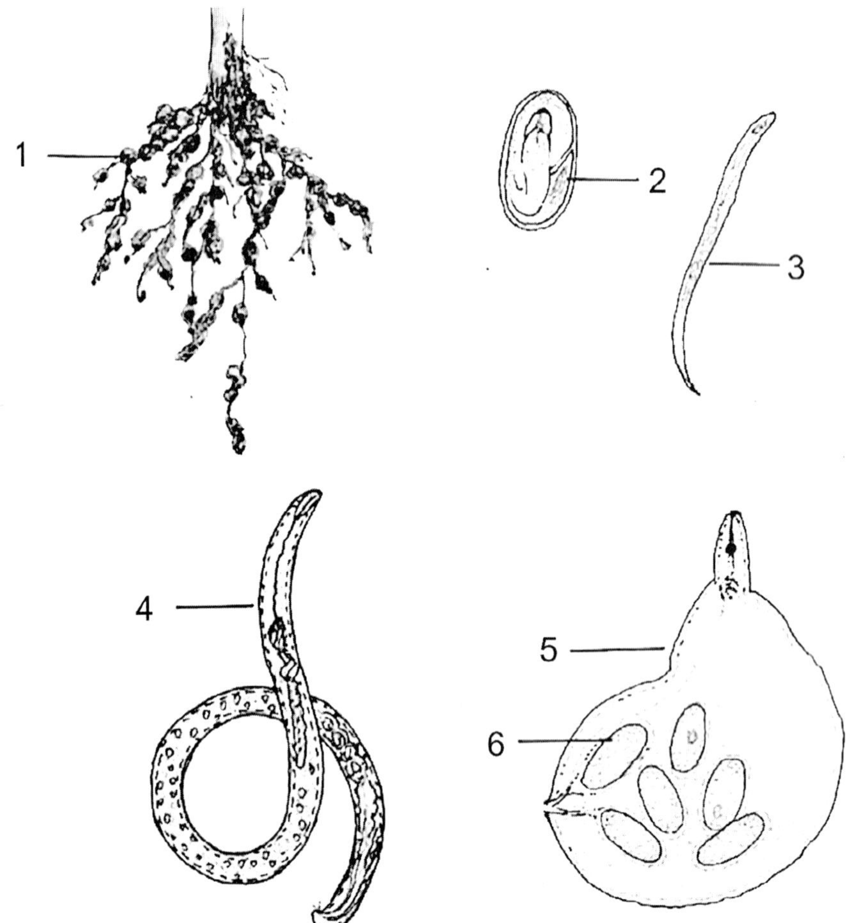

Fig. 112. Tomato root knot nematode-*Meloidogyne incognita*

1. Root knots 2. Larva inside the egg 3. Young nematode 4. Male nematode 5. Mature female nematode 6. Eggs inside the female nematode

Control measures

Cultural methods (i) During the summer season, the fields should be ploughed deep, so that the eggs are brought to the surface of the soil and destroyed by direct exposure to the heat of the sun (ii) After the harvest of the crop, the stubble along with the roots should be removed and burnt (iii) During the off season, when water is available, the field should be flooded and the water impounded in the field for a few days, thereby the larvae are destroyed (iv) In the crop rotation sequence, paddy may be included as one of the crops and the continuous inundation of water in the paddy crop helps to destroy the larvae (v) Natural manure such as, farm yard manure, compost or green leaf manure may be applied to the fields in large quantities, as a result the naturally occurring soil microorganisms, especially the actinomycetes may multiply in large numbers and destroy the nematodes. Further, such natural manure during the process of decomposition produces toxic substances such as, butyric acid, which destroy the nematodes. Application of natural manure also encourages the young plants to put forth healthier and stronger roots, which may resist nematode infestation (vi) Application of neem seed cake or saw dust at 100 kg./ acre also helps to reduce the nematode population in the soil.

Chemical control (i) Application of soil fumigants, such as dichloropropane-dichloropropene (DD mixture), vapam or nemagon is effective in controlling root knot nematode. But the cost of soil treatment is very high, as more quantity of chemical is required. DD mixture should be applied to the soil at the rate of 80-100 lit./acre with a soil injector at a depth of 15.0-22.5 cm. and at points 30.0 cm. apart. In a similar manner vapam at 1.25 lit., mixed with 100 lit. of water or nemagon at 12.0 lit./acre should be applied to the soil. These chemicals are highly toxic to the plants also, especially to the young plants and hence should be applied to the soil one month prior to planting (ii) If pest infestation is noticed after planting, nemagon may be applied at 8.0 lit./acre mixed with irrigation water (iii) Application of aldecarb 10 G-4.0 kg. or carbofuran 3 G-13 kg. or phorate 10 G-4.0 kg., mixed with 10 kg. of sand per acre uniformly over the field and irrigating the field immediately afterwards is also effective in controlling the nematodes.

4. Burrowing nematode of banana

Radopholus similis

Order - Tylenchida

Family - Pratylenchidae

Nature of damage. This migratory, endoparasite is a serious pest of banana. The pest is mostly prevalent in the tropical and subtropical regions. The infested plants are generally stunted, show leaf chlorosis, have thinner pseudostem, produce small bunches and may lodge prematurely as a result of extensive damage to the root system. The juveniles and females essentially burrow and damage the cortex of the young roots causing dark red lesions on the cortical or outer part of the roots. The roots show numerous depressed and open lesions, which cause the roots to break from the cortical region.

Such plants with short, necrotic and broken roots cannot support the weight of the bunch or they cannot withstand the force of strong winds and so they topple over. The nematodes penetrate the young roots at any point, feed on the cell contents and migrate freely through the tissues of the root cortex. In advanced stages, the infested corms also show characteristic lesions with discoloration and rotting.

The pest has a wide host range including several field, horticultural and plantation crops. Black pepper, citrus, sugarcane, coffee, coconut, tea etc. are the other economically important crops infested by the pest. Besides causing direct damage to banana, the pest paves the way for attack by the fungal pathogen-*Fusarium oxysporum* f.sp. *cubense,* which causes '*Fusarium* wilt' or **'Panama disease'**. The bacterial pathogen-*Pseudomonas solanacearum* causing 'bacterial wilt' or **'Moko wilt'** of banana may also enter the plants through the lesions caused by the nematodes. In pepper, the nematodes cause **'pepper yellows'** disease and in citrus they cause **'citrus decline'**.

Life cycle of the pest. The adult female nematode lays 4-5 eggs per day along its trail in the cortical parenchyma tissues. The eggs hatch within the root tissues. The life cycle consists of an egg stage, four juvenile stages and an adult stage and the cycle from egg to egg is completed in 20-25 days. All stages of development of the nematodes may be found within the cortex region. More than one generation of the pest can occur in the host. After deterioration and death of the infested roots, the nematodes migrate to the soil and attack nearby healthy roots. A soil temperature of about 24°C. is optimum for the survival and multiplication of the nematodes. They are found in the soil even at depths of 8 feet or more below the soil surface and can move from one plant root to another. They can survive in the soil for about 30 days in the absence of a host.

Control measures

Cultural methods (i) During the off-season, when water is available the field should be flooded and the water impounded in the field for a few days, thereby the larvae are destroyed (ii) Suckers free from nematode infestation selected from fields free from the pest infestation should be used for planting (iii) In fields infested by the pest, planting of banana should be avoided (iv) Ratooning of affected crops should be avoided (v) Hot water treatment of pared corms at 65°C. for 5 minutes is effective in eliminating the nematodes present inside the corms

Chemical control (i) The roots around the corms of suckers selected for planting are cut and removed (pared). A fungicidal solution of carbendazim (1.0 gm./lit. of water) is prepared and a small quantity of fine clay is added to the solution and mixed well. The pared corm of the sucker is dipped in the fungicidal-clay mixture and taken out. A quantity of 40 gm. of carbofuran 3 G is sprinkled uniformly over the corm. The carbofuran granules will stick on to the fungicidal-clay mixture. Then the treated suckers are planted. This treatment will afford protection against both nematode infestation and 'Panama disease'. (ii) The corms are pared and then dipped in neem oil-5 % emulsion for 30 minutes and then planted (iii) Application of aldicarb 10 G-5.0 kg. or carbofuran 3 G-13 kg., mixed with 10 kg. of sand per acre, 3-4 months after planting in the planting furrows followed by irrigation also provides protection against the nematode infestation.

Other nematode pests. Besides the nematode pests described above in detail, several other phytonematodes attack a number of other crops and cause damage to varying extent. The root knot nematodes viz., *Meloidogyne hapla* infests almost all vegetable crops, linseed etc. and produces galls and interferes with the photosynthetic activity and water absorbing capacity of the infested plants; *M. arenaria* and *M. graminicola* attack the roots of rice, millets and several other crops and weeds and produce small galls; *M. hapla* and *M. arenaria* cause root galls and coarse roots in groundnut, grapevine etc.; *M. hapla, M. arenaria* and *M. thamesi* attack potato and produce scab-like warts on the roots and tubers; *M. hapla* and *M. coffeicola* produce galls in the roots of coffee plants; *M. africana* and *M. indica* infest roots of citrus plants; *M. thamesi* infests sugarcane roots. The cyst nematodes viz., *Heterodera avenae* infests the roots of wheat and barley. The bacterial parasite-*Bacillus penetrans* is a biocontrol agent of this nematode; *H. oryzae* infests rice roots. The wheat seed gall nematode-*Anguina tritici* infests wheat seeds and produces galls. The root lesion nematodes viz., *Pratylenchus coffeae, P. brachyurus, P. penetrans, P. thornei, P. pratensis* and *P. vulnus* cause root lesions in many vegetable crops; *P. coffeae* causes root lesions in coffee, betelvine, banana, citrus, groundnut etc.; *P. vulnus* infests roots of grapevine; *P. thornei* infests wheat roots; *P. zeae* and *P. brachyurus* produce dark necrotic lesions on the roots of sugarcane plants. The stem nematodes viz., *Ditylenchus dipsaci* attacks the stem region of wheat, beans, peas, carrot, potato and stems, leaves and bulbs of onion and garlic; *D. destructor* infests the stem region of potato, cowpea, carrot, tomato etc.; *D. angustus* feeds on all aboveground parts of rice plants and causes 'ufra' disease. The stunt nematodes viz., *Tylenchorhynchus vulgaris, T. nudus, T. eremieolus* and *T. agri* cause stunted growth of wheat plants; *T. brevilineatus* causes yellow lesions on the pegs, pod stalks and on young pods; *T. brassicae* infests cabbage and cauliflower; *T. claytori* infests the roots of tobacco and causes stunting. The spiral nematodes viz., *Helicotylenchus dihystera* attacks the roots of wheat plants; *H. dihystera, H. multicinctus* and *H. erythrinae* attack sugarcane roots; *H. multicinctus* infests the roots of banana plants; *H. incisus* attacks betelvine roots. The lance nematodes viz., *Hoplolaimus galeatus, H. indicus* and *H. pararobustus* infest the roots of sugarcane; *H. indicus* infests the roots of wheat, betelvine and citrus. The white tip nematode of rice-*Aphelenchoides besseyi* attacks chrysanthemum, onion, soybean, sweet potato, sugarcane etc.; the testa nematode of groundnut-*A. arachidis* infests the inner tissues of testa, hypocotyl, roots and pods. The rice root nematode-*Hirschmaniella gracilis* attacks the roots of rice, vegetable crops, maize, sugarcane and several other hosts; *H. oryzae* attacks the roots of rice plants. The reniform nematode-*Rotylenchulus reniformis* feeds on the roots of a large number of vegetable crops, such as beans, cabbage, carrot, cauliflower, cucumber, eggplant, peas, cowpea, radish, lady's finger, lettuce, coffee, betelvine, banana, grapevine etc. The needle nematode-*Longidorus siddiqui* infests the roots of groundnut and also transmits the virus causing 'clump' disease in groundnut. The stubby root nematode-*Trichodorus christiei,* a cosmopolitan pest attacks the roots of many crop species. The dagger nematode-*Xiphinema index* infests the roots of grapevine. The citrus nematode-*Tylenchulus semipenetrans* infests the roots of citrus plants and grapevine.

PHYTOPHAGOUS MITES

The **'mites'** belong to the **Class-Arachnida, Order-Acarina**. Most of the phytopathogenic mites are classified under the two **Families-Tetranychidae** and **Eriophyidae**. The mites belonging to the Family-Tetranychidae are generally known as **'Tetranychids'**. They are very minute, and less than 100 microns in size. The body is non-segmented and the head, thorax and abdomen are not distinct, but rather fused together. The life cycle of a tetranychid mite consists of an egg stage, three nymphal stages viz., protonymph, deutonymph and tritonymph and an adult stage. In the protonymphal stage, which is also called the larval stage, the mites have only 3 pairs of legs. In all the other nymphal stages and the adult stage they have 4 pairs of legs. The mites are roundish or oval in shape, with biting and chewing, as well as piercing and sucking type of mouthparts (Fig. 113).

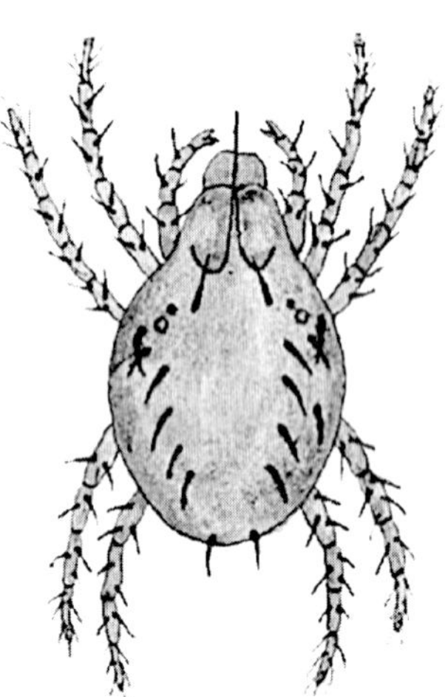

Fig. 113. Tetranychid mite

The mites belonging to the Family-Eriophyidae generally known as **'Eriophyids'** differ from the mites of other families in that they have long, worm-like or carrot-shaped body. The head and thorax are fused together, while the abdomen is long and becoming narrow at the anal end. They have two pairs of legs and biting, chewing and sucking type of mouthparts. The eriophyid mites have only two nymphal stages. The young mites resemble the full-grown adult mites, but are smaller than the adult mites. The males are smaller in size than the females (Fig. 114).

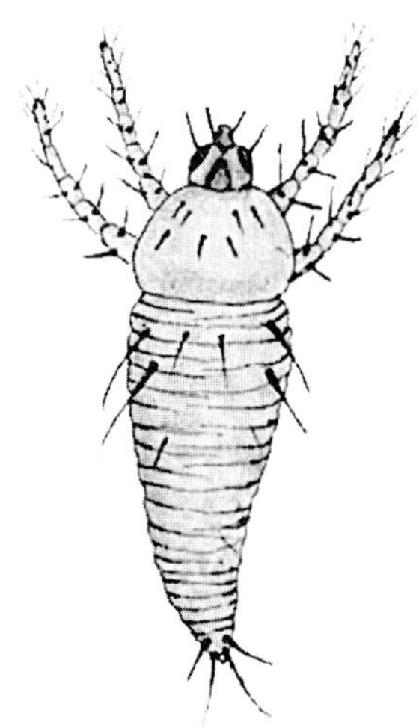

Fig. 114. Eriophyid mite

Mites pierce the plant tissues with their sharp, needle-like stylet, suck and feed on the plant sap. Whitish or reddish spots or patches may be formed on the affected parts and the plants become stunted and unhealthy. Some mites may cause galls and malformations on the attacked plant parts. Some mites are vectors of viruses and other plant parasitic microorganisms.

Some of the mites, which are responsible for causing considerable damage to economically important crop plants are discussed below: -

1. Coconut eriophyid mite

Eriophyes (Aceria) guerreronis

Class	-	Arachnida
Order	-	Acarina
Family	-	Eriophyidae

Among the non-insect pests of coconut, the eriophyid mite causes extensive damage to coconut. The pest has been reported from India, Africa, South and Central America and along the Pacific Coasts. In India, the pest is found in Tamil Nadu, Kerala and Karnataka. The mite was reported as a serious pest of coconut in Tamil Nadu and Kerala in 1998.

Nature of damage. Both the nymphs and adult mites congregate in colonies in the perianth of young, developing nuts and live in the white, tender portion covered by the inner bract of perianth and suck the sap from the tender meristematic mesocarp tissues. The feeding results in warts and numerous longitudinal fissures on the husk of developing nuts and a viscous, gummy exudate may also be seen on the surface of the affected portions. In case of severe infestation the developing nuts drop off prematurely.

Life cycle of the pest. A female mite lays about 18 eggs in the perianth of young, developing nuts. The eggs hatch in two days. The life cycle consists of an egg, two nymphal and an adult stage. The nymphal instars are not very different from the adult, except for the smaller size of the nymphs. The mites are very minute creatures, having a vermiform body, distinctly divided into a cephalothorax and a long, tapering abdomen, with only two pairs of legs situated near the anterior end of the body both in the adult and in the immature stages. The head and thorax are fused together to form the cephalothorax. The males are smaller in size than the females. The mouthparts are adapted for biting, piercing and sucking. The whole life cycle is completed in 9-12 days and there may be several overlapping generations in a year. The pest infestation is more severe during the hotter months when the humidity is less.

Control measures

Chemical control (i) Spraying with wettable sulfur-0.4 % (4.0 gm./lit. of water) is effective in controlling the pest (ii) Many of the synthetic organic insecticides introduced for the control of insect pests are effective in controlling mites also. However, dicofol-0.04 % (2.0 ml./ lit. of water) and phosalone-0.1 % (3.0 ml./ lit. of water) are widely used for the control of the pest. A single application of the pesticide is not sufficient to control the pest. Repeated application of the pesticide, once in every month may be required to effect good control. While spraying, the spathes should be thoroughly covered with the pesticide (iii) Root feeding with monocrotophos at 5.0 ml. in 20 ml. of water has also been found to be effective, but this treatment should also be repeated.

Biological control (i) The thrips-*Scolothrips indicus* and the Coccinellid bug-*Brumus suturalis* are predatory on the nymphs and adult mites (ii) The fungal parasite-*Hirsutella thompsonii* is capable of parasitizing the mites.

2. Jowar mite

Oligonychus indicus

Order - Acarina

Family - Tetranychidae

Nature of damage. It is one of the most serious and destructive pests of sorghum and is commonly found in all parts of Tamil Nadu. The mites of different growth stages remain on the under surface of leaves in large colonies within fine, delicate silken webs, suck and feed on the sap from the tissues and cause severe damage. All stages of the pest are capable of spinning webs, however usually only the females web the nest. While the nymphs and females feed on the sap, the males rarely feed. The injury caused by the pest and the continuous drainage of sap by the nymphs and females lead to development of reddish spots, which gradually extend over the entire leaf area. In case of severe infestation, the leaves and stem dry up resulting in considerable yield loss. The pest attacks sugarcane, banana and a few other crops, besides jowar.

Life cycle of the pest. The adult female mite lays spherical, whitish eggs on the under surface of the leaves. One female lays a few eggs in a day and on an average about 87 eggs are laid over several days during its life span. The egg stage lasts for 3-4 days. The larvae emerging from the eggs are six-legged, somewhat spherical and light amber-colored, which later becomes elongate and greenish in color. The nymphs become grayish-green in color.and after one or two moults become adults. The larval and nymphal stages last for 3-8 days. The entire life cycle is completed in 6-12 days. The adults soon after emergence mate and the males die soon after. One or two days after mating egg laying commences and continues for about a fortnight. Up to 40 generations may appear in a year. The mites are disseminated through wind and through contact of infested leaves with adjacent leaves.

Control measures

Chemical control (i) Spraying with wettable sulfur-0.4 % (4.0 gm./lit. of water) is effective in controlling the pest (ii) Many of the synthetic organic insecticides introduced for the control of insect pests are effective in controlling mites also. However, dicofol-0.04 % (2.0 ml./lit. of water) and phosalone-0.1% (3.0 ml./lit. of water) are widely used for the control of the pest.

3. Red spider mite

Tetranychus neocaledonicus (telarius)

Order - Acarina

Family - Tetranychidae

Nature of damage. The pest is of very common occurrence and has a very wide host range. The nymphs and adults harbor on the under surface of leaves in large colonies, puncture the leaf tissues, suck and feed on the sap oozing out. In case of severe infestation, the mites are found to infest both the surfaces of leaves. The injury caused by the pest and the continuous drainage of cell sap with chlorophyll and other pigments results in characteristic reddish-brown blotches, which may spread over the entire leaf area. The hosts of this pest includes cotton, castor, citrus plants, jasmine, rose, tomato, lady's finger, eggplant, cucurbits, grapes, papaya, sunnhemp, mulberry, tea, soybean and many other crops and weed plants.

Life cycle of the pest. The adult female mite lays on an average 100, reddish, spherical eggs. The newly hatched larvae are pale white in color. The sexes can be differentiated at the deutonymphal stage. The males are smaller, while the females are bigger and darker in color. The entire life cycle is completed in 5-12 days and about 40 generations may appear in a year.

Control measures

Chemical control. Foliar application of dicofol, phosalone, tetradifon or wettable sulfur is effective in controlling the pest. Some of the commonly used synthetic organic insecticides, such as phosphamidon, methyl demeton, dimethoate and formothion are also effective in controlling the pest.

Other mites. Besides the mites discussed above in detail, a number of other phytophagous mites attack several other crops and cause severe damage to the crops, resulting in appreciable yield loss. The rice mite-*Oligonychus oryzae* infests rice crop; the sugarcane mite-*Schizotetranychus andropogoni* infests sugarcane and sorghum; the citrus leaf mite-*Eutetranychus banksi* infests castor, papaya, moringa, gliricidia, ipomoea etc., besides citrus plants; the tea red spider mite-*Oligonychus coffeae* attacks tea and coffee plants; the citrus rust mite-*Phyllocoptruta oleivora* infests citrus, the jasmine eriophyid mite-*Aceria jasmini* infests the leaves of jasmine plants; the spider mite-*Aceria mangiferae* infests the inflorescence of mango trees leading to excessive and dense flower bud formation and premature shedding of flower buds in large numbers. The mite is also responsible for the formation of galls in the unripe mango fruits; the spider mite-*Aceria gossypii* infests cotton plants resulting in abnormal leaf growth and excessive outgrowth of hairs on the leaves; the spider mite-*Aceria tulipae* besides attacking and causing damage to leaves of wheat plants serves as the vector of the virus that causes 'stripe mosaic virus disease'; the sweet potato rust mite-*Oxypleurites convolvuli* infests sweet potato; the coconut mite-*Raoiella indica* infests coconut leaves; the tea purple mite-*Calacarus carinatus* infests tea leaves; the tea pink mite-*Acaphylla theae* infests coffee and some other weed plants, besides tea; The tea scarlet mite-*Brevipalpus australis* besides attacking tea, infests coffee, guava and some weed hosts; the tea yellow mite-*Phytotarsonemus latus* attacks tea, jute and cotton.

VERTEBRATE PESTS

MAMMALIAN PESTS

Rats and Mice

Class - Mammalia

Order - Rodentia

Family - Muridae

Among the mammalian pests, 'rats' and 'mice' are the most serious and destructive pests. They attack and cause severe damage not only to standing crops, but also cause extensive damage to stored grains and stored products of all kinds. The field rats cut down the nursery plants and feed upon the succulent fleshy stems of several crops, such as rice, wheat, sugarcane, millets, pulses and many vegetable crops. In rice crop, the damage caused by rats is estimated to be 5-10 per cent. They attack rice crop at all the growth stages and cut the plants at the base of the stem with a characteristic slanting cut. Rat damage can be seen in patches even deep inside the field and the cut plants are seen floating on the water. After grain formation, they cut off earheads and carry them to their burrows. Up to 4.0-5.0 kg. of grains are stored in each of the burrow. Presence of standing water in the field does not prevent them from causing damage, as most of them are good swimmers. Rats attack groundnut crop at the harvest stage, dig out the pods and feed on the kernels. Rats also attack immature cotton bolls, burrow them and feed on the developing seeds. The common rat, which lives and breeds in nest built at the crown region of coconut trees, bites holes in tender coconuts and drinks the sweet coconut water.

Rats cause considerable damage to stored grains and stored products in warehouses and store rooms. They bite and damage much more food materials than they can actually consume. They tear off bags filled with grains and other products, burrow holes in the walls and floors of warehouses and godowns, spill and scatter grains and other products. They pollute and contaminate grains and stored products with their hairs, urine and faecal matter and make them unfit for consumption.

Rats, which are nocturnal in habit are found everywhere and can withstand any unfavorable conditions and carry on their lives. They live in burrows in the fields, field bunds and embankments. The house rats live in roofs of buildings, crevices and holes in the walls and floors of houses and surrounding places, in drains and in concealed places. They can squeeze through holes half the size of their body and can jump up to a height of 60 cm. According to an estimate there are about 2,500 million rats in India. Rats are very suspicious and cunning creatures. They do not have good eyesight, but have exceptionally strong senses of touch, smell, taste and hearing. Though rats generally feed on vegetarian diet, they eat non-vegetarian foods, such as fish, meat etc.

Rats have very sharp, chisel-like front teeth with which they bite, cut, scrape and grind the food. On an average a rat consumes 10-15 gm. of solid food, such as grains, flour, vegetables, fruits etc. and drinks about 10 ml. of water in one day. Worldwide, the damage and loss caused by rats is estimated to be about 20 million tons in the field and about 33 million tons in the store houses. Rats are known to breed and multiply very rapidly. They start breeding when they are 3-4 months old. They produce about 10 litters in a year and in each litter there may be about 10 young ones. The gestation period is about 10 weeks.

Besides causing such direct damage and loss, they make burrows and long tunnels in irrigation channels and field bunds leading to wastage and loss of water. Rats are also responsible for spreading dangerous diseases, such as plague, rabies, rat fever and food poisoning.

Rats are generally classified into two main categories viz., field rats and house rats. The field rats live in burrows in the field, field bunds and embankments, while the house rats live around dwelling places. Four species of field rats, the house rat and the common Indian rat or coconut rat occur quite commonly and inflict serious damage and heavy loss.

FIELD RATS

1. The mole rat or Field rat

Bandicota bengalensis

The rats of this species are dark grayish-brown in color and the under surface of the body is grayish-white in color. The tail is distinctly segmented and hairless. They are 30-41 cm. long from head to tail and the tail alone measures 15-18 cm. They are quite stout, with roundish ears and short, broad face. The young ones at birth are reddish in color, with eyes and ears closed. They feed on the mother's milk for about 15 days and afterwards start feeding on the available food by themselves. They become full-grown in 4-6 months, attain maturity and start breeding. One rat may give birth to 2-18 young ones at a time. In a year a rat may give birth to young ones thrice and thus the rats multiply very rapidly. The rats live for 27-37 months.

These field rats live in long, winding burrows, which are 9-12 meters long and 1.0-1.5 meters deep in the field bunds or embankments or in the field. For their protection they provide many barriers inside the burrows to ward off snakes and other natural enemies and there may be 4-5 exit holes in each burrow for escape in case of emergency. However, the rats generally use only one exit and loose, fresh soil mound is always found at the mouth of the exit hole. Inside the burrow there are a few wide chambers, which are used as bed chambers, mating chambers, chambers for the litters and chambers for storing food materials. In one burrow only one rat lives along with its young ones.

The field rats feed on grasses, grass roots, tubers, grains, vegetables, fruits and such other food materials. In Tamil Nadu, rats cause enormous damage to rice and pulse crops. When the crop is nearing maturity, the damage caused by them is much more severe. At the booting stage, the rats cut the rice plants from the base and cause severe damage. In the attacked areas, round patches of cut plants are seen. Before harvest, the rats cut the earheads, carry them to their burrows and store them in the food chambers. About 2.0 kg. of grains may be stored inside each burrow. The rats consume the stored food material, when availability of food is scarce.

During the hot summer months and during continuous heavy rains many young ones may die inside the burrows and the rat population may decrease drastically during such periods. Concerted and joint efforts to control the rats during such periods give good results (Fig. 115).

Fig. 115. The mole rat-*Bandicota bengalensis*

2. The grass rat

Millardia meltada

The rats of this species are smaller in size than the mole rats and measure about 23 cm. from the head to the tail end and the tail alone measures about 10 cm. The body is dark brownish-gray in color, while the under surface of the body is light brown in color. The body is covered with fine, soft hairs and the legs are white in color. They live in burrows in the field and field bunds. The burrows resemble those of the mole rat, but are shorter and narrower. In a single burrow more than one rat live together. One female rat may give birth to 1-12 young ones at a time.

The grass rats feed on fleshy plants, stem portions, grains, grass roots, rhizomes and such other food materials. They attack and damage rice crop at all growth stages. They feed on the germinating rice seeds, cut and feed on the young seedlings and plants at the booting stage. But these rats never store food materials inside the burrows. They feed on the immature cotton bolls also (Fig.116).

Fig. 116. The grass rat-*Millardia meltada*

3. The gerbil rat

Tatera indica

The rats of this species are reddish-gray in color and the under surface of the body is whitish in color. They are almost of the same size as the common house rats. They measure about 18 cm. from the head to the end of the body, while the tail is longer than the rest of the portion. They have big eyes and the tail is covered with fine, soft, short hairs, with a thick bunch of long, black hairs at the tip of the tail. The hindlegs are longer and adapted for jumping. They can jump to a height of 1.2-1.5 meters easily.

They feed on grains, soft and fleshy stem portions, roots, leaves, fruits, vegetables, grasses etc. They attack and cause damage to the plants at all growing stages. They live in burrows near hedges and bushes. More than one rat live together in each burrow. These rats also do not store food materials inside the burrows (Fig. 117).

Fig. 117. The gerbil rat-*Tatera indica*

4. The field mouse

Mus booduga

The rats of this species are much smaller. They measure about 13 cm. from the head to the tail end. The tail is short, measuring about 5.0 cm., thin and without hairs. They are brown-colored and the under surface of the body is white in color and have small legs. They live in narrow burrows in the field bunds, causing severe damage to the field bunds leading to wastage of irrigation water. They are found mostly in fields and gardens, but sometimes found in houses also. A pair of mice lives in a burrow. In paddy fields, they cut the earheads and cause severe damage (Fig. 118).

HOUSE RAT AND COCONUT RAT

The common house rat

Rattus rattus rufescens

The rats of this species are brownish-gray in color and the under surface of the body is blackish in color. The head and body together measures about

11.5 cm., while the tail is long, measuring 11.5-28.3 cm. and thin. A coat of fine, soft hairs covers the body. They live in burrows under the ground, on rooftops, on lofts and such other secluded and concealed places. They feed on all types of vegetables, animal feeds, poultry feeds, grains and also non-vegetarian foods. They cause extensive damage to stored grains and stored products in granaries, godowns, storehouses, grocery shops etc. They also cause damage to household articles, such as wooden articles, clothing etc. (Fig. 131).

Fig. 118. The field mouse-*Mus booduga*

The common Indian rat or **Coconut rat**

Rattus rattus wroughtoni

The rats of this species are reddish or yellowish-brown in color and the under surface of the body is white in color. They burrow the tender coconuts in the trees and drink the sweet coconut water. The burrowed coconuts drop down and are seen scattered below the trees. They live and breed in nests built at the crown of the trees between the fronds with fibrous materials from the coconut trees. In closely planted coconut gardens, the rats jump from one tree to another easily.

Control of rats

Natural control (i) Rats have many natural enemies, such as cats, owls, snakes, mongooses etc. These predators catch large number of rats from the fields and houses and prey on them (ii) Rain is also a natural agent that destroys large number of young rats. After heavy and continuous spells of rain, drowning in the inundated water destroys many young rats.

Physical and mechanical methods. (i) Places around store houses, houses, gardens, road sides etc. should be kept clean. Waste materials, unwanted things, rubbish and useless articles should not be allowed to accumulate around these places for long periods, as these may provide breeding grounds for rats (ii) After the harvest of crops, live burrows in the fields and bunds may be dug and the rats caught and killed (iii) Grains and other stored products should not be dumped on the floor of store houses. They should be packed in gunny bags and stacked on lofts at least 45 cm. above the floor (iv) Consumable things should be kept in metal or plastic closed containers (v) Burrows, holes, damaged areas etc. found on the floors, walls and roofs of store houses and houses should be patched and plastered with cement. Holes, gaps etc. in doors, windows and ventilators should be blocked with metal sheet (vi) Many types of rat traps, such as box-type traps, snap traps, bamboo bow traps, mud pot traps etc. are being used to trap and kill rats in the fields and in houses. The Thanjavur bamboo bow traps are widely used in the paddy fields to trap and destroy field rats (vii) A type of sticky trap to catch rats, especially in houses is available in the market. It consists of a thick, folded cardboard with a thick, glue-like, sticky substance in-between the fold. This simple contraption is spread out and placed in a few places frequented by rats. The rat, which sticks on to the glue cannot escape and it can be killed easily.

Biological control (i) Rats constitute 90 per cent of food of barn owls. They prey on all types of rats and mice. It has been estimated that a pair of barn owls with their six chicks consumes 500-600 rodents during their nesting period. To encourage the owls to sit and prey on rats, 'owl torches' or 'owl perches' may be provided in the fields. White pieces of cloth are wound round the tip of poles of 1.0-1.5 m. height and planted in the fields at the rate of 15-40 per acre, at a distance of about 4.0 m. apart. These are known as owl torches or owl perches. During the nighttime, the owls come and perch on these torches conveniently, catch and prey on the prowling rats. Instead of cloth, small, inverted mud pots with white spots on the surface may be placed on the pole top or 'T'-shaped wooden perches with straw lining may be planted for the owls to sit on them comfortably (ii) Indiscriminate killing of snakes, especially non-poisonous snakes found in the fields should be avoided, as snakes are one of the chief biological agents, which can catch, eat and destroy large number of rats.

Chemical control. Rat poisons, such as Zinc phosphide, warfarin, tomorin, Strichnine hydrochloride, Barium carbonate, Calcium cyanide dust, bromodiolone etc. are being used for the control of rats. Fumigation of burrows with cyanogas is advocated for the control of rats, especially in big godowns. Use of Zinc phosphide is one of the most effective methods of controlling rats both in the fields, as well as in houses, warehouses, godowns etc. One part of Zinc phosphide is mixed with 19 parts of bait material, such as popped corn or rice or dry fish and a few drops of coconut or gingelly oil. This mixture is placed in places frequented by rats. Prebaiting without the poison for one or two nights is necessary. Instead of Zinc phosphide, warfarin can be used. For this five parts of warfarin mixed with 95 parts of bait material and a small quantity of oil is used.

OTHER MAMMALIAN PESTS

Besides rats, several other mammalian pests, such as monkeys, jackals, porcupines, rabbits, squirrels, flying foxes, wild pigs, deers and even wild elephants cause extensive damage to crops resulting in considerable yield loss.

Monkeys *(Macaca radiata).* Monkeys visit fields and gardens and cause damage. They bite and damage crops like sugarcane. They also bite and eat fruits, such as papaya, banana, mango etc. and cause extensive damage. They also venture into human dwellings in some localities, eat and carry away food and other consumable things and cause annoyance to people. They also carry the dreaded *'rabies virus'*.

Deers *(Axis axis).* In some of the villages around sanctuaries, deers come in herds, trample down cultivated crops, feed on the vegetative parts and cause severe damage.

Jackals *(Canis aureus)* and **Foxes** *(Vulpes bengalensis).* These canines usually come in small packs and cause enormous damage to sugarcane and maize crops by biting and cutting the stem portion of plants. Besides damaging crops, they act as potential carriers of the *'rabies virus'* and are regarded as a serious danger to pet animals, cattle and human beings.

Porcupines *(Hystrix indica)* and **Wild pigs** or **Wild boars** *(Sus cristatus).* These wild animals occasionally cause considerable damage to tuber crops by digging and feeding on the tubers of potato, sweet potato, tapioca, carrot etc. and also groundnut.

Flying foxes *(Pteropus giganteus).* These flying mammals come during the nighttime, bite and feed on fruits, such as guava, mango, papaya etc. They also cause severe damage to arecanuts by biting, chewing and feeding on the juice from the outer fleshy epicarp of ripe arecanuts.

Rabbits *(Lepus nigricollis).* Rabbits feed on the foliage of vegetables, tubers, cereals etc.

Squirrels *(Funambulus palmarum).* Squirrels eat fruits, nuts etc. and also nibble and damage flowers of crops.

Elephants *(Elephas indicus).* Herds of elephants sometimes wander down from their territories in the forest areas and hilly terrains into neighboring villages in search of food and water and cause damage to all types of cultivated crops. They trample, pull down and eat crops, such as sugarcane, banana, coconut etc.

Control

Monkeys can be trapped by using nets or different types of traps with the help of Forest Department Officials and released into their natural habitats. Trained dogs may be used to scare and drive away monkeys. In the human dwellings, wire nets may be provided in the windows, ventilators etc. to prevent entry of monkeys into the houses. Religious sentiments stand in the way of killing monkeys by any means.

Provision of wire fencing around cultivated fields, setting up powerful electric lights in the fields, engaging laborers to watch the crops during the nights and driving away the deers that venture into the fields help to ward off crop damages by deers. Hunting of deers or catching them by any means or poisoning them is prohibited according to law, as they are considered to be endangered species.

Jackals can be killed by using poison baits. Pieces of meat, each weighing about 0.5 kg. are taken. A small cavity is made in each piece by scooping out a small piece from the center. A few grains of Strychnine hydrochloride or a small quantity of Zinc phosphide is put inside the cavity and the cavity closed with the piece of meat scooped out. The pieces of poisoned meat are placed on a big leaf and placed in locations frequented by the jackals in the evening hours. The jackals, which eat the meat die of poisoning. The uneaten pieces of meat should be collected on the next morning and stored in a safe place. The bait once prepared may be used for 2 or 3 days and the remnants if any after this period should be disposed off by burying deep in the soil.

Porcupines can also be destroyed by using poison baits. For this Strychnine hydrochloride or Zinc phosphide is placed inside pieces of guava fruits and placed in locations frequented by these animals. The porcupines, which are attracted by the smell of guava fruits, eat the poisoned guava pieces and die of poisoning. Introducing cyanide gas into burrows occupied by porcupines is also effective in killing them

Wild pigs and Flying foxes can be destroyed by hunting. Explosives are also used inside baits to kill wild pigs.

Rabbits can also be destroyed by either hunting or by introducing cyanide gas into the rabbit burrows.

Squirrels can be destroyed by poison baiting with Zinc phosphide or by hunting.

Elephants can be prevented from intruding into cultivated fields by erecting low voltage electric fence around the fields. They can also be driven away by using firecrackers, beating of drums or by setting up bonfires

NON-MAMMALIAN PESTS

BIRDS

A few species of birds are also responsible for causing much damage to field and horticultural crops, especially the produce. They generally destroy much more than they can actually consume. Among them the omnivorous crows, the cosmopolitan house sparrows, the weaver birds, pigeons, peacocks, parrots, mynahs and quails are more important.

Common crows *(Corvus splendens)*. Besides feeding on the grains of rice, sorghum, wheat, lesser millets, groundnut etc., crows peck, tear and feed on the fleshy pulp of fruits, such as mango, papaya, guava, mulberry etc.

House sparrows *(Passer domesticus)*. Sparrows feed on sorghum, millets, wheat, rice, small fruits etc. in the fields. They also pick and feed on any kind of grains from threshing floors, drying yards and from houses.

Weaver birds *(Ploceus philippinus)*. Weaver birds come in large numbers during the nesting period and build artistic, hanging nests at the top of palmirah and other tall roadside trees roundabout cultivated fields. They cut and tear off long strips of leaves from rice and millet crops for weaving their nests and thus cause damage to the crops. At the time of crop maturity they feed on the grains of rice and millets, thus causing appreciable yield loss.

Pigeons *(Columba livia)* and **Peacocks** *(Pavo aristatus)*. These birds feed on the grains from matured crops and also pick and feed on the grains, such as groundnut, pulses etc. sown in the fields.

Parrots *(Psitacula krameri)*. Green parrots alight in large numbers on cultivated crops, cut the earheads, as well as feed on the grains of sorghum, maize, pearl millet, wheat, barley etc. They also peck and feed on the fleshy pulp of fruits such as, mango, guava, papaya, pomegranate etc., thus cause damage.

Mynahs *(Acridotheres tristis)* and **Koels** *(Eudynamys scolopacea)*. These birds also feed on various kinds of fruits and cause extensive damage in orchards and in kitchen gardens. The partially eaten fruits may either remain attached to the trees or may fall down to the ground.and become unfit for marketing

Control

(i) The bird nests, as well as the resting places of birds around the fields may be destroyed (ii) Nets may be used by experienced persons to trap and capture birds, which come in large numbers to feed on the grains (iii) Beating of drums or empty tins or cans and making loud noise, using firecrackers to produce cracking noise, using sling shots to throw stones, setting up scare crows and adoption of such other tactics may help to scare and drive away the birds (iv) Commercially available mechanical bird scarrers, which produce very loud explosive noise automatically at regular intervals of time ranging from 10-15 minutes may be set up at the rate of one device for every 8-12 acres to scare away birds (v) Poison baits may be used to kill the bird pests. A poison bait with sorghum grains as bait and malathion 5 D as poison may be used to destroy sparrows. Poison bait with sorghum grains as bait and carbofuran 3 G as poison in the ratio 100 : 5 may be used to kill sparrows and pigeons, which pick and eat grains sown in the fields.

OTHER NON-INSECT INVERTEBRATE PESTS

Rice field crab - *Paratelphusa hydrodromus* (Decapoda, Crustacea)

Crabs make holes in the sides of field bunds, irrigation channels and at corners of fields, which are slightly raised and higher than the water level in the fields and live within the holes. They protect the holes from water inundation by heaping mounds of soil around the opening of the holes. Though crabs are generally carnivorous, plant-feeding habit has been noticed in some of the fresh water crabs. They cut seedlings in the nurseries and young transplanted crop mostly within a fortnight after planting. They cut the young plants at the ground level into small bits and carry the pieces to the holes to be used as food. In the older plants, they cut open the outer leaf sheaths and feed on the inner tender portions. In the attacked fields, bits of stems and leaves are seen floating on the water. In severely attacked fields, the damaged areas appear in patches near about the bunds and irrigation channels. The pest generally does the damage only during the dusk or at night. In water inundated and in fields with poor drainage, they cause more damage. Besides causing such direct damage to plants, the crabs cause damage to the bunds by burrowing holes, which leads to breaches and loss of irrigation water.

The crabs multiply mostly during the dry periods from April-June. The adult female lays about 200 eggs. The eggs are carried in a pouch-like abdominal flap on the ventral side of the abdomen till they hatch. On hatching, the young ones come out of the pouch, separate out, make their own holes and start living independently. Crabs have an oval body and the abdomen is tucked under the thoracic region and has a hard, chitinous dorsal covering on the body. They have 10 conspicuous legs and eyes.

The measures for the control of crabs include physical, chemical and biological methods:-(i) Stagnating excess of water in the fields should be avoided and excess water should be drained off. In the drained fields, the crabs coming out from their holes are easily caught and eaten by rats and predatory birds such as, pond herons, storks etc. (ii) The crabs moving about in the fields may be hand picked and destroyed (iii) The crab holes may be dug and the crabs caught and destroyed (iv) Moistened rice bran kneaded and made into large lumps are kept as bait inside wide-mouthed earthen pots and are buried in the soil near about the bunds and irrigation channels with their rim just above the water level. The crabs, which are attracted by the bait fall into the pots. They can then be collected and destroyed. Such pots may be placed at the rate of four pots per acre (v) Poison bait with cooked rice as a bait and malathion EC as a poison may be places at the mouth of crab burrows in the evening time to attract and kill the crabs (vi) Poison bait with popped rice mixed with fried onion and dry fish bits as bait and warfarin 0.025 % as a poison is also effective in killing the crabs (vii) A few Copper sulfate crystals may be dropped inside the crab holes to destroy them (viii) Cyanide gas may be introduced into the crab holes using a cyanogas pump to destroy the crabs. But this treatment requires trained personals and is also very costly

Snails and slugs (Gastropoda, Mollusca)

Snails and slugs live on land and are often found feeding on vegetation. Snails have an external hard, spiral shell covering their soft body, while the slugs do not generally have such a conspicuous hard shell, but have a smooth shell. Slugs are found sometimes in large numbers in betelvine gardens and nibble betelvine and agathi leaves. Leaves of paddy seedlings and vegetable crops are sometimes nibbled, eaten and damaged by small snails. The giant African snail-*Achatina fulica,* an exotic pest has established well in many countries including India. These snails nibble, eat and destroy many vegetable crops, ornamental plants, plantation and horticultural crops. Snails and slugs are more active during the rainy season and occur more in moist and damp areas. They are usually active during the nights and take shelter under fallen leaves or underneath shades of leaves of plants, such as banana, papaya etc. during the daytime.

The giant African snail is large in size, about 7.0-15.0 cm. in length, with shells having 6-7 whorls. They are chestnut brown in color, with light brown and yellow longitudinal stripes. The female lays about 200 eggs at a time in batches on the soil surface or just beneath the soil surface. They live for about 3-4 years

Cultural, physical and chemical methods can be adopted to destroy the snails and slugs:- (i) In rice fields excess water should be drained off to expose them, so that they may be devoured by predators (ii) During rainy season, moist gunny bags or moist fallen leaves may be heaped in a few places near about the fence of cropped areas. The snails and slugs that take shelter underneath the heaps can be collected the next morning and destroyed (iii) Triphenyl tin hydroxide, a fungicide available under the trade name Du-ter acts as a molluscicide also. Du-ter may be applied uniformly over the field at the rate of 200-400 gm. ai./acre to kill the snails and slugs (iv) Poison bait containing wheat bran as a bait and Metaldehyde or Paris green as a poison may be used to attract and destroy the snails and slugs (v) A few hermit crabs and a predatory millipede eat and destroy the snails and slugs.

❑❑❑

Chapter-3

Household and Livestock Pests

Insect and non-insect pests cause extensive damage to standing crops, stored grains and stored products resulting in huge economic loss. Besides these, several pests cause annoyance, as well as attack livestock such as, cattle, poultry and pet animals. Some others damage household articles such as, clothing, upholstery, carpets, paper, books, leather articles, wooden and bamboo furniture and all sorts of wooden structures. These obnoxious creatures besides consuming any kind of food materials, spoil them by contaminating and polluting them with their faecal matter and urine and impart a repulsive, bad odor and make them unfit for consumption by humans and livestock. The damages caused by them pave the way for easy access and infection by some of the saprophytic microorganisms, which hasten the deterioration process. A few of these microorganisms produce mycotoxins, which are poisonous and may even be fatal to the consumers.

Among the household pests, cockroaches, crickets, silver fish, ants, white ants, carpet and powder post beetles and clothes moths are quite important. These insects not only cause annoyance but different types of flies, mosquitoes, bugs, fleas and lice either suck and feed on the blood from humans, cattle, poultry etc. but are also responsible for disseminating several disease causing microorganisms.

HOUSEHOLD PESTS

1. Cockroach (*Periplaneta americana*)-Dictyoptera, Blattidae

Cockroaches are very familiar in houses and are found running about in storerooms, kitchens, bathrooms and toilets. They are also quite common in restaurants, bakeries, food stores etc., especially in dark basements and dirty surroundings. They are active during nights and are fond of warm and moist atmosphere. They touch, feel, taste and eat any type of food products and prepared food, either vegetarian or non-vegetarian. Further, they pollute the food materials with their excreta spread allover the food and leave a foul, unpleasant odor. They also scrape, feed and damage bindings, cardboard boxes, magazines and other materials dumped in dark places.

The adult cockroach-*Periplaneta americana* (Fig.119) is a fairly large, flattish insect with a uniform dark brown color and 3.50-4.00 cm. long and has two dark, triangular patches on the prothorax. Two other smaller ones, the German cockroach-*Blatella germanica* (Fig.120), which has two characteristic dark brown streaks on the prothorax and the slightly bigger Australian cockroach-*Periplaneta australasioe* (Fig. 121), which has a circular yellowish band on the prothorax and a narrow, yellowish patch on each of the forewings are also found in some localities.

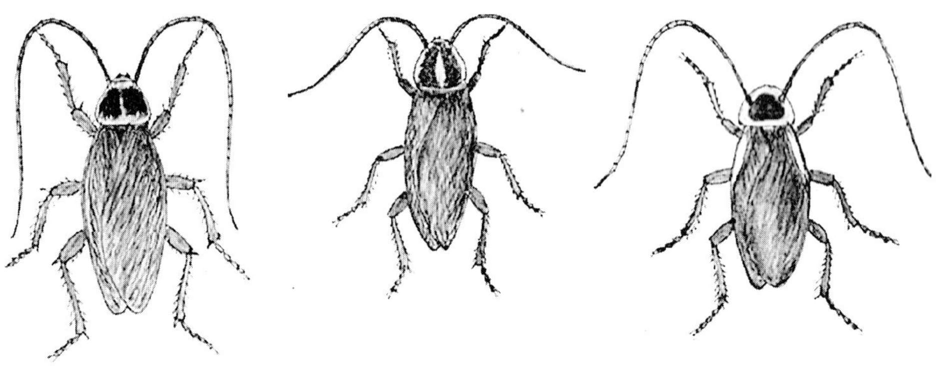

Fig. 119. Common cockroach-*Periplaneta americana*

Fig. 120. German Cockroach-*Blatella germanica*

Fig. 121. Australian cockroach-*Periplaneta australasioe*

Cockroaches lay eggs in a characteristic seed-like case or egg capsule called **'ootheca'**. These egg capsules are often found sticking out of the abdomen of the female as it moves about. The female produces an ootheca once in 4 or 5 days and a female may lay 15-90 oothecae. Each ootheca contains 14-16, cigar-shaped eggs. The incubation period ranges from 15-100 days. The young ones reach adulthood in about 3 months and during that period they moult their outer skin 6 times.

(i) Keeping the houses and surrounding areas clean and dry, keeping foodstuffs covered, proper disposal of food remnants etc. help to ward off the insects (ii) Dusting with propoxur (Baygon) 1 D or malathion 4 D or spraying with propoxur 20 EC at 0.04% ai (2.0 ml./lit. of water) or malathion 50 EC at 0.1% ai. (2.0 ml./lit. of water) is effective in controlling the pest

2. **Silver fish** (*Lepisma saccharina*)-Thysanura, Lepismantidae

The adult insects are quite active and fast runners. They are glistening, grayish in color and are commonly found behind pictures, wall hangings, calendars, in books, shelves and in-between washed clothing that have not been disturbed for a long time. They are nocturnal in habit and shun light. The adults are about 2.0 cm. in length and wingless. They are spindle-shaped, compressed dorso-ventrally, soft-bodied and scaly with three characteristic long anal processes (Fig.122). The adult females deposit eggs loosely in concealed places and the incubation period is about 7 days. The adult stage is reached in 3-24 months. There are many instars during the growth stage and some of them moult several times throughout their lives. The young ones do not have scales and anal processes.

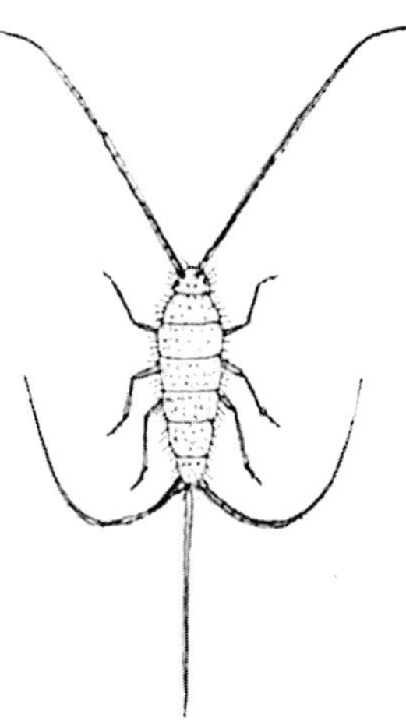

Fig. 122. Silver fish *Lepisma saccharina*

These insects feed on a variety of materials such as, starched clothes, rayon fabrics, bindings of books, papers on which paste or glue has been used and many other materials that contain cellulose

(i) These insects can be controlled by using poison bait consisting of oat meal, sodium fluoride or white arsenic, sugar and common salt (ii) Spraying phenyl solution or propoxur or malathion also controls the pest (iii) Keeping naphthalene balls or camphor in cup boards, shelves, almirahs, card board boxes etc. wards off the pest.

3. **Crickets**-Orthoptera, Gryllidae

Common house cricket *(Gryllus domesticus)*

Mole cricket *(Gryllotalpa africana)*

Crickets have almost the same habits as cockroaches. They also feed on various food materials and contaminate them. They usually harbor inside cracks and crevices in walls, door frames and in such other hiding places during the daytime and come out during the nights. They are often found in damp, dark corners in the house. The house crickets are grayish-brown in color with characteristic long anal processes (Fig. 124). The mole crickets are yellowish-brown in color and are somewhat fleshy (Fig. 123). These nocturnal creatures, mostly the males make a disturbing, shrill and annoying noise at night by rubbing the wings together. Crickets are capable of leaping long distances. Crickets live in holes, cracks etc. under moist soil in orchards, backyards, lawns, coconut gardens etc., catch hold of small insects and feed on them. The female crickets lay eggs singly in such underground dwellings.

Crickets can be spotted out easily and destroyed.

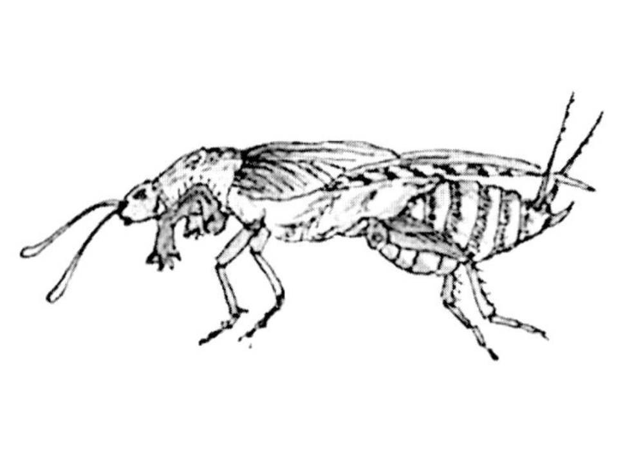

Fig. 123. Mole cricket-*Gryllotalpa africana*

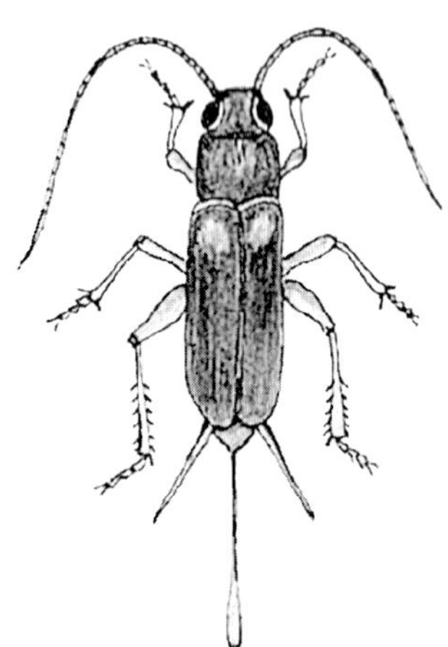

Fig. 124. House cricket-*Gryllus domesticus*

4. Carpet beetles *(Anthrenus vorax* and *A. pimpinella)*-Coleoptera, Dermestidae

These are small beetles, the grubs and adults of which bite small holes and are often destructive to carpets, bed spreads, curtains, upholstery, silk, woolen goods and even museum specimens of insects, stuffed birds and animals, dried meat etc. that may contain wool, fur, feather and hairs. The adult female deposits the eggs in dark, secluded places or on woolen and other fabrics. The grubs are dark brown or black and wooly and hence are commonly known as **'wooly bears'**. The pupa is also covered with long, wooly hairs (Fig. 128).

(i) Systematic cleaning of stored fabrics and other materials helps to ward off the pests (ii) Godowns where fabrics and other materials are stored should be fumigated with fumigants such as methyl bromide (iii) Fabrics and other preserved materials in small enclosures and boxes can be protected by using paradichloro benzene crystals, naphthalene balls or camphor.

5. **Powder post beetle** or **Furniture beetle** *(Sinoxylon sudanicum)*-Coleoptera, Bostrychidae

These small, reddish and dark brown beetles attack cane, bamboo and other wooden structures and furniture. The beetles bore small holes and feed on the hard wood. The grubs cut the hard and dry wood and make zigzag tunnels in the wooden structures. The wood is reduced to powder, which is pushed out through the holes made by them in the form of fine powder or dust. Numerous small, round, shot-holes are visible on the infested wooden structures (Fig.125)

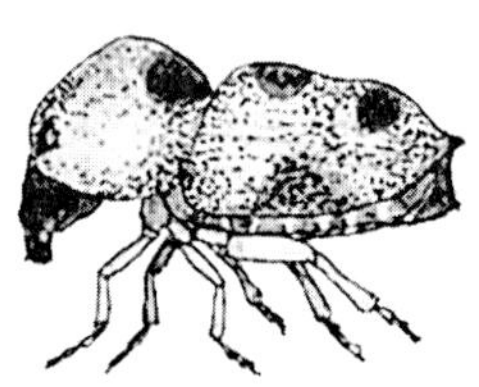

Fig. 125. Furniture beetle *Sinoxylon sudanicum*

(i) Painting the wooden structures periodically with varnish, paint, tar or creosote deters the pest from attacking the wood (ii) Treating wood with copper sulfate solution or zinc chloride solution preserves the wood from attack by this pest.

6. **Clothe moths-**Lepidoptera, Tineidae

The black and white moth *(Trichophaga abruptella)*

The brown moth *(Tinea pachyspila)*

The slender caterpillars of these small moth pests attack wool, fur, hair, feather and sometimes damage dead insects, dry, dead animals, animal and fish feed, milk powder, leather, trophies etc. The caterpillars generally move about, their body enclosed in small cases of cotton, wool or silk and feed on the fabrics. The moths of *Tinea pachyspila* are light grayish in color (Fig.129), while those of *Trichophaga abruptella* are black and white-winged (Fig.130).

The methods suggested for the control of carpet beetle is applicable for the control of these moth pests also.

7. **Ants-**Hymenoptera, Formicidae

The small house ants *(Monomorium destructor* and *M. criniceps)*

The household red ant *(Solenopsis geminata)*

The black ant *(Camponotus compressus)*

Besides causing annoyance, ants of different kinds, which are commonly found in households attack all kinds of foods and provisions. Ants are social insects and they assemble in large numbers and carry bits of food materials to their nests as they march in rows. Some species nest in the sills and woodwork of houses and cause damage. Some species that are found in the lawns, gardens and floors build nests under the ground and throw mounds of earth all round the entrance of their nests. They carry whatever food materials available to their nests. Some species feed on grains, seeds, dead insects, fats etc. They cause annoyance and pain by their stinging bite (Fig.126, 127)

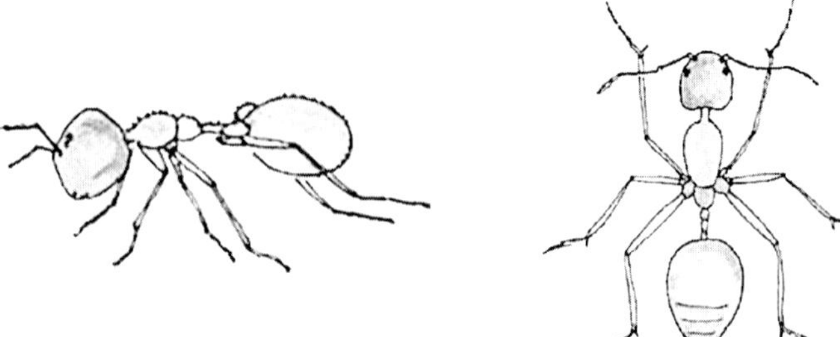

Fig. 126. Household red ant *Monomorium destructor*

Fig. 127. Small house ant *Solenopsis geminata*

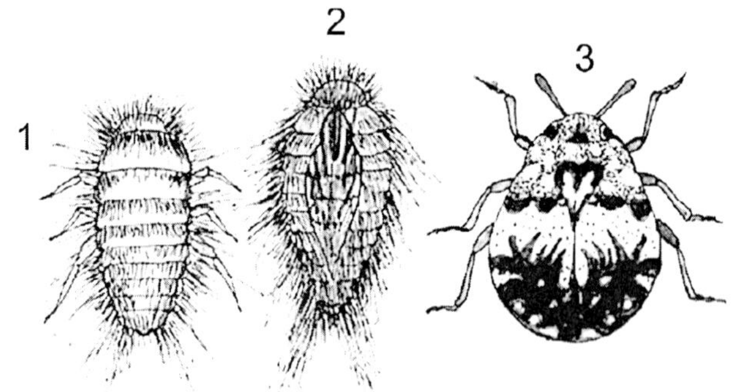

Fig. 128. Carpet beetle-*Anthrenus voras*
1. Grub 2. Pupa 3. Adult beetle

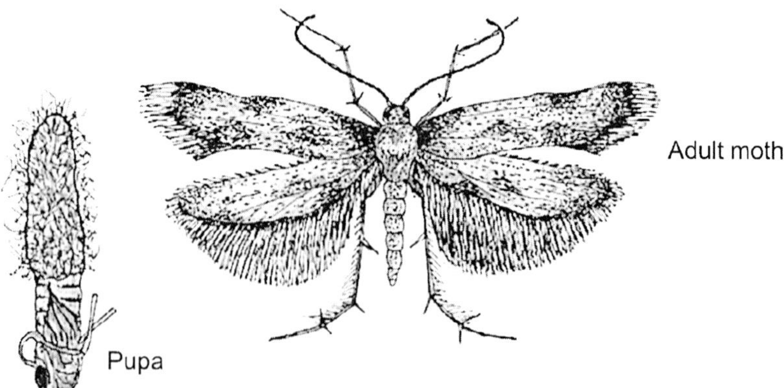

Fig. 129. Brown Cloth moth-*Tinea pachyspila*

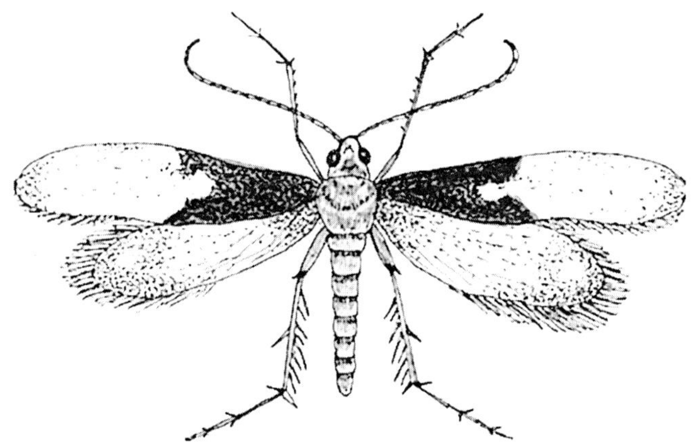

Fig. 130. Black and while cloth moth-*Trichophaga abruptella*

(i) Dusting on the ant nests and surrounding places with carbaryl 5 D or methyl parathion 2 D eliminates the pests. However repeated application whenever they reappear is necessary (ii) Spraying with chlorpyrifos 20 EC (0.04% ai) or lindane 20 EC (0.04% ai) or propoxur 20 EC (0.04% ai) in kerosene oil is more effective in controlling ants.

8. White ants or Termites

Wood dwelling termites *(Cryptotermes domesticus* and *Kalotermes indicus)*- Isoptera, Kalotermitidae

Soil dwelling termites *(Odontotermes obesus* and *Microtermes obesi)*- Isoptera, Termitidae

Among the various forms of termites, the wood inhabiting forms and the soil inhabiting or ground dwelling forms cause extensive damage to various household articles. The wood inhabiting termites damage woodwork of all kinds, furniture, picture and photo frames, wooden poles etc. by boring through them and feeding on the cellulose material. They form a series of galleries with their excreta, earth and wood dust without any external opening, remain inside and feed on the wood. Besides wooden materials they destroy books, pictures, clothing, dried plant and animal products and anything that contains cellulose.

The soil inhabiting termites live in the ground and infest wood through the soil and damage woodwork of buildings and wooden structures in contact with the ground. They emerge aboveground to get access to other building materials and for this purpose they construct covered galleries with their faecal matter, earth and wood dust (Fig. 102).

(i) Wooden structures can be protected by coating the surface with coal tar or crude oil (ii) Application of creosote, Mercuric chloride, Sodium fluosilicate, chlorpyrifos or lindane is also effective in controlling the pest. Application of chlorpyrifos or lindane in kerosene oil is more effective (iii) Soil application of lindane 20 EC or chlorpyrifos 20 EC at 250 ml. diluted in 5.0 liters of water at the time of construction of buildings is effective in protecting the wooden structures against termite infestation.

9. The common house rat *(Rattus rattus rufescens)*-Mammalia, Rodentia, Muridae

Among the mammalian pests, the common house rat is the most serious and destructive pest and is responsible for causing considerable damage to foodgrains, foodstuffs and feeds, household articles, wooden structures etc. Rats cause much more damage to foodstuff than they can actually consume. They spoil the quality of foodstuffs, feeds etc. by polluting with their hairs, urine and faecal matter and make them unfit for consumption and marketing. They burrow through wooden doors, windows, partitions and even concrete floors and walls with their chisel-like teeth and make their way into human dwellings and consume whatever is available, either vegetarian or non-vegetarian. A single rat consumes 10-15 gm. of solid food such as, grains, flour, vegetables, fruits etc. and drinks about 10 ml. of water in one day.

House rats build nests under roof tops, in rafters where unwanted things are dumped, in card board boxes and in such concealed places or may live in burrows underground. They carry pieces of clothing, cotton, furs, woolen materials, fibers, paper etc. to their hiding places and use them for building their nests for their young ones. When there are young ones, they eat and damage much more (Fig. 131).

Fig. 131. House rat-*Rattus rattus rufescens*

(i) Rats can be caught and destroyed by using wooden box-type traps, wire mesh cage-type traps, spring loaded metal traps etc. (ii) Poison baits with Zinc phosphide as poison and pop corn or fried fish or boiled tapioca as bait material can be used to attract and kill rats (iii) Readymade poison bait with Bromodiolone as a poison and other food materials as attractant available in the market as cakes may be used to attract and kill rats (iv) Putting 20-25 gm. of phorate 10 G into the rat burrows and covering the bore immediately with moist soil is also effective in killing the rats.

LIVESTOCK PESTS

Besides attacking standing crops, harvested and stored produce and other household articles, several pests are known to attack livestock such as, cattle, sheep and goat, horses, poultry, pet animals etc. These pests, not only cause direct injury by sucking and feeding on the blood, cause and transmit some serious diseases. They are also a source of perpetual annoyance and disturbance to livestock. These pests are specially adapted for continuing their lives as parasites on warm-blooded livestock. They include insect pests such as, different types of flies, gnats, maggots, fleas, lice etc. and a few non-insect pests such as, ticks and mites. The injury and annoyance caused by these pests result not only in certain functional disorders, but also leads to reduction in productivity of milch animals and lowering the working efficiency of farm animals.

1. The gadfly *(Tabanus striatus* and *Chrysops dispar)*-Diptera, Tabanidae

These flies are also called as **Horse flies**, **Marsh flies**, **Breeze flies** or **Green-head flies**. Only the females of this fly pest are bloodthirsty and attack all animals such as, cattle, buffaloes, horses and mules and sometimes, even men. They are seen mostly in summer and are very fond of sunlight. They are strong fliers and mostly diurnal in habit and are seen during the hottest part of the day. They suck blood from the naval area, neck, withers, abdomen, back and hind legs. They are virulent biters and bite intermittently. They cannot be disturbed and driven away easily after starting a bite. Even after the fly has left after feeding, blood continues to ooze out from the punctured wound. The female flies are found hovering about in cattle yards, stables and grazing grounds, visit the hosts frequently, pierce the skin of animals, suck and feed on the blood. However, the males are not parasitic and they feed only the nectar from flowers.

The flies breed in marshy places around the cattle yard. The female fly lays hundreds of dark, elongated eggs, which are glued to aquatic plants. The maggots that emerge from the eggs in 4-7 days fall down to the moist soil and are found wriggling around. They are cylindrical, elongated, and maggot-like with the body having transverse ridges, pointed at both ends, apodous and hemicephalus. They are carnivorous and feed on small organisms such as, earthworms, crustacea, fly larvae, naiads etc. found in the moist soil or water and continue to grow. The larval period lasts for a month or two and there may be 7-11 larval instars during that period. The respiratory siphon is found on the eighth abdominal segment. The full-grown maggot pupates in the same surrounding. Pupa is obtectate. Respiration of pupa is through ear-shaped spiracles. The flies that emerge out from the pupae are stout-built and are very active. They resemble ordinary houseflies but are larger in size and have prominent compound eyes, which in the males almost meet at the median line and have three rows of whitish marks on the abdomen. The entire life cycle is completed in 5 weeks to 5 months and there may be only 1 or 2 broods in a year. The adult fly is large and robust, about 1.2-1.8 cm. long, with piercing and sucking type of mouthparts and has a tooth-like projection (spur) on the basal part of the antenna (Fig. 132).

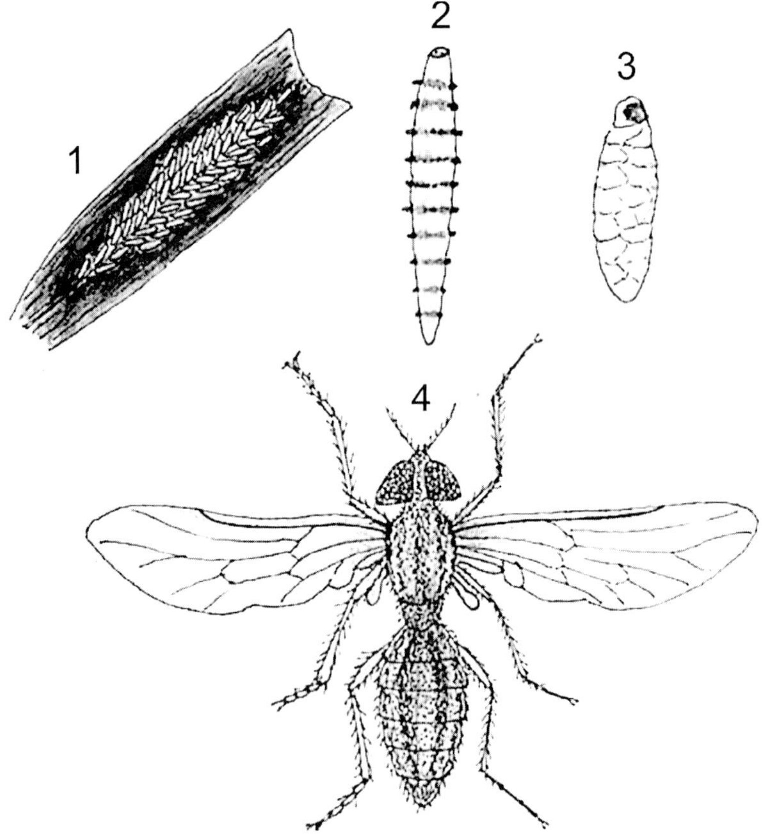

Fig. 132. The common gadfly-*Tabanus striatus*

1. Eggs 2. Larva 3. Pupa 4. Adult fly

The bites of tabanid flies are painful and irritating and may give rise to itching and rashes in soft-skinned animals. Animals become restless and unmanageable. There will be blood loss in animals. The fly acts as a mechanical vector for *Trypanosoma evansi* that causes **'Surra'** in cattle. It also transmits *Anaplasma marginale* that causes **'Anaplasmosis'**, *Bacillus anthracis* that causes 'Anthrax', *Francisella tularensis* that causes **'Tularaemia'** and a few virus diseases mechanically.

Several other blood-sucking flies are found attacking different hosts. The hill gad fly-*Pangonia (Corizoneura) taprobane,* which has a remarkably long and pointed proboscis, pierces the skin of wild animals such as, elephants, bison etc., sucks and feeds on the blood (Fig. 134).

The walls, barricades etc. of cattle sheds, horse pens etc. may be sprayed with malathion, fenitrothion or synthetic pyrethroids to control the flies. The flies have a habit of skimming over water and occasionally dipping their bodies into it. So, pouring kerosene oil on to water kills the flies when they dip into it. A few Hymenopterous parasitoids such as, *Telenomus* spp. parasitize and destroy the eggs.

2. Horse flies *(Hippobosca maculata* and *H. capensis)*-Diptera, Hippoboschidae

The flies are also known as **'forest flies'** and **'louse flies'**. These flies attack horses, cattle, dogs etc. Only the adult flies are parasitic and suck blood from the hosts. The flies, which have strong claws, attach themselves to their host firmly while sucking blood. Sometimes a number of flies are found attached to the sides of the neck in cattle. The flies though provided with wings, fly only short distances. When disturbed, these flies, which remain clinging on to the skin of their host fly out from one part and again settle on some other part of the body of the host. They are active in summer, especially in sunny weather and remain on their host for long periods.

The life history of these insects is rather peculiar. Instead of laying eggs like many other species of flies, here the female fly deposits fairly matured larva one at a time in sheltered spots where there is dry soil or humus. The sub-globular larva almost immediately changes into a short, roundish, seed-like puparium. The length of the pupal period is greatly influenced by temperature. Such flies that exhibit peculiar feature are known as **'pupipara'**. The flies are about 1.0 cm. long, somewhat leathery and flattened and have wings and well-developed, strong tarsal claws, which help them to cling on to their host. They have yellowish or reddish- brown body, often marked with spots and bands (Fig.133, 135).

The flies are known to transmit **trypanosoma** in cattle. The flies can be controlled by spraying synthetic pyrethroids, dichlorvos etc.

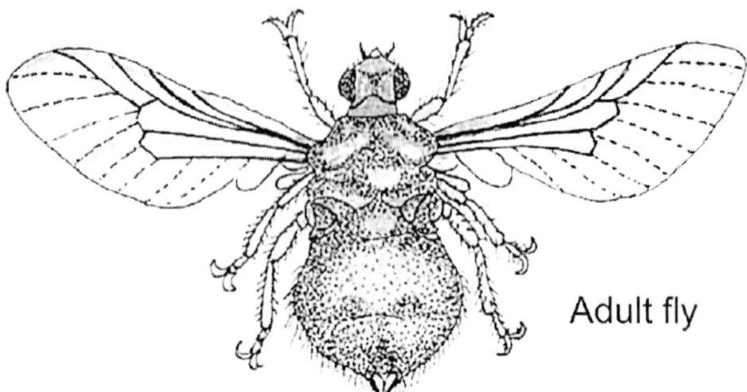

Fig. 133. Horse fly-*Hippobosca maculata*

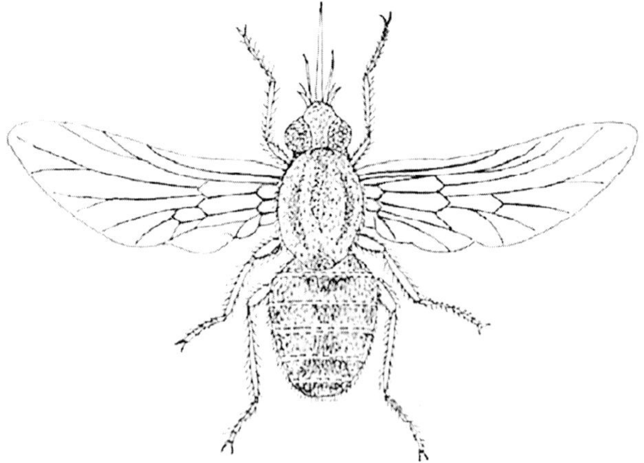

Fig. 134. Hill gad fly-*Pangonia taprobane*

Fig. 135. Horse fly-*Hippobosca capensis*

3. The stable fly *(Stomoxys calcitrans)*-Diptera, Muscidae

These biting flies are cosmopolitan in nature and are found throughout the world. Both the female and male flies suck and feed on the blood from cattle, horses and other animals including humans, mostly in the ankle region. The fly bites intermittently like *Tabanus* and feeds more than once a day. The maggots are not parasitic. They are closely allied to the common housefly in general features and life cycle, but are slightly smaller in size and are light gray in color. The thorax has 4 longitudinal stripes. The abdomen is short and broad and three dark spots are found on the second and third segments. The wings are kept at an divergent angle when at rest. Unlike houseflies, these flies possess mouthparts with a distinctly pointed proboscis for piercing the skin and suck blood from the host. They are diurnal, prefer bright sunlight and are mostly seen in summer. They are very commonly found in cowsheds and stables. Though they are fast fliers, they do not fly long distances.

The life cycle of the flies is quite similar to that of the housefly. They breed in moist situations, in wet straw, grass and other waste materials along with cattle or horse dung. A female fly lays 600-800 eggs and the immature stages are found in cattle and horse manure. The eggs are white in color and they hatch in 1-4 days. The larva is acephalus and apodus. There are three larval instars, which last for 20 days. The pupal stage lasts for about 8 days. The life cycle is completed in 30-35 days and there may be several broods during a year. After the rains they are found in very large numbers (Fig. 138).

The flies act as mechanical vector for *Trypanasoma evansi* that causes **'Surra'** in cattle. They also transmit anthrax bacilli, rinderpest virus, tetanus bacteria and poliomyelitis virus mechanically. The control measures suggested for the control of houseflies may be adopted.

4. The blood sucking fly *(Siphona exigua)*-Diptera, Muscidae

These parasitic, blood-sucking flies are quite common and they attack cattle, buffaloes and dogs. Both female and male flies are blood-suckers and they are usually attracted by the odor, warmth and sweat of their hosts. The female fly lays about 20 eggs on fresh dung of cattle or buffalo. The eggs hatch in about 10 hours. The maggots live on the dung and become full-grown in about 7 days. The grown-up maggot pupates in the soil near about the dung and emerges as an adult fly in 4-5 days. Several broods may appear in a year.

5. The ox bot fly or Warble fly *(Hypoderma lineatum)*-Diptera, Oestridae

The fly is also known as **'Heel fly'**, **'Ox warble fly'** or **'Common cattle grub fly'**. The maggots of the flies, which are commonly known as warbles, generally attack cattle. The females oviposit while walking on the animals. The eggs laid by the female fly are glued to the hairs on the body, mostly around the lower regions of the legs of the host.. The eggs hatch in 2-3 days and the young maggots penetrate the skin causing typical swellings and inflammations called warbles on the skin of the host. They live in the pouches thus made and feed on the skin tissues and grow into fleshy maggots. The larval stage may last for several weeks. When full-grown and about to pupate, the larva pierces the skin of the host and through the hole drops to the ground and changes into the pupal stage. The adult fly emerges from the pupa after about 3 weeks time. The entire life cycle lasts for about a year. Adult flies are hairy, resembling bumble bees in appearance, somber-colored forms with heavy body and have rudimentary, non-functional mouthparts and large head, but are rarely seen. They are active during summer, but do not bite or sting. They live only for a few days, rarely for about 2 weeks. The warbles on the body make the ox, cow or goat loose condition and become unhealthy. Their productive capacity and working efficiency are adversely affected. The hides of these animals show warbled spots with holes and are known as warbled hides and fetch very low prices in the market.

Matured larvae may be squeezed out of the warble swelling. Ruptured skin during extraction may lead to localized inflammation and abscess formation, which should be treated separately. Insecticides with low dermal toxicity such as, fenthion, fenitrothion, famphur, phosmet may be applied dermally.

6. The horse bot fly *(Gastrophilus equi* or *G. intestinalis)*-Diptera, Oestridae

The female fly hovers above the animal with its ovipositor extended and oviposits the eggs singly while in flight and attaches the eggs along the distal half of the hairs. Large number of eggs are found mainly around the fetlocks of the forelimbs and also higher up on the legs. Hatching takes place only by the application of moisture and friction, which is provided by rubbing and licking of the host. The maggots hatching out of the eggs are not swallowed directly into the mouth, but penetrate the mucosa of the mouth and gradually move down the mucosa as far as the pharynx. Afterwards, the larvae penetrate the dorsal mucosa of the anterior region of the tongue and burrow slowly towards the posterior end. The larvae remain in the tongue for 24-28 days. Subsequently the larvae move into the pharynx and then pass into the stomach. In this manner several maggots may find their way into the stomach of the horse. They feed on the mucous secretions in the stomach and grow, at the same time they cause trouble to the host. The larvae are known as **'bots'** or **'stomach bots'**. The larvae remain inside the host for about 10 months. Before pupation, the full-grown maggot escapes to the soil along with the horse dung and pupates under the soil. The pupal period lasts for 3-5 weeks. The total duration of the life cycle is about one year. The heavily built adult fly is somber-colored, somewhat resembling a bee, hairy, about 1.8 cm. long and has rudimentary mouthparts and is not parasitic. The female fly has a long and crooked ovipositor.

The fly causes annoyance to the animals. The maggots cause ulceration in the oesophageal region of the stomach. Suitable treatments have to be given by a qualified veterinarian to get rid of the pest.

7. **The sheep bot, sheep nasal fly** or **Head maggot of sheep** *(Oestrus ovis)*-Diptera, Oestidae

The larvae occur in the nasal cavity and adjoining sinuses in sheep and rarely in goats. The female fly lays the first stage larvae as in the case of horse flies (larviparous) inside the nasal chambers of the sheep. The maggots find their way into the other chambers of the head such as, mouth, frontal sinuses etc. The larvae are white in the early stages and turn slightly yellow later on. The full-grown larva is about 3.0 cm. long, tapering anteriorly and has a flat posterior end. The larvae possess large, black, oral hooks. They feed on the mucus secretions of the mouth and damage the mucus membranes resulting in irritation, constant nasal discharge and sometimes obstruction of air passages into the lungs in case of severe infestation. Full-grown maggots either crawl out or are expelled out forcibly as a result of sneezing by the sheep and then pupate under the soil. The flies emerging out from the pupa are dark gray colored, heavily built and rough. They have a big head and rudimentary mouthparts. The duration of the life cycle is 7-12 weeks (Fig. 136).

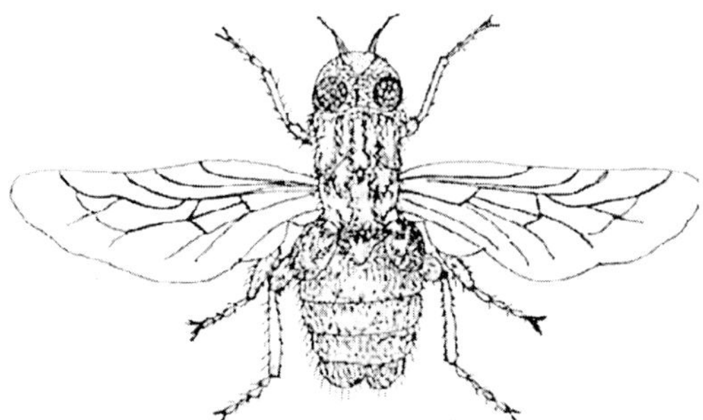

Fig. 136. Sheep bot fly-*Oestrus ovis*

The flies cause great annoyance when they attack the sheep to deposit larvae. The animals stop feeding and become restless. They shake their heads or press their noses against the ground or rub between other sheep. The larvae irritate the mucosa with their oral hooks and spines, causing the secretion of a viscid, mucous exudate on which they feed. Sometimes the flies deposit their larvae in the eyes, nostrils and on the lips of man, which may cause serious trouble. The larvae most often affect shepherds.

The sheep may be allowed to feed in narrow troughs, the edges of which are smeared with tar. The animals automatically tar themselves and this acts as a repellent and prevents the flies from laying their larvae on the nostrils of the host. Suitable treatments may be given with the help of a veterinarian.

8. The sheep ked *(Melophagus ovonus)*-Diptera, Hippoboschidae

The fly is also known as **'sheep tick fly'**. Keds are permanent ectoparasites. Females attach larvae to wool of sheep by means of a sticky substance. The larva is immobile and soon turns into a chestnut-brown pupa. The pupa is ovoid in shape with broad ends and the pupal stage lasts for 19-23 days. Female ked lives for 4-5 months in sheep. A female may produce 10-15 larvae during each gestation period lasting 10-12 days. They usually spread from sheep to sheep by contact. The flies are degenerate, apterous, leathery, hairy and pupiparous. The head of the ked is short and broad and not freely movable. The proboscis is well developed. Thorax is small, brown and armed with number of markings.

Abdomen is large and broad. Legs are strong and armed with stout claws. The adult flies, which are parasitic, live amidst the wool, suck and feed on the blood. The severe irritation caused prompt the sheep to bite, rub and scratch the area, thus damaging the wool. Faeces of the keds produce stains in the wool, which reduce the market value of the wool.

Shearing the wool markedly reduces the ked population. The insects are very susceptible to organophosphorous insecticides. Ronnel, diazinon and ciodrin are effective in controlling the keds.

A few other pests such as, the biting louse-*Bovicola caprae* and *B. ovis* are ectoparasites of goats and sheep respectively.

9. The blow flies or Blue bots *(Lucilia serinissima* and *Pycnosoma flaviceps)*-Diptera, Calliphoridae

Unlike the blood-sucking flies, here only the maggots of the bottle flies are parasitic. The maggots attack exposed flesh or open wounds and sores and feed on the decaying tissues. In case of severe infestations, the maggots cause cutaneous diseases. The flies may sometimes be found attacking exposed and neglected wounds in humans also.

The fly lays eggs in carcasses, foul smelling wound, urine or faeces of animals and human beings. There are three larval stages, each lasting 5 days. The pupal stage lasts for 4 days. The duration of life cycle is 10-12 days. The flies, which are bright blue and green colored, are commonly found in dirty locations, especially on night soils and near about slaughterhouses. The fly is a primary myiasis producer (Fig. 137).

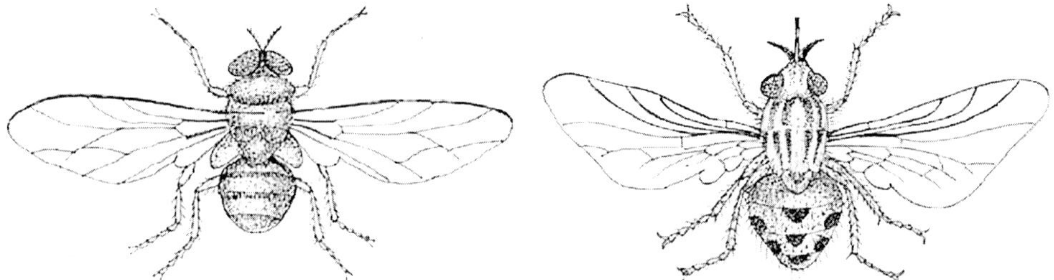

Fig. 137. Blue bottle fly-*Lucilia serinissima* **Fig. 138.** Stable fly-*Stomoxys calcitrans*

10. The eye fly *(Siphunculina funicola)*-Diptera, Choloropidae

This is a minute, bluish-black fly, quite similar to a small house fly in general appearance. These flies have the habit of clinging on to ropes, strings or threads hanging from roofs etc. especially cattle sheds and in damp, warm places in swarms. They hover in front of the eyes of animals, settle around the corners of eyes and feed upon the secretions from the eyes. They breed in moist soil and faecal matter. They are a perpetual nuisance and cause much annoyance to almost all kinds of animals, both domesticated and wild (Fig.142)

11. The house flies *(Musca* spp.*)*-Diptera, Muscidae

Of the many species of *Musca*, a few are bloodsuckers, while the rest are non-bloodsuckers. However, these non-biting ones cause considerable annoyance and a source of perpetual nuisance to various livestock and humans. *Musca domestica, M. nebula, M. vicina* and *M. sorbens* are the important non-biting flies. They breed in any kind of filth, decomposing organic matter including cattle dung and night soil. Even their presence is quite annoying and the sense of their touch is also repulsive. The houseflies are carriers and in some cases intermediate hosts of many serious diseases in cattle and humans (Fig.143).

12. The buffalo louse *(Haematopinus tuberculatus)*-Siphunculata, Haematopinidae

This minute, flattish, sucking insect pest attacks buffaloes, oxen and other animals. The insects, which are generally found in large numbers inside the ear lobes and other protected parts, suck and feed on the blood (Fig. 140).

The moths-*Arcyophora icterica* and *Pionea damastesalis* settle on the eyelids of cattle, buffalo, horse, donkey, cat etc. and feed on the lachrymation and secretions from the eyes.

13. **The bird louse** or **Shaft louse** *(Menopon pallidum)*-Mallophaga, Menoponidae

Among the different species of lice, which attack birds, the bird louse is quite important and a serious pest of many bird species. This very common and permanent ectoparasite of fowl is distributed all over the world. This louse differs from the typical lice in that they are not blood-suckers. The mouthparts of these lice are mandibulate and as such they cut, bite and scratch the skin, feathers, body scales etc. of their host. The adult lice are very small, grayish-brown in color, quite active and have hard and horny bodies, which are flattish and are wingless. The female louse fastens minute, whitish eggs on to the basal barbs in large numbers.

The nymphs that hatch out from the eggs in about 2 weeks time are yellowish in color and they become adults in 10-12 days. The entire life cycle is spent on the body of the host bird. The annoyance and irritation caused to fowls and chicken by this parasite is often quite serious and the infested birds can be seen rubbing their bodies in loose soil or ash pits to get rid of the lice or to get relief from the irritation caused by the lice (Fig. 139).

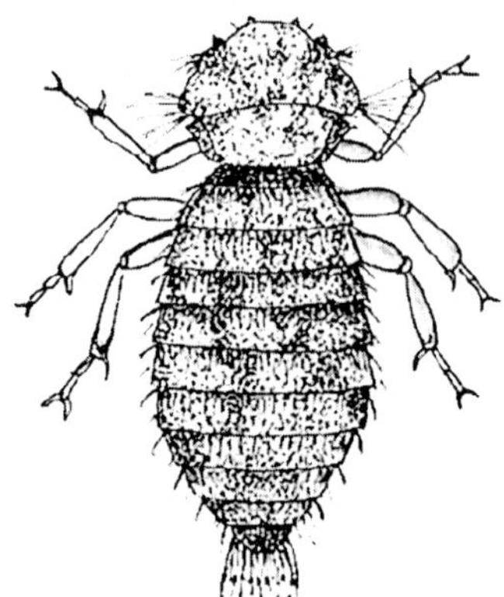

Fig. 139. Bird louse-*Menopon pallidum*

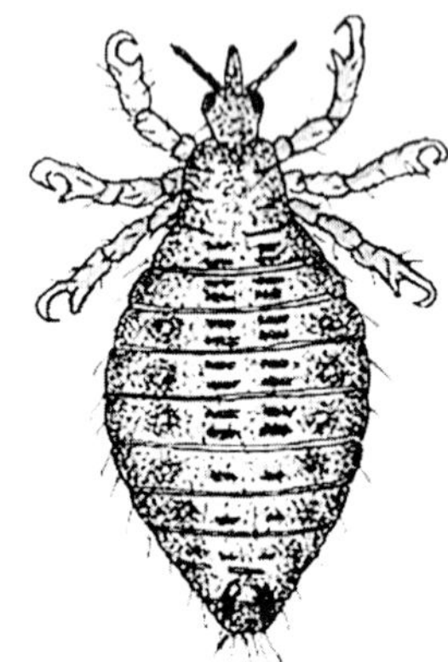

Fig. 140. Buffalo louse-*Haematopinus tuberculatus*

14. **The hen flea** or **Chicken flea** *(Echidnophaga gallinaceae)*-Siphonaptera, Tungidae

These minute, dark brown, hard and horny, blood sucking creatures are found in large numbers adhering to the face, comb, wattles and the area around the eyes. They are wingless and their bodies are compressed laterally. Hundreds of these insects may be found on a single bird. Only the adult fleas are parasitic. The flea remains attached to the host with its head embedded into the skin of the host, sucks and feeds on the blood. The fleas cannot be dislodged or brushed off easily. Heavily infested birds become weak and emaciated. The female flea lays eggs loosely, which drop to the ground and the larvae emerging from the eggs grow in the filth. The grown-up larva pupates in a cocoon and the adult flea emerges from the pupa. The fleas are provided with strong hind legs fit for jumping. Sometimes these fleas may attack man also.

Besides the above, a few other insect pests viz., the biting body louse-*Menacanthus stramineus* and the wing louse-*Lipeurus caponis* also attack poultry birds; the silver-spotted mosquito-*Aedes aegypti* attacks poultry birds and transmits-*Plasmodium gallinaceum* that causes 'malaria'; both the nymphs and adults of the bed bug-*Cimex hemipterus* suck and feed on the blood from poultry birds.

15. Ticks and mites-Arachnida, Acarina

Besides insect pests, a few non-insect pests are also known to attack cattle and other livestock and cause serious consequences. All of them feed on the blood from their hosts. Ticks and mites belong to this category. They are classified under Class-Arachnida, Order-Acarina. They are like some spiders and their body is not distinctly divided into two or three separate regions as in the case of insects and they have 4 pairs of legs as against 3 pairs of legs found in insects. The mouthparts are adapted for piercing and sucking or biting and sucking. The ticks and mites that attack livestocks are all ectoparasites. The ticks are fairly large creatures, dark brown to black in color, smooth and quite plumpy. They attach themselves strongly to their hosts, bite and suck the blood from the hosts and it is quite difficult to dislodge them from the hosts. The chief forms of ticks found commonly are the spinose ear ticks-*Ormithodorus megnini*, the poultry tick-*Argas persicus*, the dog tick-*Rhipicephalus haemaphysaloides, Boophilus* spp. etc. and the cattle and horse tick-*Hyalomma aegyptium* (Fig. 141).

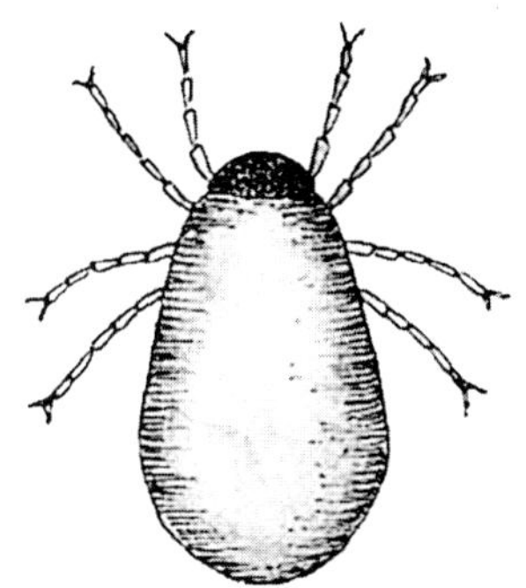

Fig. 141. Poultry tick *Argas persicus*

The mites are very minute creatures and they pierce the skin of the host. suck and feed on the blood. The common and well known mite-*Psoroptes communis*, besides sucking blood from animals such as, dogs, sheep etc., cause diseases such as, **'mange'** and **'scabies'**.

Fig. 142. Eye fly-*Siphunculata funicola*

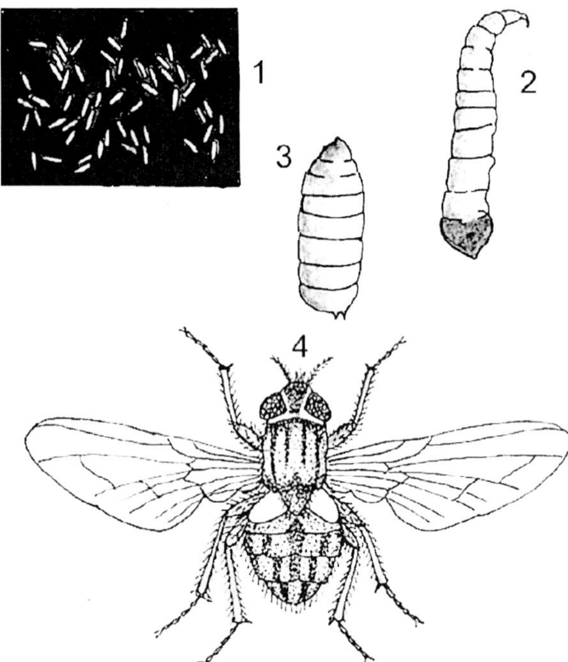

Fig. 143. House fly-*Musca domestica*
1. Eggs 2. Larva 3. Pupa 4. Adult fly

Dermal application of the synthetic pyrethroid, acaricide-flumethrin 1% solution available under the trade names, 'Poron', 'Tikkil power' etc. is very effective in controlling all tick species including strains resistant to carbamates, organo phosphate and organo chlorine groups of insecticides. Besides having direct lethal effect on the ticks, it inhibits the production of viable eggs by the ticks. The residual toxicity of the chemical may last for 20-70 days. It is an ectoparasiticide and is applied at the rate of 1.0 ml./10.0 kg. body weight, evenly along the mid line of the back from the front of shoulder to tail setting area and it spreads all over the body surface, but does not penetrate the skin in significant amounts. It can be used safely on cattle and other animal pets.

Dermal application of the ectoparasiticide-amitraz 12.5% W/V belonging to the class-farmamidines at 2.0-3.0 ml. in 1.0 liter of water is also very effective in controlling ticks, mites, fleas and other ectoparasites attacking cattle, dogs, cats etc. It is available under the trade name 'Ridd'.

DISEASE CARRYING INSECTS OF HUMANS

Besides causing direct damage to livestock and humans in various ways, some insects are known to transmit very serious diseases, which may even be lethal at times. Among them, the most important insects are houseflies, mosquitoes, fleas, lice, bed bugs, eye flies, sand flies and the cone-nose bug. Some mites are also capable of transmitting a few diseases.

1. The housefly *(Musca domestica)*-Diptera, Muscidae

Houseflies are one of the chief agents that are responsible for mechanical transmission of several bacterial diseases such as, typhoid fever, paratyphoid, cholera, diarrhea caused by *Escherichia coli*, *Salmonella* dysentery, tuberculosis, leprosy, anthrax etc., viral diseases such as, Poliomyelitis and jaundice, protozoan diseases such as, Amoebic dysentery caused by *Entamoeba histolytica* and a few Helminthic diseases. To some extent they are also responsible for transmitting other diseases such as, ophthalmia, typanosomiasis etc. The causative microorganisms are picked up by the flies from human excrements, sputum, and carcasses of diseased animals, manure and other filth.

The female fly lays small, cigar-shaped, pearl white eggs in batches of over 100 in fresh manure, night soil or in other decomposing organic matter. Each female fly may lay up to 600 eggs during its life span. The eggs hatch in 12 to 24 hours depending upon weather conditions. The maggots coming out of the eggs are very small, creamy white in color, smooth and the head indistinct and pointed. The hind portion of the larva is broader. The larvae feed on bacteria, organic matter etc and become full-grown in 3-7 days. They wriggle their way to drier parts bordering the manure heap or nearby places and pupate. The pupae are small, oval-shaped, smooth, brownish and are seed-like. The adult fly emerges from the puparium in about 3 days. The life cycle is completed in 7-12 days. It takes a shorter time for the completion of the life cycle in the hot weather than in the colder season. During summer, the flies breed rapidly and multiply in abundance. The adult fly is about

6.0 mm. in length, dirty gray in color with bilaterally plumose arista and have some dark stripes on the thorax and thin hairs all over the body. The flies have sponging or lapping type of mouthparts known as muscoid-type. The mouthparts are not adapted for piercing and sucking blood from the host as in the case of blood-sucking flies. The flies start laying eggs within 4 days after fertilization (Fig. 143).

The houseflies cause considerable annoyance by settling upon our bodies and our foodstuffs and contaminating them with filthy particles from various sources carried in their mouthparts, body hairs, wings, feet etc. They also make the foodstuffs unwholesome by dropping 'flyspecks' (faecal matter) on them. The flies vomit fluid drops known as 'vomit drops' on liquefiable solid food to liquefy them before it is sucked up. This is important in connection with the disease transmitting capacity of the fly. They may also fall into liquids like milk, coffee etc. and make them unsuitable for drinking. The flies may contaminate any kind of food on which they alight after having fed upon infective substances containing bacteria, viruses or any other infective microorganisms. Further, the intestine of the fly is also charged with infective material in concentrated form, which may be discharged upon any food it seeks. *Musca nebula* another housefly, the life cycle of which is almost similar to that of *Musca domestica* is also commonly found in houses.

Because flies are found everywhere, especially in all dirty surroundings, it is very difficult to control them. However, by taking concerted measures by every individual, as well as by the Public Health Department, their population can be brought down. The first and foremost effective control measures against houseflies are proper attention to their breeding sites and prompt disposal of household garbage and rubbish, manure and night soil, carcasses of animals, birds etc. Food materials should always be kept covered with housefly-proof nets made of nylon or metal wire. Providing fly-proof screens to windows, doors etc. prevents entry of houseflies into human dwellings. Other methods include use of various devices such as fly-paper, fly and maggot traps, poison baits, insecticidal sprays etc. Dusting manure heaps with methyl parathion dust, diazinon dust or carbaryl dust destroys the developing maggots and prevents flies from laying eggs on the manure heaps. Spray application of diazinon or dimethoate in the breeding area is effective in destroying the developing maggots. Poison baits with curd or sugar syrup as bait and propoxur (Baygon) as poison or milk as bait and formalin as bait are effective in killing adult flies. Surface spray application of propoxur 20 EC (0.04 % ai.) effectively controls flies. Sprinkling powdered common salt on the surface of dining tables, kitchen tables and such surfaces deters houseflies from alighting on the surfaces. A few insect growth regulators such as, cyromazine, which prevent moulting and pupation are being used with some amount of success. Spores of *Bacillus thuringiensis* parasitize the larvae.

2. **Mosquitoes**-Diptera, Culicidae

Mosquitoes are quite familiar insects and are much more annoying than houseflies as they pierce the skin and suck blood, thus causing pain, irritation and loss of blood. Mosquitoes are small, delicate, long-limbed and quite active insects. A characteristic feature of these insects is that their wings, especially along the veins and the body are covered with fine, minute scales as in butterflies. Their mouthparts are adapted for sucking and both the sexes feed on various juices of fruits and nectar from flowers. However, the mouthparts of the females are especially well-developed for piercing the skin and sucking blood. The larvae, known as wrigglers develop in water and feed on algae, bacteria, and minute insects found in water. The pupae, which are also aquatic, move about actively, but they do not feed. The entire life cycle is completed in 9-10 days. Mosquitoes are generally divided into two groups viz., Culicines and Anophelines. They differ in their general features, habits and certain variations in the developing stages. The Culicine group includes the Genera-*Culex* and *Aedes*, while the Anopheline group is represented by the Genus-*Anopheles.*

Anopheles mosquitoes are the vectors that transmit human 'malaria fever' caused by the protozoa-*Plasmodium falicifarum, P. vivax, P. malariae* and *P. ovale* and *Anopheles culicifacies, A. fluviatilis* and *A. stephensi* are the major vectors. *Anopheles fluviatilis* is the main vector of 'malaria' in hilly areas and *A. stephensi* in urban areas. 'Filariasis' in man caused by *Wuchereria bancrofti* and *Brugia malayi* is also transmitted by *Anopheles stephensi.*

The Anophelines lay eggs singly. The larvae are very active and they feed on algae, bacteria, fungi and decaying vegetation. The pupae are also very active, but they do not feed. Both the larvae and pupae are dependent on free oxygen from the atmosphere for respiration. The larvae have an abdominal breathing tube, which pierces the water surface for a supply of air. The pupae have ear trumpets on the thorax for the same purpose. They prefer clean, still water in shady areas. The larvae feed in the sun and rest in the shade. The larvae remain parallel to and just under the surface of water. The pupae remain just under water with the thorax just under the surface of water. Adult mosquitoes are 4.0-5.0 mm. in length, dark brown to nearly black in color. Palpi are long and about the same length as the proboscis and the wings have black scales with white spots. When at rest the head and body are in a straight line and the hind portion elevated. No hairs are present on the scutellum (Fig. 146).

The mosquitoes belonging to the Genus-*Culex* lay eggs in rafts, each containing several eggs. The larvae remain under water with the head downward and the anal tip towards the surface. The breathing tube or siphon tube is rather long and narrow. The pupa remains just under water with the thorax just under the surface of water, while the breathing tube pierces the water surface to take in air supply for respiration. The adult mosquitoes have plain wings. The palpi are short and shorter than the proboscis. When at rest, the head and body forms an angle and the body remains almost parallel to the resting surface. The scutellum has three groups of hairs. In the Genus-*Culex, Culex fatigans* is quite commonly found in households and it is the main intermediate host that transmits 'filariasis' in man caused by a filarial plasmodium-*Wuchereria bancrofti* and *Brugia malayi* and 'elephantiasis'. *Encephalitis virus,* causing 'brain fever' in children is also transmitted by species of *Culex.* Fowl malaria caused by *Plasmodium gallinaceum* is also transmitted by some species of *Culex.*

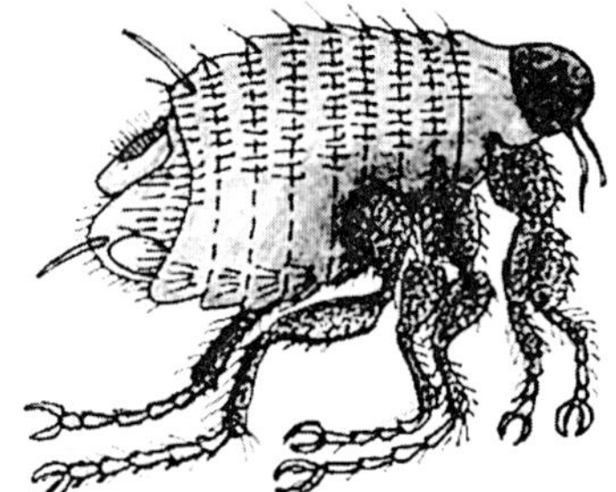

Fig. 144. Rat flea-*Xenopsilla cheopis*

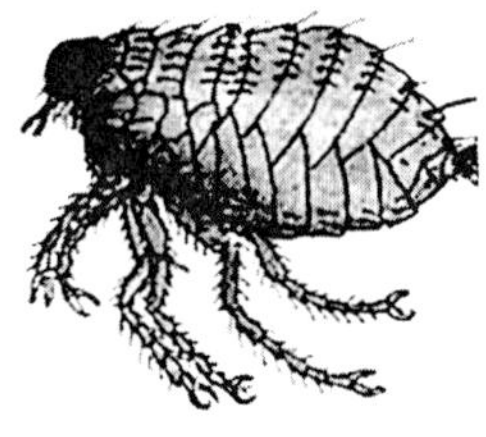

Fig. 145. Cat flea-*Ctenocepalides felis*

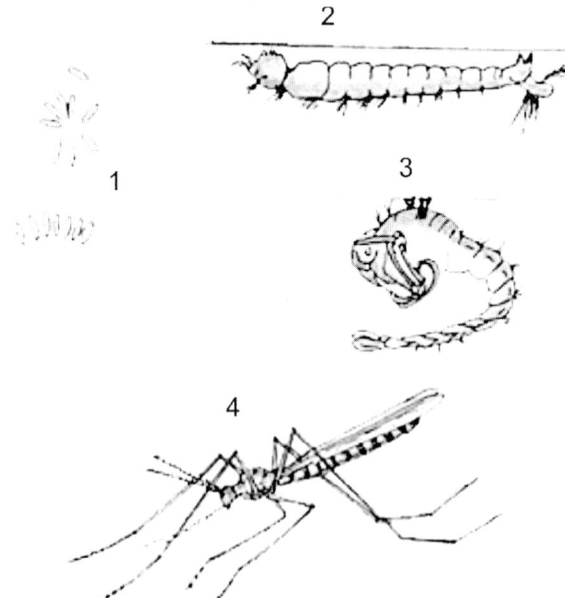

Fig. 146. Anopheles mosquito *Anopheles stephensi*

1. Eggs
2. Larva
3. Pupa
4. Adult mosquito

The egg raft of *Culex fatigans* is boat-shaped, with about 200 eggs in each raft. The mosquitoes breed heavily in waters contaminated with sewage. The entire life cycle is completed in 8-10 days. The larvae of the mosquitoes are associated with aquatic plants such as, pistia that are found commonly in ponds. The mosquitoes breed very rapidly during the monsoon seasons when there is water stagnation in many places and these seasons are favorable for the transmission of the diseases also (Fig. 147).

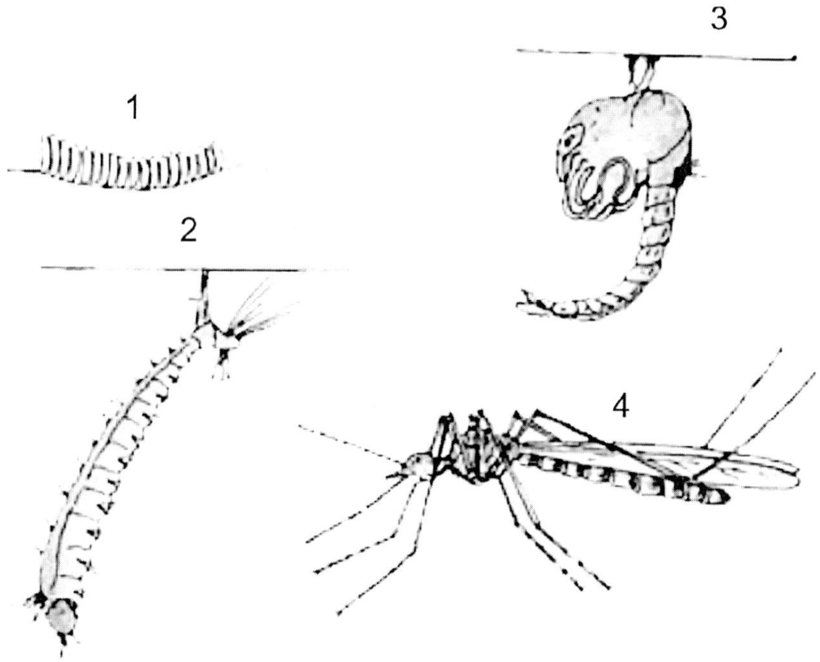

Fig. 147. Culex mosquito-*Culex fatigans*
1. Eggs 2. Larva 3. Pupa 4. Adult mosquito

Mosquitoes of the Genus-*Aedes* are responsible for the transmission of important diseases such as 'dengue' and 'yellow fever' caused by viruses in humans. 'Encephalitis' caused by a virus is also transmitted by species of *Aedes*. These mosquitoes lay eggs singly and the larvae have a short and broad breathing tube. The pest breeds in marshy places and the adult mosquitoes are carried over long distances to towns and cities by wind. *Aedes aegypti* and *A. fasciata*, the silver spotted or tiger mosquitoes are the most important carriers of these virus diseases. After acquisition of the viruses, an incubation period of 11-14 days is required for the mosquitoes before they can transmit the viruses. Once the viruses are acquired, the mosquitoes remain viruliferous for the rest of their lives (Fig. 149).

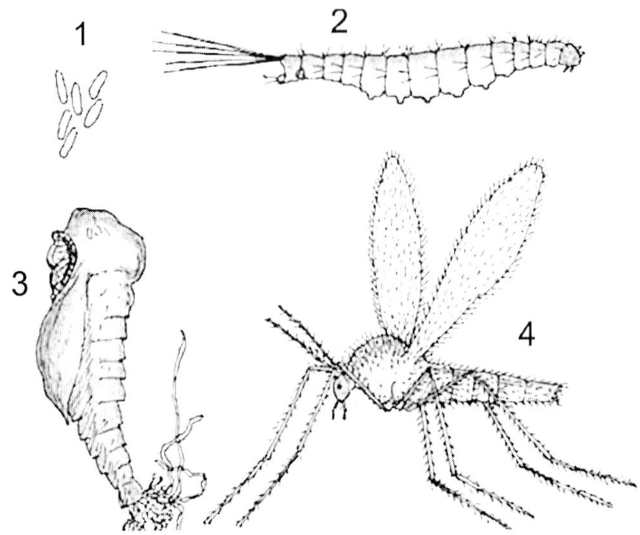

Fig. 148. Sand fly-*Phlebotomus minutus*
1. Eggs 2. Larva 3. Pupa 4. Adult fly

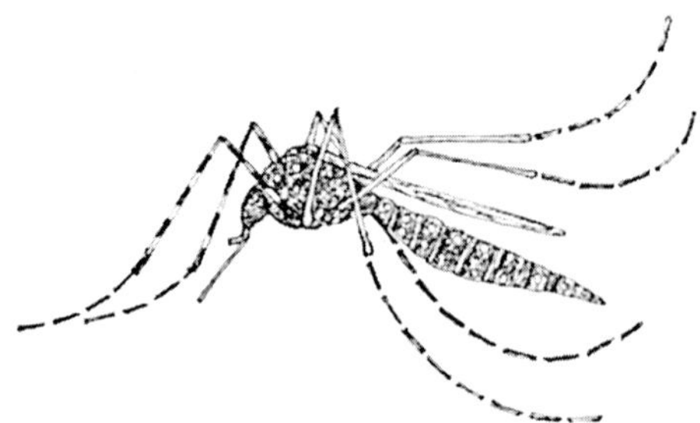

Fig. 149. Aedes mosquito-*Aedes aegyptii*

As in the case of houseflies, it is very difficult to control mosquitoes and concerted efforts have to be taken by every individual, as well as the Public Health Department to reduce the population of mosquitoes. Proper attention should be bestowed to mosquito breeding sites such as, ponds and streams of stagnant water, ill-drained swamps, gutters and all the places in houses, latrines, gardens etc., where water stagnates in various ways. Mosquitoes being very light may be carried over long distances by wind currents from far off places away from local breeding sites also. Small ponds and swamps may be filled up with earth to prevent water stagnation. Arrangements should be made to provide suitable drainage facilities in the case of stagnating sewers and channels. A small quantity of mineral oil may be poured on the surface of the stagnant water, which spreads over the entire area of the water surface and prevents the larvae and pupae from coming to the surface of the water to take in air from the atmosphere and due to suffocation they are killed. As a temporary measure of relief against mosquito bites, use of mosquito nets, provision of wire mesh screens to windows, doors and ventilators etc., application of mosquito repellent creams such as, odomos, citronella oil etc. are being followed. In places where there is stagnation of water for long periods and when the water cannot be drained off, insecticides may be applied to kill the larvae and pupae. Application of fenitrothion at 0.5-1.0 gm. ai./ sq.m. surface area of water is effective in eliminating the larvae and pupae for about 6 weeks. Fenthion and malathion have also been found to be effective. Surface spraying with fenitrothion, fenthion, propoxur, deltamethrin, dichlorvos etc. kills adult mosquitoes, which come and rest on the treated surface. Nowadays mosquito mats, mosquito coils, mosquito vaporizing liquids containing different synthetic pyrethroids as the active material are being marketed widely, which can be used in houses. Battery operated mosquito bats have also been introduced to kill flying mosquitoes. Some species of fresh water fishes have also been introduced, which feed on the larvae and pupae. A water dispersible formulation of a specific strain of *Bacillus thuringiensis* var. *israelensis* has been reported to be highly effective in parasitizing and killing early instar larvae. Some water inhabiting species of nematodes have also been found to be predaceous on the mosquito larvae.

3. Bed bug *(Cimex hemipterus)*-Hemiptera, Heteroptera, Cimicidae

Bed bugs are ectoparasites that pierce, suck and feed on the blood from man, mammals and birds. They are present in large numbers in dirty, ill-maintained houses, lodging houses, public conveyances etc. They are nocturnal in habit and hide in cracks, crevices and joints in walls, floors, woodwork, furniture etc., in-between folds of mattresses, pillows etc. during the daytime. However, even during daytime, when there is a chance, they bite and suck blood. Bed bugs often migrate from room to room. The venom injected during the bed bug bite causes itching, burning sensation and inflammation. The insects possess stink glands and a disgusting odor emanates from them. The bugs live long and can survive long periods of starvation. Bed bugs attack man, animals and poultry.

The female bug lays about 150-200 yellowish-white, flask-shaped eggs singly in their haunts. The eggs hatch in 3-14 days during the summer months and take a longer time to hatch during the winter season. The young nymphs hatching out from the eggs resemble the parent except in their size and color. There are 4 nymphal stages, each lasting for 5-7 days. The life cycle is completed in 8-13 weeks. Both the nymphs and adults are bloodsuckers.

The adult bugs are dull reddish-brown in color, broadly oval in shape, about 4.0-5.0 mm. in length, flattened and with pad-like rudimentary hemelytra. The rostrum lies in a ventral groove at the anterior end of the head. The whole body is covered with characteristic spinose bristles and some hairs. While *Cimex hemipterus* is found mostly in the tropical countries (Fig. 150), *C. lectularius* is found in the temperate countries. Bed bugs are disseminated through the agency of personal clothing, travel articles such as, luggages, beddings, holdall etc.

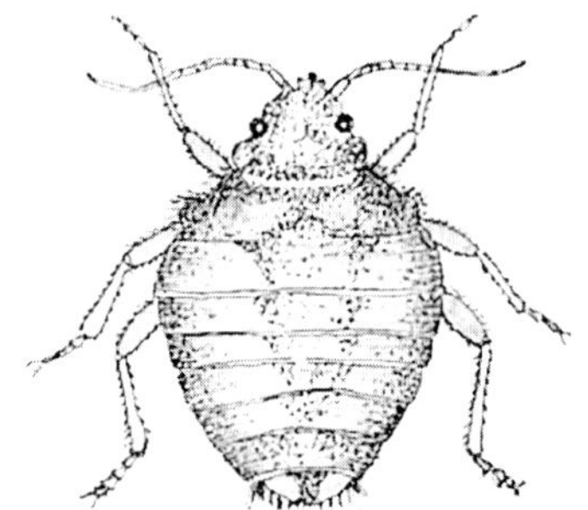

Fig. 150. Bed bug *Cimex hemipterus*

Bed bugs are the transmitting agents of several diseases such as, 'Kala azar', caused by protozoa, 'Relapsing fever', caused by a spirochete, 'typhoid fever', 'plague', 'leprosy' etc. Cleanliness of rooms, mattresses, pillows, furniture etc. plays a vital role in controlling bed bugs. Cracks, crevices, holes, joints etc. in the walls and floors should be patched up and the walls white washed. Joints in wood works and furniture should be patched up. The mattresses, pillows etc. should be dried in open sun. Care should be taken not to introduce bed bugs from outside into the houses along with bedding and luggage after a journey. Application of boiling water to carts and wooden furniture is a common practice followed in houses. Spray application of fenitrothion, propoxur, dichlorvos etc. is also effective in controlling the pest. Spray application of synthetic pyrethroids such as, permethrin has also been found to be very effective.

4. Human lice - Siphunculata (Anoplura), Pediculidae

Human head louse *(Pediculus humanus capites)*

Human body louse *(Pediculus humanus corporis)*

Human pubis louse or **Crab louse** *(Phthirius pubis)*

'Human lice' are ectoparasites that live on the blood of humans. Both the nymphs and adults are bloodsuckers. While the head louse is found mostly on the head among the hairs (Fig. 151), the body louse is found on the clothing and body among body hairs (Fig. 152) and the pubis louse among the hairs in the pubic and anal regions (Fig. 154). The louse bite causes a lot of irritation and itching. The skin around the bitten area becomes scarred, thickened and bronze-colored with brownish spots.

The adult female lays about 300 dirty white, slightly elongated eggs within a span of about 10 days, which are cemented to the hairs in rows. The nymphs that hatch out of the eggs have 3 nymphal instars, each lasting over a few days. The nymphs are quite similar to the adults in all respects, except that they are smaller and light-colored. The adult insects are apterous, grayish in color and the body flattened dorsoventrally. The head is conical and pointed with short, 3-5 segmented antennae. The mouthparts are modified for piercing and sucking blood. The legs are adapted for clinging to the host. The single-segmented tarsus has a powerful claw with which the louse clings firmly to the hair of the host.

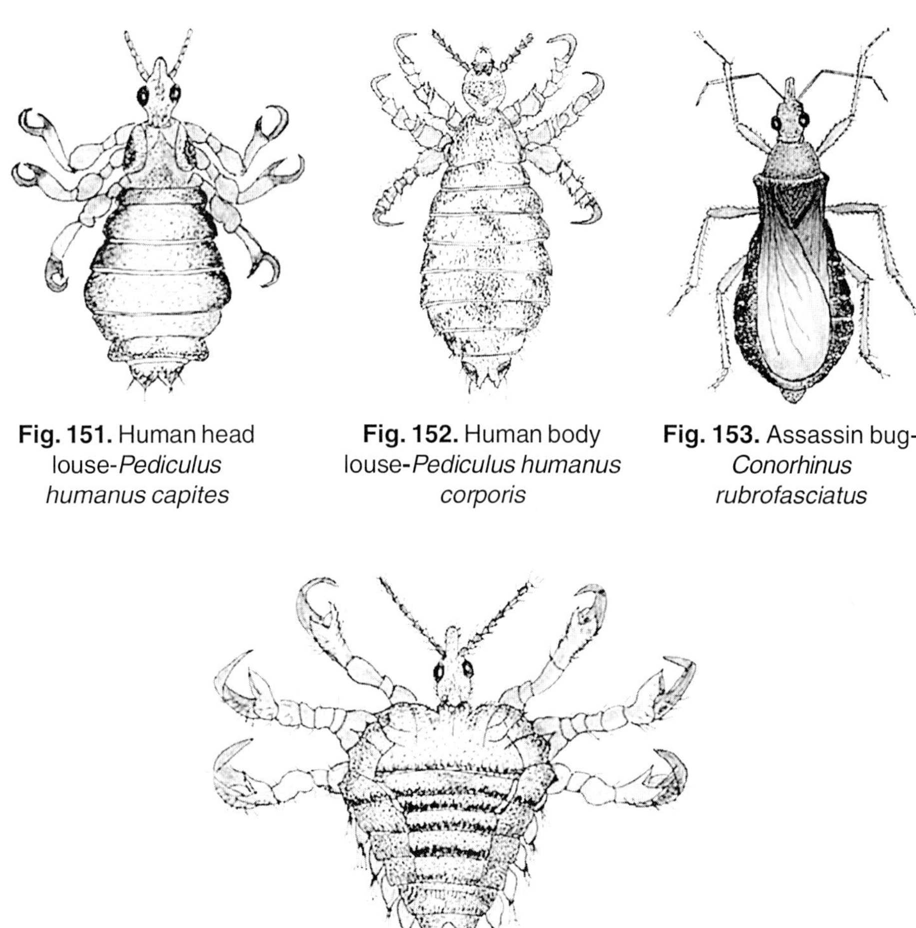

Fig. 151. Human head louse-*Pediculus humanus capites*

Fig. 152. Human body louse-*Pediculus humanus corporis*

Fig. 153. Assassin bug-*Conorhinus rubrofasciatus*

Fig. 154. Human pubic louse-*Phthirius pubis*

The human lice transmit several diseases such as, the 'epidemic typhus fever' caused by *Rickettsia prowaseki*, 'relapsing fever', 'trench fever' etc. Thorough cleanliness of the body by regular baths, proper cleaning of clothing and wearing of clean cloths, avoiding contact with infested persons, use of clean, pest-free combs etc. help to prevent lice infestation. The hairs should be combed with fine-toothed comb and the lice collected and killed. The infested head may be treated with commercial preparations such as, 'kenz' to destroy the lice.

5. Fleas - Siphonaptera, Pulicidae

Human flea *(Pulex irritans)*

Rat flea *(Xenopsylla cheopis)*

Dog flea *(Ctenocephalides canis)*

Cat flea *(Ctenocephalides felis)*

'Fleas' are ectoparasites of warm-blooded hosts only in their adult stage and they infect rats, dogs, cats, pigs, birds and humans. They siphon out blood from the host. Some species of fleas attack only a single host, while some others attack more than one host. However, most of the fleas attack man. They like warmth and so once the host dies and its body turns cold, the fleas abandon the dead body and seek other living hosts. Fleas are voracious bloodsuckers and their bite is not felt immediately. But the irritation, reddening, itching, rashes and other such allergic reactions occur sometime after the bite and may last for a long time.

The adult female flea lays ovoid, whitish or cream-colored eggs loosely on dust, excrement or on the body of the host, but are not gummed to the hair of the host. These eggs fall off and are found in the haunts of the host. After each blood meal, 6-8 fertile eggs are laid and in all about 200-500 eggs are laid by a female in 2-4 days time. The larvae that hatch out from the eggs are whitish, slender, vermiform maggots, which live on a variety of organic material generally found under dirty carpets, matting, unclean corners of rooms and in the sheds of domestic animals such as, dogs, cats, rabbits, pigs etc. The larvae possess well-developed head and mastigatory-type mouthparts but lack eyes. Full-grown larva pupates inside a pale silken cocoon and in about 15 days the adult flea emerges out. The entire life cycle is completed in 4-5 weeks. The adult fleas are small, active, wingless insects, about 2.2-2.5 mm. in length and laterally compressed. They have large simple eyes and possess piercing and sucking type of mouthparts for sucking blood and have stout and strong hindlegs adapted for leaping. The characteristic shape of the body is well suited for gliding easily and freely among hairs, furs or feathers of mammals and birds, which form their natural hosts. The heavily chitinized body is usually covered with spines and bristles, which are sharply inclined backwards. Often fleas leave the host between feeds and hide in cracks of floors and walls, in the litter found in grain godowns, pet sheds etc.

Fleas are notorious transmitters of 'Bubonic plague' caused by *Pasteurella (Bacillus) pestis* and the vector of the disease is the domestic rat flea-*Xenopsylla cheopis* (Fig.144). Rat fleas are also responsible for transmitting endemic or murine 'exanthematous typhus' *(Rickettsia mooseri)* in humans. The rickettsia multiply in the gut of the flea and excreted in faeces of fleas. *Pulex irritans* is another human flea, but its favorite host is the pig. *Xenopsylla brasiliensis* and *X. astia* also attack humans. The chief host of *Ctenocepalides canis* is dogs and of *C. felis* is cats (Fig.145) and these fleas attack humans also. *Ctenocephalides canis* and *C. felis* transmit 'tapeworm' of dog *(Dipylidium caninum)*. They also transmit 'filarial worm' of dog *(Dipetalonema reconditum)*. Fleas also spread 'Tularaemia' *(Francisella tularensis)* mechanically

Destruction of rats, proper attention to the healthy condition of pets such as, dogs, cats, rabbits, birds etc., proper dusting and cleaning of carpets and matting and thorough washing of pet sheds and rooms haunted by fleas with sanitary fluids such as, Lysol, Dettol etc. will control fleas effectively. Dead rats should be disposed off immediately by burying them deep in the soil or by burning after pouring kerosene. The breeding sites should be sprayed with propoxur or dichlorvos for controlling the fleas. Carbaryl formulated shampoo is effective for dogs and cats. Combination of dichlorvos and fenitrothion is effective for dogs and cats. Hexachlorohexane (HCH) dust or wash is useful for dogs only. Commercial preparations of coumaphos, ronnel, dichlorvos or synthetic pyrethroids such as, allethrin, resmethrin etc. can be used.

6. **Eye-fly** *(Siphunculina funicola)* - Diptera, Choloropidae

'Eye-flies' settle on the face, especially on the corners of the eyes of persons with eye complaints and cause considerable annoyance. They breed in damp, warm places under thatched sheds and such concealed places. The adult flies are minute, bluish-black in color and in general appear like small houseflies. The flies have a tendency of attaching themselves in swarms on ropes and threads hanging from roofs etc. (Fig. 142).

The flies transmit some common eye complaints including 'conjunctivitis'. Surface spray application of propoxur, endosulfan, fenthion etc. in places haunted by the flies gives effective control.

7. **Sand fly** or **Owl midge** *(Phlebotomus minutus)* - Diptera, Psychodidae

Due to their sand color, the flies are called 'sand flies'. The female sand flies are haematophagus (bloodsuckers), nocturnal in habit and can even penetrate through mosquito curtains. They prefer to attack toes, knuckles and joints and cause severe irritation. The adult sand fly is closely allied to mosquitoes, but are smaller in size with the body and wings thickly covered with hairs. The wings are held erect like a roof over the abdomen. They are weak fliers and can fly only short distances (Fig.148).

The adult female sand fly lays 40-80 eggs at a time singly in moist, dark places, especially in rock crevices and between stones. The eggs are white in color when laid, become dark in a few hours, They are torpedo-shaped with reticulated surface. The larva resembles small caterpillars (eucephalus) and feed on faeces of lizards, bats and other animals and on dried leaves. The head is quite prominent and well chitinized, but eyes are absent. There are 4 larval instars. Pupa is obtectate and curved like a 'comma', resembling a miniature ram's head. The entire life cycle is completed in about 35 days.

Phlebotomus minutus is the vector that causes 'Sand fly fever'.

P. argentipes transmits 'Kala azar' a disease caused by a flagellate-*Leishmania donovani. Phlebotomus serpenti* transmits 'cutaneous Leishmaniasis' caused by *Leishmania tropica* in man and dogs.

Dense vegetation around human dwellings should be removed. The control measures recommended for the control of mosquitoes may be adopted.

8. Assassin bug *(Conorhinus rubrofasciatus)*-Hemiptera, Heteroptera, Reduviidae

The bug is also known as 'cone nose bug' or 'kissing bug'. The bugs with piercing and sucking type of mouthparts attack both man and animals and are predacious on other insects. Sometimes the bugs find their way into human dwellings and attack humans. The bugs usually bite man at night for blood meal and the bite causes intense pain and swelling. The adult bug is fairly large, 20-24 mm. long, dark brown in color and the abdomen is broader at the middle exposing the margins of the abdominal segments beyond the wings. Some species serve as the carriers of *Trypanosoma cruzi,* which causes 'human Trypanosomiasis'. The bugs being large can be collected and destroyed (Fig. 153).

The methods of dissemination of different diseases vary with the causal organism and the vector. In cases such as, mosquitoes, bed bugs, fever ticks etc. the causal organism responsible for the disease actually gets into the body of the insect vector, undergoes certain changes inside the vector, multiplies and is then transmitted to the body of human beings and higher animals as in the case of Malaria, Trypanosomiasis etc. On the other hand, in the case of insects such as, the housefly, the disease causing organism is merely carried on the mouthparts, on or in the body of the insect and deposited on the human being or on food material without the disease organism undergoing any change in the vector as in Cholera, Typhoid, Tuberculosis etc. This latter phenomenon, where the insect merely serves as a porter or messenger is termed **'Myiasis'**. This term is applicable to the spread of bacteria, eggs of intestinal worms and larvae associated with open wounds, sores etc. in humans and cattle. Zumpt (1965) has defined Myiasis as 'the infestation of live human and vertebrate animals with dipterous larvae, which at least for a certain period, feed on the host's dead or living tissue, liquid body substances or ingested food'.

9. **Mites** - Arachnida, Acarina

Certain species of mites, which are non-insect, ectoparasites also attack humans and cause complaints known as 'Acariasis'. The mites are very minute creatures, less than 100 microns in size and possess biting and chewing, as well as piercing and sucking type of mouthparts.

The most important mites attacking the human skin include the 'itch mite' or 'dust mite'-*Sarcoptes scabici* var. *hominis*, which mainly attacks the skin of fingers, knee and elbow and causes insufferable itching, thickening of the skin, death of skin tissues and acute pain; the follicle mite-*Demodex folliculorum* found attached to the hair follicles, giving rise to itching and inflammation; the straw and grain mite-*Peduculoides ventricosus*, which attacks the skin of people handling grains, straw etc. often causing itching and Dermatitis and the tiny mite-*Tyroglyphus longior* affecting persons handling copra, dried nuts, cereals etc. and causing 'copra itch' or 'grocer's itch'.

Appendices

Appendix-I : List of insecticide formulations approved by the Government for plant protection purposes and dosage to be applied.

Chemical name	Trade name	Dosage to be applied for Spraying	
		Active (% a.i.)	ml./gm. per lit. of water
(1)	(2)	(3)	(4)
Acephate 75 WSP	Orthene, Lancer, Starthene, Acemil, Agrophate, Acelate, Acephate.	0.15%	2.0 gm.
Carbaryl 50 WP	Sevin, Sevimol, Sevidol, Carvin, Carbamate, Carbaryl, Corovin, Devicarb, Hexavin, Ryltex.	0.10-0.20%	2.0-4.0 gm
Cartap 50 WSP	Padan, Cartap, Thiobel, Vegetox.	0.10%	2.0 gm.
Chlorfenvinphos 50 EC	Birlane, Supona, Heptarex, Apachlor.	0.05%	1.0 ml.

Contd...

Chlorpyriphos 20 EC	Coroban, Dursban, Agrofos, Chlorpyrifos, Radar, Ruban, Lorsban, Spannit.	0.04%	2.0 ml.
Cypermethrin 10 EC	Cymbush, Ripcord, Polytrin, Challenger, Bullet, Cymet, Cyperguard, Cyperkil, Ninja, Paraban, Mortal, Cypermar, Polytrin, Lacer, Cypermethrin.	0.005%	0.5 ml.
Deltamethrin 2.8 EC	Decis, Cislin, Decathrin, Deltamethrin, Deltaphos, Glossinex, K.Othrine.	0.0015%	0.5 ml.
Diazinon 20 EC	Basudin, Agroziron, Diacap, Bazanon, Dazinol, Diazol, Disonex, Paraban, Neocidol, Diazinon.	0.04%	2.0 ml
Dichlorvos 76 WSC	DDVP, Nuvan, Vapona, Bangvos, Dichlorvos, Divipan, Luvan, Paradeep, Nuvanex, Nuvasul, Piran.	0.076%	1.0 ml
Dicofol 12.5 EC	Kelthane, Acarin, Banmite, Dicofol, Decofol, Delcofol, Hexakel, Micothane.	0.02%	1.0 ml
Dieldrin 18 EC	Quintox, Dieldrix, Dieldrin, Octabox, Alvit, Celatox.	0.02%	1.0 ml
Dimethoate 30 EC	Rogor, Cygon, Roxion, Tara 909, Dimethoate, Hexagor, Devigon, Agromat, Corothiate, Hexaban, Paragor, Parrydimate.	0.06%	2.0 ml
Dinacap 48 EC	Karathane, Crotothane, Arathane, Capryl, Mildex.	0.06-0.075%	1.25-1.5 ml
Endosulfan 35 EC	Thiodan, Agrosulfan, Endocel, Endocid, Marvel, Hexasulfan, Endostar, Endosulfan, Thiotox, Parasulfan, Pesticel, Thiokil.	0.035-0.07%	1.0-2.0 ml
Ethion 50 EC	Mit, Ethion, Fosmite, Force, Lazor, Rathion.	0.05%	1.0 ml
Fenitrothion 50 EC	Sumithion, Folithion, Demise, Fenitrothion, Gilmore.	0.05-0.10%	1.0-2.0 ml
Fenthion 82.5 EC	Lebaycid, Baycid, Baytex, Fenthion, Queletox.	0.08-0.16%	1.0-2.0 ml

Contd...

Fenvalerate 20 EC	Sumicidin, Agrofen, Fenicidib, Devifenvalerate, Fenhil, Fenvel, Fenvalerate, Gilten, Grofen, Monocil, Parafen, Parryfen, Starfen, Sumitox, Tagfen, Valour.	0.0075%	0.375 ml
Formothion 25 EC	Anthio, Anthiomix, Aflix, Sandothion.	0.05%	2.0 ml
Lindane 20 EC	Agrodon, Canon, Lindone, Gamma BHC, Lindane, Marsoline, Micodane, Paraban.	0.04%	2.0 ml
Malathion 50 EC	Cythion, Corothion, Agrolmol, Taimal, Bangmal, Bugtax, Chemathion, Entomol, Hilthion, Maltox, Malathion, Sandoz 50.	0.05-0.10%	1.0-2.0 ml
Methyl demeton 25 EC	Metasystox, Hexasystox, Himax, Knockout, Parasystox.	0.025-0.05%	1.0-2.0 ml
Methyl parathion 50 EC	Agrimethyl, Corocid, Milon, Devithion, Metacid, Luthion, Folidol M, Methyl parathion, Paramar, M.Devithion, Paradol, Paracrop, Paramet, Metaphos, Ramthion, Tagpar, Vicomethyl..	0.05-0.10%	1.0-2.0 ml
Monocrotophos 36 SL	Azodrin, Croton, Nuvacron, Luphos, Macrophos, Parryfos, Corophos, Entofos, Hexaphos, Gilmore Monocrotophos, Micophos, Indophos, Monocid, Monocil, Monocron, Monostar, Triphos, Monovip.	0.04%	1.0 ml
Nicotine sulfate 40 S	Nicotine sulfate, Black leaf, Destruxol, Shreds.	0.02%	0.50 ml
Permethrin 25 EC	Ambush, Dragnet, Duracide P, Gilmore Permethrin, Pestop, Permethrin, Multispray, Perthrins, Tornade, Torpedo, Trinex, Permasect, Perigen.	0.01%	0.40 ml
Phenthoate 50 EC	Agrophen, Aimsan, Cidial, Gilphenthoate, Delsan, Deslan, Elsan, Phenthoate, Phendal, Tagsan, Tanone.	0.05-0.10%	1.0-2.0 ml

Contd...

Phosalone 35 EC	Zolone, Agrosalone, Phosal, Gilmorephosalone, Micozone, Sugalone.	0.07-0.105%	2.0-3.0 ml
Phosphamidon 85 WSC	Dimecron, Agromidon, Daron, Agrophos, Bangdon, Cildon, Phamidon, Phoskil, Sumidon, Vimidon, Phosphamidon, Sudon.	0.05%	0.50 ml
Propoxur 20 EC	Baygon, Blattanex 20, Domexil, Domeel, Pilargon, Propogon, Propoxur, Propyon, Suncide, Unden.	0.04%	2.0 ml
Quinalphos 25 EC	Ekalux, Agriphos, Quinal, Quinalphos, Tagquin, Suquin, Starlux, Quinavip, Quinaltaf, Sicophos, Solux, Quinatox, Rambux.	0.05%	2.0 ml
Profenofos 50 EC	Polytrin C, Fenom C, Selecron, Curcacron.	0.05-0.10%	1.0-2.0 ml
Propetamphos 20 EC	Biotic, Panik B, Safrotin, Seraphos.	0.04%	2.0 ml
Sulfur 80 WP	Thiovit, Cosan, Microsul, Sulfex, Elosan, Earmasul, Wettable sulfur.	0.32%	4.0 gm
Thiodicarb 75 WP	Larvinbrand, Securex, Semivin.	0.15%	2.0 gm
Thiometon 25 EC	Ekatin, Hexatin, Thiotox, Tombel.	0.05%	2.0 ml
Triazophos 20 EC	Deltaphos, Hostathion.	0.02%	1.0 ml
Trichlorphon 50 EC	Dipterex, Tugon, Dylox, Danex, Denkaphon, Trichlorphon.	0.05%	1.0 ml

EC - Emulsifiable concentrate
WP - Wettable powder
SL - Soluble liquid
WSC - Water soluble concentrate
WSP - Water soluble powder
S - Solution

Appendix II. List of granular insecticide / nematicide formulations and dosage to be applied.

Granular formulation (1)	Quantity of granules to be applied per acre in kg. (2)
Aldicarb 10 G	5.0 kg.
Carbaryl 4 G.	12.0 kg
Carbofuran 3 G.	13.0 kg.
Chlorfenvinphos 10 G.	5.0 kg.
Chlorpyriphos 10 G.	5.0 kg.
Diazinon 10 G.	5.0 kg.
Disulfotan 5 G.	10.0 kg.
Endosulfan 4 G.	12.0 kg.
Fenitrothion 5 G.	10.0 kg.
Fenthion 5 G.	10.0 kg.
Lindane 6 G.	8.0 kg.
Phenthoate 5 G.	10.0 kg.
Phorate 10 G.	5.0 kg.
Quinalphos 5 G.	10.0 kg.
Sevidol 4 G.	12.0 kg.

The granular formulation is mixed with sufficient quantity of sand to make the quantity to 20 kg. and then applied uniformly over the soil, incorporated into the soil and then irrigated.

Appendix III. Quantity of spray fluid that may be required to spray one acre of crop by different types of sprayers.

Crop (1)	High volume sprayers (2)	Semi-low volume sprayers (3)	Low volume sprayers (4)
Rice	200 - 250 lit.	50 - 100 lit.	5.0 - 50 lit.
Groundnut	250 - 300 lit.	65 - 120 lit.	6.0 - 60 lit.
Cotton	300 - 500 lit.	75 - 200 lit.	7.5 - 100 lit.
Millets	200 - 350 lit.	50 - 140 lit.	5.0 - 70 lit.
Potato	300 - 350 lit.	75 - 140 lit.	7.5 - 70 lit.
Pulse crops	250 - 350 lit.	65 - 140 lit.	6.0 - 70 lit.
Sugarcane	350 - 500 lit.	85 - 200 lit.	9.0 - 100 lit.
Wheat	250 - 350 lit.	65 - 140 lit.	6.0 - 70 lit.

Contd...

Chillies	200 - 350 lit.	50 - 140 lit.	5.0 - 70 lit.
Tomato	200 - 350 lit.	50 - 140 lit.	5.0 - 70 lit.
Eggplant	200 - 350 lit.	50 - 140 lit.	5.0 - 70 lit.
Tobacco	200 - 350 lit.	50 - 140 lit.	5.0 - 70 lit.
Grapevine	500 - 600 lit.	125 - 240 lit.	13.0 - 120 lit.
Banana	250 - 600 lit.	65 - 240 lit.	6.0 - 120 lit.
Coconut	3.0 lit. / tree	0.75 lit. / tree	0.6 lit. / tree
Arecanut	3.0 lit. / tree	0.75 lit. / tree	0.6 lit. / tree
Mango	5.0 - 10 lit. / tree	1.25 - 4.0 lit. / tree	0.125 - 2.0 lit. / tree
Guava	5.0 - 10 lit. / tree	1.25 - 4.0 lit. / tree	0.125 - 2.0 lit. / tree
Sapota	5.0 - 10 lit. / tree	1.25 - 4.0 lit. / tree	0.125 - 2.0 lit. / tree
Citrus plants	5.0 - 10 lit. / tree	1.25 - 4.0 lit. / tree	0.125 - 2.0 lit. / tree

While using semi-low volume sprayers or low volume sprayers, the quantity of the chemical formulation, which is required to spray the crop with a high volume sprayer should be used, although the quantity of water required to prepare the spray fluid is much less. Because a lesser quantity of water is used for preparing the spray fluid, the chemical formulation should not be reduced proportionately.

Appendix IV. Classification of pesticides on Acute Toxicity levels.

Particulars (1)	**Category I Extremely Toxic** (2)	**Category II Highly toxic** (3)	**Category III Moderately Toxic** (4)	**Category IV Slightly toxic** (5)
LD_{50}				
Oral (mg./kg.)	0.0 - 50	51-500	501-5000	> 5000
Dermal (mg./kg.)	0.0-200	201-2000	2001-20000	> 20000
Color of the lower triangle on the label of the container.	Red	Yellow	Blue	Green
Signal in the upper triangle	'poison', skull and cross bones	'Poison'	'Danger'	'Caution'
Few examples	Aldicarb, Aluminium phosphide, Monocrotophos.	Cypermethrin, Dichlorvos, BHC.	Carbaryl, Malathion	Delfin, Nimbecidine, Wettable sulfur.

References

Anon. 1997. Handbook of Agriculture. *Director, Director of Publications and information on Agriculture, Krishi Anusandhan Bhavan, Pusa, New Delhi* - 110 012. pp. 1303.

Ashok Kumar and Prem Mohan Nigam. 1989. Economic and Applied Entomology. *Emkay Publications, B-19, East Krishna Nagar, Delhi* - 110 051. pp. 349.

Edward, J.C. and Misra, S.L. 1974. An Introduction to Plant Nematology. *Central Book Depot., Allahabad.* pp. 82.

Gailce Leo Justin, C., Johnson Thangaraj Edward, Y.S. and Swamiappan, M. 1995. A compendium of Entomology. *Butterfly Publications, Coimbatore, Tamil Nadu, India.* pp. 543.

Ganapathy, T., Rabindran, R. and Sabitha Doraiswamy. 2002. An Illustrated Glossary of Plant Pathology. *AE Publication, Coimbatore*- 41. pp. 251.

George N. Agrios. 1997. Plant Pathology. *Academic Press, California, USA.* pp. 635

Grist, D.H. and Lever, R.J.A.W. 1969. Pests of rice. *Longmans Green & Co., Ltd., London.* pp. 520.

Hem Singh Pruthi. 1969. Textbook on Agricultural Entomology. *Indian Council of Agricultural Research, New Delhi.* pp. 977.

Kanna, S.S. 1983. Agricultural Entomology. *Adarsh Prakashan, Agra.* pp. 304.

Krishnan, N.T. 1993, Economic Entomology. *J.J.Publications, 32-A, H.A.K.Road, Madurak.* pp. 265.

Lewin Devasahayam, H. 1997. Insect and Non-insect Crop Pests. *Zion Printers and Publishers, 91, Arunachala Sreet, Chintadripet, Chennai* - 600 002. pp. 478 (In Tamil)

Pradhan, s. 2002. Agricultural Entomology and Pest control. *Indian Council of Agricultural Research, Krishi Anusandhan Bhavan, Pusa, New Delhi* - 110 012. pp.267.

Prem Mohan Nigam and Asok Kumar. 1991. Plant Protection - Insect control. *Emkay Publications, B-! 9, East Krishna Nagar, Delhi* - 110 051. pp. 191.

Ramakrishna Ayyar, T.V. 1963. Handbook of Economic Entomology for South India. *Controller of Stationary and Printing, Government of Madras, Madras.* pp. 516.

Ross H. Arnett and Richard L. Jacques. 1981. Simon and Schulter's Guide to Insects. *Simon and Schulter Inc., New York.* pp. 512.

Saxena, R.C. and Srivasta, R.C. 2009. Entomology at a Glance. *Agrotech Publishing Academy, Udaipur* 313002. pp. 464.

Upadhyay, K.D. and Kusum Dwivedi. 1996. A Textbook of Plant Nematology. *Aman Publishing House, Madhu Market, Meerut* - 250 002. pp. 148.

Vasantharaj David, B. 1981. Indian Pesticide Industry- Facts and Figures. *Vishwas Publications, No.1, Jeevan Tara Co-op. Hsg. Soc. Ltd., Vile Parle (East), Bombay* - 400 057. pp. 378.

Vasantharaj David, B. 2005. Elements of Economic Entomology. *Popular Book Depot., 29, Potters Street, Saidapet, Cennai* - 600 015. pp. 590.

Vasantharaj David, B and Kumaraswami, T. 1975. Elements of economic Entomology. *Popular Book Depot., 29, Potters Street, Saidapet, Chennai* - 600 015. pp. 507.

Glossary of Scientific Terms

Abdominal legs. Fleshy legs found in the abdomen of caterpillars (Lepidoptera, saw flies), bearing characteristic arrangement of crochets (See prolegs).

Abiotic factors. Non-living factors or environmental factors that influence the rate of multiplication of any organism and is independent of population density. Abiotic factors include topography, soil type, relative humidity, temperature, light, moisture and wind.

Acariasis. Any skin disease resulting from infestation with acarids or mites.

Acaricide. A chemical or chemical formulation employed to control mites and ticks; alternate name miticide.

Acephalus larva. Larva with a reduced or no head capsule and appendages. The larva is also called a maggot (e-g) Larva of housefly and many other dipterous insects.

Acquired immunity. Resistance to a microbial or other antigenic substance acquired either actively or passively by a naturally susceptible individual.

Active ingredient (a.i.). Toxic component present in a formulated pesticide.

Adhesive. A material added to a formulation to increase its sticking power or adhesiveness.

Aerosol. A pesticide formulation that is dispersed in the form of spray droplets of diameter 0.1-5.0 μm. with the help of compressed air, usually from a canister.

Aestivation. Dormancy during a hot or dry season.

Chrysalis. A naked pupa; an open pupa, not enclosed or covered; the naked pupal stage (e-g) pupa of butterfly.

Classification. The systematic arrangement of insects in series showing their relation or agreement of structure, life habits or other characters forming the basis of classification.

Cocoon. A silken case woven by a full-grown larva or a fibrous case made by a larva or an earthen case in the soil made by a larva, in which it pupates.

Collateral hosts. Host plants belonging to species other than the species of the primary host, but within the same Genus.

Copulation. Sexual act between the male and female, resulting in fertilization.

Crawlers. The early stage nymphs of mealy bugs, which have functional legs and can crawl about short distances.

Cysts. The oxidized cuticle of dead adult females of the cyst nematodes that protects the eggs contained within; the carcass of dead female adult nematodes of the Genus - *Heterodera* or *Globodera,* which may contain eggs.

Dead heart. Wilting and eventual death of the central shoot in the case of monocots, which come off quite easily when pulled and wilting and death of the main and side shoots in the case of dicots, as young caterpillar bores into the stem of the plant either from the top or side, eats the inner tissue and makes tunnels inside (see white ears).

Defoliator. Any biting and chewing insect that destroys the leaves of plants extensively.

Dengue. A virus disease of man marked by fever and severe pain in head, eyes, muscles and joints, and transmitted by certain mosquitoes.

Determinate growth. With a distinctive final growth stage, adult instar.

Deterrent. A chemical substance that deters feeding or oviposition of an insect.

Detritivorous. Feeding on organic debris of plant or animal origin.

Deutonymph. The third instar stage in the development of mites and ticks.

Dioecious. Having the male and female sex organs in different individuals; any one individual being either male or female.

Disinfectant. A physical or chemical agent that frees a plant, organ or tissue from infestation; a substance, which eliminates contaminant organisms from surface.

Disinfestant. An agent that kills or inactivates pests in the environment or on the surface of a plant or plant organ before infestation takes place.

Diurnal. Active during the daytime; habitually flying by day only.

Dormancy. A state of quiescence or inactivity.

Earthen cocoon. A type of cocoon constructed by the larva with soil and saliva (e-g) *Helicoverpa armigera.*

Ecdysis or Moulting. The process of shedding the skin (exoskeleton) undertaken periodically by insects and other arthropods.

Ecology. The study of living organisms in an area and their physical environment.

Economic control. Control of a pest in which the expenditure on control measures is more than compensated by increased value of yield.

Economic injury level (EIL). The infestation level of a particular pest at which the damage caused by the pest will result in economic loss; the lowest population density of a pest that causes economic damage.

Economic pest. A pest causing a crop loss of 5-10 per cent.

Economic threshold level (ETL). The infestation level of a particular pest, beyond which level there will be economic loss; pest population level at which control measures have to be initiated to prevent the pest population rising to the economic injury level.

Ecosystem. A self-containing habitat in which living organisms and the physiochemical environment interact in an exchange of energy and matter to form a continuing cycle.

Ectoparasite. A parasite that lives externally and feeds from the host by sending in specialized organs or mouthparts to ingest nourishment from the host, but does not kill the host; a parasite feeding on host from the exterior.

Egg case. The case or covering formed or secreted by an insect to contain or hold together the egg mass as a whole (e-g) Praying mantis.

Egg mass. A group of eggs deposited by the female insect, which are arranged in a characteristic manner adjacent to each other as in the case of butterflies and many moths or in rows as in rice bug or overlapping as in the case of many moths.

Egg parasitoids. The parasitoids deposit their eggs inside or outside the host eggs and the progeny of the parasitoid emerging from the eggs, parasitize the host eggs and destroy them (e-g) *Trichogramma* spp. on lepidopteran eggs.

Egg pod. Eggs laid in a compact cluster arranged in a vertical, spiral manner inside a capsule; a capsule, enclosing the egg mass, which is formed through the cementing of soil particles together by secretions of the ovipositing female (e-g) eggs of grasshoppers and locusts.

Egg raft. A compact mass of floating eggs consisting of 200 - 300 eggs, which are laid in water (e-g) Culex mosquito.

Elytra. The hardened, modified, veinless, leathery or chitinous forewings of coleopteran beetles and weevils that serve as protective covering to the membranous hindwings and commonly meeting in a straight line down the middle dorsum in repose.

Embryo. The young one before emerging from the egg.

Emulsifiable concentrate (EC). An insecticide formulation in which the active ingredient is dissolved in a solvent and when the formulation is mixed with water, the active ingredient is held in a suspended state in water.

Endemic pest. A pest permanently established in a moderate or severe form in a defined area, commonly a country or part of a country.

Endoparasite. A parasite, which enters a host, lives within the host and feeds from the host; a parasite, which enters a host and feeds from within.

Entomophagous. Organisms, which feed on insects or their parts; insectivorous.

Entognathous. Having the mandibles and maxillae retracted into pouches in the head as in the case of Protura, Collembola and Diplura.

Enzyme. A protein produced by living cells that can catalyze a specific biochemical reaction.such as, digestion of food etc.

Ephemeral. Short-lived.

Epidemic (Epiphytotic). A sudden and rapid flare up of a pest from its endemic area; a widespread and severe, but temporary flare up in the incidence of a pest, particularly within a season.

Epidemiology. The study of disease development and spread throughout a host population.

Eruciform larva (Polypod larva). A larva with a more or less cylindrical body, a well-developed head and with both thoracic legs and abdominal prolegs; a caterpillar (e-g) larvae of moths and butterflies.

Eucephalous larva. Larva with a distinct, well-developed, sclerotized head capsule, mouthparts and antennae (e-g) Larva of mosquito.

Exarate pupa (Free pupa, pupa libera or decticous pupa). Pupa remains uncovered, without puparium. All the appendages such as, the wings and legs are free from the rest of the body and the pupa resembles somewhat the adult condition (e-g) Beetles.

Excretion. Elimination of waste products of metabolism such as, faecal matter, urine etc.

Exotic peat. Pest introduced from a foreign country; not native.

Exuviae. The cast off skin of nymphs or larvae at the time of metamorphosis.

Fecundity. The number of eggs produced, regardless of their fertility.

Femur. The single-piece part of the leg in-between the trochanter and tibia. In jumping legs as in the case of grasshoppers, the femur is much enlarged and very strong and helps the insect to jump.

Fibrous cocoon. A type of cocoon constructed by the larva with fibres (e-g) Red palm weevil.

Field resistance. The resistance observed under field conditions due to the exposure of plants to natural population of insects; moderate resistance.

Filariasis. A human disease caused by the filarial nematode - *Wuchereria bancrofti* transmitted by the mosquito - *Culex fatigans*; also called as Elephantiasis.

Forma specialis **(f.sp.).** A group of races and biotypes of a pathogen species differing in some physiologic properties that can infect only plants within a certain host Genus or species; a special form.

Frass. The wet or dry, saw dust-like excrement of borers usually evident at their exit holes in fruits and other plant parts.

Frassy cocoon. A type of cocoon constructed by the larva with frass and saliva (e-g) *Opisina arenosella.*

Free pupa. A pupa in which the appendages and limbs are not fused to the outer covering.

Fringed wings. Wing margins fringed with long setae or hairs (e-g) Wings of thrips.

Frond. The entire leaf of a coconut palm. The fleshy midrib of the frond consists of 100 - 120 leaflets on either side.

Fumigant. A substance or mixture of substances, which produce gas, vapor, fume or smoke intended to destroy insects or other pests (e-g) carbon disulfide, carbon tetrachloride, chloropicrin, dibromo chloropropane, EDCT, ethylene dichloride, hydrogen cyanide, methyl bromide, phostoxin etc.

Fumigation. The application of a fumigant for disinfestation of an area.

Furrow application. Placement of pesticide with seed or seed material in the furrow at the time of sowing.

Gall. A swelling or outgrowth produced by a plant as a result of attack by an insect, mite or microorganisms.

Gelatin capsules. Capsules filled with pesticide for use as root, stem or zone application.

Giant cells. A host response in which many adjacent cells at the feeding site of a nematode coalesce to form large, multinucleate cells with mass of protoplasm.

Granulosis viruses (GV). Viruses causing insect diseases characterized by the presence of large number of very small, but microscopically dissembled granular inclusion bodies in infected cells.

Gravid female. Adult female that is capable of laying eggs; adult female insect containing fertilized eggs.

Gregarious. Forming aggregations.

Grubs. The larvae of Coleoptera and Hymenoptera are generally called as grubs.

Haematophagus. Blood suckers; Feeding or subsisting on blood.

Hairy cocoon. A type of cocoon constructed with body hairs (e-g) Wooly bear.

Halteres. Knob-shaped, modified hindwings of Diptera, which act as balancing organs (see balancers).

Hemelytra. Wings in which the basal half is thick and leathery, while the distal half is membranous (e-g) Forewings of heteropteran bugs.

Hemicephalus larva. The head capsule and appendages are considerably reduced and the head is retracted into the thorax and the head capsule is in-between that of eucephalous and acephalous larva (e-g) Brachycera, Tipulidae.

Heterogenesis. A form of reproduction in some insects, which has sexual and asexual or parthenogenetic forms or some cyclic form of alternation of generations.

Hibernation. State of inactivity; passing the winter in sleep or inactivity.

Histological changes. Changes in the cellular structure of tissues.

Honeydew. A sugary exudate secreted by sucking insects, such as aphids, white flies, scale insects and plant hoppers; excess sugary plant sap secreted from the hind end of the gut by sap-sucking insects.

Hopper-burn. Burnt-up appearance of rice crop in patches, as a result of severe infestation by rice brown plant hopper; a toxicogenic symptom produced in crops like rice, lady's finger, cotton, potato etc. as a result of feeding by toxicogenic insects such as leafhoppers.

Horizontal resistance. The type of resistance, which is effective against all the known biotypes of the insect. It is quantitative, as the degree of resistance depends on the number of minor genes each contributing a small effect. It is also called non-specific resistance.

Hormone. An internally secreted chemical messenger substance produced by endocrine tissues (glands), which influences or activates other organs or physiological processes within the same organism; a chemical secreted in one part of the body that regulates the function of some other part of the body.

Host plant. Living plant attacked by or harboring a pest or parasite and from which the invader obtains part or entire requirement of its nourishment.

Host range. The various kinds of host plants that may be attacked by a pest.

Husk. The outermost, fibrous structure of a coconut.

Hyperparasitoid. An insect parasite of another primary parasitoid; also known as secondary parasitoid.

Hyperplasia. A plant overgrowth due to increased cell division; an abnormal increase in the number of cells of a tissue.

Hypertrophy. A plant overgrowth due to abnormal cell enlargement; an abnormal increase in the size of cells of a tissue.

Hypognathous. Having the head vertical and the mouth directed downward (e-g) Orthoptera.

Hypoplasia. A defective or incomplete development of an organ system, organ or tissue

Imago. The adult, sexually matured form of an insect.

Immature stage. Any stage of an insect prior to the imago (adult form).

Immune. Cannot be infested by a given pest.

Indigenous. Native to an area; native to a particular region or country.

Infest. To occupy and cause injury to either a plant, animal, stored product or soil.

Ingestion. Taking in food; eating.

Injury. The effect of pest activity on the host physiology, which is usually injurious or deleterious to the host.

Insect growth regulator (IGR). A general class of natural and synthetic chemical compounds associated with the control of growth and metamorphosis in insects; chemicals that interfere with normal insect growth or development, usually by affecting the growth hormone systems.

Insecticide. A toxic chemical substance or compound employed to kill and control insects; chemicals prepared with the express purpose of controlling insect pests, the active ingredient of which may be in the form of gas, liquid or solid and formulated as liquefied gases, smokes, dusts, water dispersible powders, emulsions or solutions.

Insect pest management. An ecologically based strategy of maintaining insect pest population below the economic injury level by the use of any or all control techniques that are economically, ecologically and socially acceptable.

Instars. Stages between moults.

Integrated control. An approach that attempts to use all available methods of control of a pest or of all the pests and diseases of a crop plant for best control results, but with least cost and least damage to the environment.

Integrated pest management (IPM). An approach utilizing all available methods of pest suppression including mechanical, biological, chemical and natural control in a systematic way with the primary goal of safe, effective and economical pest population reduction and to maintain the pest population at a level below that causing economic damage or loss. It may be directed at a single important target pest species or at a pest complex, integrating the individual control measures applied against each, so as not to interfere one with the other.

Intermediate host. The host that harbors the immature stages or the asexual stages of a parasite (e-g) Mosquito for the malarial parasite.

Juvenile hormone (JH). A multi carbon chain hormone secreted by the corpora allata into the haemolymph that maintains the immature form of an insect during early moults.

Juvenile stage. Any or every post-embryonic stage in the development of an organism that precedes the sexually mature adult stage such as, egg, larval and pupal stages of insects.

Key pest. An important major pest species in the pest complex attacking a crop and causing economic damage, with a dominating effect on control practices.

Knockdown effect. An insecticide that has a power to knock down flying insects rapidly but may not kill them immediately.

Lacerate. Tearing off the tissues and causing injury.

Larva. The immature form of holometabolous insects, greatly different in form from the adult; young insect that comes out of an egg in the early stage of morphological development and differs fundamentally in form from the adult.

Larva aculeata. Larva with dense, fur-like hairs.

Larva cornuta. Larva with fleshy horns or processes.

Larva furcifera. Larva with furcate processes.

Larval parasitoid. Parasitoid that deposits its eggs in or on the host larvae. The progeny of the parasitoid emerges from the host larvae, parasitizes and destroys the host larvae (e-g) *Bracon brevicornis, Apanteles flavipes* etc.

Larviparous. Giving birth to larvae directly instead of laying eggs, as in certain Diptera.

LD_{50} or Median lethal dose. A value used in presenting mammalian toxicity, usually oral toxicity, expressed as mg. of toxicant per kg. of body weight; a lethal dose for 50 per cent of the test organisms; the dose of a toxicant producing 50 % mortality in a population.

Legal control. Control of pests through the enactment of legislation that enforces control measures or imposes regulations such as, quarantines to prevent introduction or spread of pests.

Lesion. A localized spot of diseased tissue.

Lethal. Fatal or deadly.

Life cycle. Definite pattern of emergence, growth, reproduction, old age and death.

Light trap. A device for collecting insects, consisting of a light source to attract insects at night and mechanism to trap the insects.

Longevity. Life span of an individual.

Longicorn. Having the antennae as long as or longer than the body as in the case of cerambycid beetle.

Looper. A caterpillar of the family - Geometridae, Order - Lepidoptera with only one pair of abdominal prolegs in addition to the terminal claspers and which moves by looping its body i.e. by moving the posterior part of the abdomen close to the thorax and then extending the anterior part of the body forward; otherwise called as a measuring worm (e-g) Daincha looper.

Maceration. Chewing without ingestion of the chewed material.

Macropterous. Winged; having wings; in insects, such as aphids, the comparatively long-winged forms, which are mostly migratory in nature.

Maggots. The larvae of a fly (Diptera) are generally called as maggots; vermiform larva devoid of true legs and without a well-developed, distinct head capsule (Diptera), the head-end of which is pointed and the posterior end is somewhat flattened and appears as if cut off at the tip.

Major pest. Insect pest that causes a loss of more than 10 percent of the produce.

Malaria. An infectious, fever causing disease caused by protozoans of the Genus - *Plasmodium*, which invade the red blood corpuscles after transmission to a human, most commonly by *Anopheles* mosquito.

Malformation. Part of a plant or parts of a plant that is deformed, badly formed or badly shaped.

Man-made pest. A species, which is a pest only because of human interference with natural control processes that normally regulate its density to non-pest levels. Most commonly generated through pest upsets i.e. the unintentional destruction of natural enemies of a non-pest species with a chemical insecticide or through monocultural practices.

Matrix. The gelatinous substance into which eggs of *Meloidogyne* and some other nematodes are deposited to form an egg mass.

Mechanical control. Any mechanical means of pest control or mechanical devices for pest control such as, insect and mammalian traps of various kinds, barriers to prevent entry or gaining access of pests to plants or other commodities.

Median valve. A valve situated posterior to the stylet, the pulsating action of which creates a vacuum and cell sap is ingested into the alimentary canal through the stylet tube.

Metabolism. The entire process or series of changes or transformation of ingested food into tissue and cell substances and of these latter into waste products.

Metamorphosis. The series of changes in the bodily form, which most insects undergo as they develop into adulthood.

Methyl eugenol. A chemical sex attractant for *Dacus dorsalis.*

Middle lamella. The cementing layer between adjacent cell walls, which generally consists of pectinacious materials, except in woody tissues, where pectin is replaced by lignin.

Migratory. Migrating from plant to plant.

Minor pests. Insect or non-insect pests that cause a loss ranging from 5-10%.

Monophagous pests. Pests that infect only a particular species of plants alone.

Mosaic disease. A disease characterized by variegated pattern of green and yellow on the foliage of plants caused by certain viruses.

Mottling. A disease symptom in the leaves comprising of numerous light and dark areas in an irregular pattern usually caused by a virus, often interchangeably with mosaic disease.

Moulting. Casting off the outer skin to facilitate further growth; the shedding or casting off of the cuticle.

Mycoplasma (pl. mycoplasmas). Obligate, plant parasitic pleomorphic mollicute (Prokaryote without cell wall) found in the phloem tissue; microorganisms intermediate in size between viruses and bacteria, possessing many virus-like properties and ultramicroscopic.

Myiasis. Infestation of live human and vertebrate animals with dipterous larvae, which at least for a certain period, feed on the host's dead or living tissue, liquid body substances or ingested food.

Naiad. The aquatic nymph of a dragonfly, mayfly or other hemimetabolous insect.

Naked pupa. A pupa not enclosed in a cocoon or other covering.

Naphthalene. Coal tar derivative used as a fumigant against clothes moth, also employed as a soil fumigant, relatively nontoxic to mammals.

Narrow spectrum insecticide. Insecticide effective against only a narrow range of insects.

Necrosis. Death of a defined area of plant tissue.

Negatively phototrophic or Photonegative. Natural instinct to shun away from artificial light source.

Nematicide. A chemical compound or physical agent that kills or inhibits nematodes.

Nematode. Generally microscopic, worm-like animals that live saprophytically in water or soil or as parasites of plants and animals.

NEPO. Nematode transmitted viruses with polyhedral (spherical) particles.

Nerve poison. Chemical formulations associated with the solubility of tissue lipid and actively block acetylchorinesterase in warm-blooded animals; chemical formulations that affect the nervous system of insects (e-g) Organochlorine, Organophosphate and Carbamate insecticides.

NETU. Nematode transmitted viruses with tubular (rod) particles.

Nocturnal. Active during the nighttime.

Non-persistent virus. A stylet-borne virus; a virus that does not undergo part of its development within the body of its vector.

Nuclear polyhedrosis virus (NPV). A virus that causes viral disease of insects, mainly the larvae of certain Lepidoptera and Hymenoptera, characterized by the formation of polyhedral inclusion bodies in the nuclei of the infected cells.

Nuptial flight or Mating flight. A flight, which may be very long or short, that precedes copulation between an adult male and female.

Nymph. An immature stage (instar), which is similar to the adult except for the absence of functional wings and sexual organs.

Obligate parasite. A parasite that can lead a parasitic mode of life and which cannot exist in any other way.

Obtect pupa. The pupa, in which the appendages and body are compactly united or fused by a hardening of the outer skin, termed the theca (e-g) Moth pupa.

Occasional pest. The pest that reaches significant levels only occasionally (e-g) Rice case worm.

Oesophagus. A tubular part of the fore gut between the mouth and the crop.

Oligopod larva. Larva has a well-marked head, thorax and abdomen. Thorax bears three pairs of thoracic legs, but no abdominal legs (e-g) Larva of beetles.

Omnivorous. Feeding on many different kinds of both plant and animal food.

Onion shoot. Refer silver shoot.

Ootheca. A hardened egg case or covering formed by the secretions of accessory genital glands or oviducts, as in cockroaches.

Opisthognathous. Having the front of the head deflected backward, so that the mouthparts project to the rear, as in certain Homoptera.

Osmaterium. Fleshy, tubular, eversible processes producing a penetrating odor, capable of being projected through a slit in the prothoracic segment of certain papilionid caterpillars or from openings elsewhere in the bodies of certain other forms.

Oviparous. Reproduction by eggs laid by the female.

Ovoviviparous. In the case of some species of insects, the eggs are retained in the genital tracts of the female parent until they are ready to hatch and so hatching occurs just before or as the eggs are being laid.

Parasite. An organism deriving all or a portion of its nutrition from another organism (the host) while permanently or temporarily attached to or within the host.

Parasitoid. Entomophagous insects, which live on or in the body of another insect (host) or in or on any stage of the host from which they get protection and nourishment at least during any stage of their life cycle.

Paris green. An arsenical insecticide, chemically copper acetoarsenite; highly toxic to mammals.

Parthenogenesis. Development of eggs without being fertilized; reproduction from eggs without fertilization by sperm cells.

Pathogen. A disease causing organism.

Peduncle. Stalk or main stem of an inflorescence, part of an inflorescence or a fructification.

Permanent pests. Pests, which have always a foothold on the host whether they have a long or short life cycle and they may have a resting period spent on the host or in the soil (e-g) Mango hopper, coccids etc.; also called as persistent pests.

Persistence. The property of a pesticide to remain effective as a residue; the property of a pesticide to remain active for long period after application.

Persistent virus. A virus that persists in a virulent state in the vector system for an extended period, often for the lifetime of the vector.

Pest. The English term 'Pest' is derived from the Latin term 'Pestis', which means causing destruction. Pests are defined as those organisms, which compete with man in his food supply, damage his possessions and attack his person; pests include insects, mites, ticks, nematodes, fungi, bacteria, weeds, rodents, birds, molluscs, crustaceans, viruses, mycoplasmas, protozoa etc.

Pesticide. A chemical formulation by virtue of its toxicity (poisonous properties) that is used to kill pest organisms. The term pesticide includes insecticides, acaricides, nematicides, fungicides, bactericides, herbicides, molluscicides, rodenticides etc.

Pesticide residue. Pesticide remaining on or in a plant, plant product, air, water or treated area following a time lapse after application.

Pest resurgence. The rapid increase in numbers of a pest following cessation of control measures or resulting from development of resistance of a particular pest to certain plant protection chemicals or due to elimination of its natural enemies as a result of plant protection measures.

Pest surveillance. A systematic monitoring of the pest population, dispersion and dynamics in different crop growth phases, so as to forewarn the farmers to take up timely crop protection measures needed.

Pharate pupa (Prepupa). The non-feeding, quiescent stage between the larval and pupal period. Usually during this period the cocoon or puparium is formed. In male coccids and thrips, the first instar in which external wing pads appear is sometimes termed prepupa.

Pheromone. Chemical substance that attracts members of the same species or one sex of that species, especially insects and nematodes.

Phloem. Food conducting and food storing tissue in the vascular system of roots, stems and leaves.

Phyllophagous. Feeding upon leaf tissues.

Physical control. Control of pests by physical means such as, heat, cold, electricity, sound waves etc.

Phytophagous. Herbivorous; plant eating.

Phytosanitation. Measures requiring the removal or destruction of infested or infected plants or plant parts, which are likely to form a source of reinfestation or reinfection.

Phytotoxicity. Substances poisonous or toxic to plants; substances or chemical compounds or formulations, such as fungicides, insecticides, nematicides etc., which may cause adverse side effects (harmful effects) to the crop.

Poison bait. An attractant foodstuff for insects, molluscs or rodents, mixed with an appropriate toxicant to attract and kill the target organisms.

Plasmodesma (pl. plasmodesmata). A fine protoplasmic thread connecting two protoplasts and passing through the wall, which separates the two protoplasts.

Polyphagous pests. Pests infecting more than one unrelated species of plants.

Polypod larva (Eruciform larva). Larva has a well-marked head, a thorax of three segments, each bearing a pair of clawed legs and an abdomen of ten segments bearing five pairs of fleshy, hooked legs called prolegs (e-g) Caterpillar of butterfly.

Positive phototrophic or Photopositive. Natural instinct that makes organisms to move towards artificial light source.

Predator. An organism that lives by preying upon other living organisms.

Proctodaeal feeding. Contents of the rectal pouch supplied as food by the workers in a termite colony to soliciting young nymphs, queen and soldiers

Prolegs. The fleshy abdominal legs found in caterpillars (Lepidoptera, saw flies), bearing a characteristic arrangement of crochets; also called as abdominal legs, pseudo legs, false legs or prop legs

Pronotum. Dorsal surface of the prothorax

Propagative virus. The circulative viruses that multiply in the bodies of their vectors (e-g) Rice tungro virus; circulative virus.

Proprietory name. Distinguishing name given by the manufacturer to his particular formulated product; trade name.

Proteolytic enzyme. Protein splitting enzyme; enzymes capable of splitting proteins into other compounds.during metabolism.

Prothoracic shield. A thick, hard, protective layer of skin covering the prothorax of larvae like a shield.

Protonymph. The first nymphal stage of a tetranychid mite. It is also called the larval stage.

Pseudocoel. A false coelome or body cavity not lined with an epithelium of mesodermal origin, containing a liquid in which various internal organs are suspended.

Pterygotes. Insects, which have wings; winged insects.

Pulverulent. Powdery; dusty; covered with very minute powder-like scales.

Pupa. A resting stage in insect metamorphosis between larva and young adult, during which movement and feeding ceases and a great change in form occurs.

Pupa adecticous (pupa complex). The appendages remain in separate sheaths and are only partly free as in the case of mosquito pupa or 'tumbler' or adhere to the body and the abdomen is capable of movement as in the case of most lepidopterous and dipterous pupa or enclosed in the last larval skin or puparium as in the case of some dipterous pupa.

Pupa coarctata. The last larval skin is not cast off but formed into a puparium. The insect inside cannot be seen. The adult insect emerges from the puparium by making a circular opening on the pupal case by means of special structures as in the case of pupa of housefly, stable fly etc.

Puparium (Pupal case). After the last moult of the larva, the larval skin is not cast off, but becomes hardened and chitinous to form a case or covering.

Pupa obtecta. The legs and wings are bound down to the body by moulting fluid and can be seen externally and the pupa is active. Adult insect emerges from the pupal case through a 'T'-shaped slit on its dorsal surface as in the case of mosquito, sand fly and horse fly pupa or the appendages are tightly oppressed to the body and are covered by an external skin and the abdomen is capable of movement. Such pupa known as a chrysalis has a posterior process or cremaster, which may be furnished with apical hooks or anchorage as in the case of some lepidopterous pupa.

Pupa nuda. A naked pupa, which lies free of any attachment.

Quarantine. All operations associated with the prevention of importation of unwanted organisms into a territory or their exportation from it; regulation forbidding sale or shipment of plants or plant parts, usually to prevent disease, insect, nematode or weed invasion of an area.

Quiescent. A state of relative biological inactivity induced by physical factors such as, low temperature or other adverse environmental conditions.

Race. Subgroup or biotype within a species or variety, which are morphologically identical but differs from other races by virulence, symptom expression or host range.

Rachis. Elongated main axis of an inflorescence.

Radicle. Part of the plant embryo that develops into the primary root.

Random pests. Casual visitors such as weevils, cockchafer beetles, blister beetles etc., which have no seasonal basis.

Recurring pests. Pests infesting specific seasonal crops such as grasses, pulses, vegetables etc. regularly. They may pass through several generations during the growth period of the respective crop, rest for a while on alternate hosts and transfer their attention again to the second season crop (e-g) rice stem borer, lady bird beetle etc. They are also known as regular pests.

Relative humidity (RH). The ratio of the amount of water present in a sample of air to that which could be contained in the same volume, if it were saturated.

Reproductive potential. The maximum possible rate of increase in population (multiplication) of an organism. This is also called 'reproductive capacity' or 'biotic potential'.

Resistance to pest attack. The inherent morphological, physiological or genetic properties present in a particular variety of the host that deters infestation by a pest.

Resistance to insecticide. The ability of some strains of insects to survive normally lethal doses of an insecticide, the ability having resulted from survival of tolerant individuals in populations exposed to the toxicant for several generations.

Respiratory poisons. Chemicals, which block cellular respiration as in the case of fumigants (e-g) Hydrogen cyanide, carbon monoxide etc.

Resurgence. An increase of a pest population after a period of decline to a level higher than its original one, especially when the decline is caused by a pest control measure.

Rhizome. Mostly horizontal, jointed, fleshy, elongated, usually underground stem.

Rhizosphere. The soil immediately surrounding the roots of living plants; the area in the soil immediately surrounding plant roots and influenced by them.

Rosette. A short, circular cluster of leaves, that gives the plant a bunchy appearance; short, bunchy habit of plant growth.

Scab. A contagious skin disease of animals caused by certain parasitic mites.

Scaly wings. Wings covered with small, colored scales, which are unicellular flattened outgrowths of the body wall and are responsible for color (e-g) wings of moths and butterflies.

Scarabeiform larva. A grub-like, stout, sub-cylindrical, 'C'-shaped, sluggish larva with a well developed head, short and diminutive thoracic legs, but no abdominal legs and has soft, fleshy body (e-g) Some families of Coleoptera.

Scientific name. A latinized name of a species internationally recognized. The scientific name of a species consists of the generic and a species name. Scientific names are always written in 'Italics'.

Scutellum. The third dorsal sclerite of the mesothorax and metathorax; the triangular piece between the elytra in Coleoptera; the triangular part of the mesothorax between the bases of the hemelytra; a hemispherical part cut off by an impressed line from the mesonotum in Diptera.

Sedentary. Stationary without any bodily movement; remaining in a fixed location.

Seasonal pests. Pests that occur during a particular season on a specific crop (e-g) red hairy caterpillar on groundnut.

Selective insecticide. An insecticide, which kills selected insects but spares many or most of the other organisms, including beneficial species either through differential toxic action or the manner in which the insecticide is used.

Semilooper. A caterpillar in which only one or two pairs of the abdominal legs are wanting and when in movement only small loops are formed (e-g) castor semilooper.

Semiochemicals. Substances that modify the behavior of insects by inducing in them the response for sex, feeding, fear, alarm etc. These are further classified into pheromones, secreted by insects for stimulating the insects of the same species (intraspecific stimulators) and allelochemicals, the substances used for interspecific communications.

Serpentine mines. Winding larval feeding tunnels in leaves or stems of plants made by some species of Lepidoptera and Diptera.

Sessile. Permanently attached.

Seta. A slender hair-like outgrowth of the integument.

Severe pest. A pest with a general equilibrium position above the economic injury level, which makes the pest a constant problem.

Sex pheromone. The substance generally released by the females to attract the males for the purpose of mating.

Sex lure. A chemical substance that attracts only members of one sex of that species, especially in insects.

Siglure. A synthetic attractant for the Mediterranean fruit fly.

Silver shoot. Formation of hollow, pink, purple, dirty white or pale green, cylindrical, tubular structure like that of the tusks of an elephant, bearing at their tips a green, reduced leaf blade with ligules and auricles.

Slug caterpillar. Thick, short, stout and fleshy caterpillar, with a short and retractile head and the body covered with poisonous spines called scoli; also called as platyform larva (e-g) *Parasa lepida.*

Social insects. Insects that live in-groups or colonies generally with differentiated castes or forms with specific functions in the colony (e-g) termites, honeybees etc.

Soil fumigant. A pesticide that will evaporate quickly under normal temperature, which when incorporated into the soil, releases a toxic gas that kills pests in the soil.

Spathes. The two, hard, boat-shaped fibrous structures that protect the inflorescence.

Sporadic pests. Pests that infest certain crops occasionally within some limited areas or pockets in a more or less severe form (e-g) rice earhead bug

Stadium (pl. Stadia). The time interval between moults in a developing insect.

Sterile male technique. A method of pest control, which involves the release of large numbers of artificially sterilized male insects in the populations.

Stomach insecticide. An insecticide that is toxic to insects when ingested, usually along with their food and takes effect through the alimentary tract. Its use is limited to surface feeding pests with biting and chewing mouthparts.

Stomodaeal feeding. A mixture of salivary secretion and regurgitated intestinal contents supplied as food by the workers to young nymphs, queen and soldiers.in the termite colony.

Stubble. Crop residue left over in the field after the harvest of the crop.

Stunting. The reduction in height of a vertical axis, resulting from a progressive reduction in the length of successive internodes or a decrease in their numbers.

Stylet. Stiff, relatively long, slender, hollow, tube-like feeding organ of plant parasitic nematodes or sap-sucking insects, such as aphids, leafhoppers etc.

Stylet-borne virus. Virus that is carried only in the stylet and is not persistent in the vector; the infectivity of the vector is rapidly lost unless there is a renewed access to a fresh virus source (e-g) banana bunchy top virus, groundnut rosette virus etc.

Stylet knob (Syn. Basal knob). Structure at the base of a nematode stylet.

Suckers. Young plants arising from the mother corm as side shoots as in the case of banana plants.

Sucking insects. Insects that suck blood, nectar, plant sap or other liquid food.

Susceptibility. The state of being readily affected or acted upon by an injurious agent.

Suspension. A state in which solid particles are kept dispersed in a liquid.

Swarms. A large number of insects moving together simultaneously.

Symptom. A visible, abnormal change in a host (including behavior) as a result of pest infestation.

Syncytium (pl. Syncytia). A multinucleate structure in root tissue formed by dissolution of common cell walls induced by secretion of certain sedentary plant parasitic nematodes (e-g) Cyst nematodes; a multinucleate mass of protoplasm surrounded by a common cell wall.

Synonym. A different name given to a species or genus previously named and described.

Systemic insecticide. An insecticide, which is absorbed through a plant surface and is translocated away from the site of application.

Taxis. A directed response involving the movement of a living organism towards or away from a stimulus.

Taxonomy. The science of classification and naming of species; arrangement of species and groups into a system, which shall exhibit their relationship to each other and their places in a natural classification.

Termitarium. The earthen structure or nest inhabited by a termite colony.

Terrestrial. Strictly living on the ground; usually living on dry land.

Thermotropism. The response or reaction of an organism to temperature changes.

Thigmotropism or Thigmotaxis. Response or reaction to touch or contact.

Tibial spur. A large spine on the tibia usually located at the distal end of the tibia.

Tissue. A group of cells of similar structure, which performs a special function.

Tolerance. A basis for resistance in which a host shows an ability to grow and reproduce or to repair injury, despite supporting a pest population equal to that, which may cause damage to a more susceptible host.

Toxicity. The capacity of a compound to produce injury; ability to poison or to interfere adversely with vital processes of the organism by physico-chemical means.

Tracheal gills. Larval stages of some species of insects, which are acquatic to some extent breathe under water through gills, which are outgrowths of the tracheae; respiratory organs of certain aquatic insects consisting of the outgrowths of the body wall containing numerous closed tracheoles into which dissolved oxygen can diffuse from the surrounding water.

Trail marking pheromone. An intraspecific chemical messenger produced by foraging ants and termites to indicate source of requisites to other colony members.

Transient phase. The stage of development, which lasts for only a short time.

Translocation. Transfer of nutrients or virus through the plant; movement of chemical substances within a plant.

Transmission. The transfer or spread of a virus or other pathogen from one plant to another.

Transovarial transmission. The ability of a virus vector to transmit the virus to its progeny by means of eggs.

Trap crop. A crop, which is preferred by a particular pest in addition to the main crop, but is less important or less valuable than the main crop.

Trench fever. A non-fatal disease of human beings characterized by sudden onset of fever, headache, dizziness and pain in muscles and bones caused by *Rickettsia quintana* transmitted by the body louse through its faeces.

Tritonymph. The third stage nymph of a tetranychid mite.

Trypanosomiasis. The disease caused by the presence in the body of a protozoan parasite of the Genus - *Trypanosoma*, marked by fever, anemia and redness of the skin, transmitted by the tse tse flies; also known as African sleeping sickness.

Tubercles. Small wart-like protuberances on the surface of the body, on which in some insect larvae hairs are found.

Tumblers. The aquatic pupal stage of mosquitoes.

Tumour. Any swelling, whether edema or a mass resulting from malformation, inflammation or repair.

Typhoid fever. An acute infectious human disease caused by a bacterium - *Salmonella typhi* characterized by continued fever, inflammation of intestine, intestinal ulcers, a rose spot on the abdomen and enlarged spleen. The disease is food and water-borne, but may be transmitted by houseflies.

Typhus fever. A human disease caused by a bacterium-like microorganism -*Rickettsia prowaseki* and transmitted by the body louse - *Pediculus humanus humanus.* High fever, backache, intense headache, bronchial disturbances, mental confusion and congested face characterize the disease. Mortality may range from 15 - 75 per cent.

Typical host. A host in which pathogenic microorganism or parasite is commonly found; synonymous with 'natural host'.

Variety. A loosely defined taxonomic category below the level of sub species, usually possessing some visibly different character.

Vascular bundle. Strand of conductive tissues, usually composed of xylem and phloem; in leaves the small vascular bundles are called veins.

Vector. A living organism, such as insect, mite, bird, higher animal, nematode, parasitic plant and human able to carry and transmit a pathogen, especially an insect, nematode etc., able to transmit a virus.

Vessel. A xylem element or series of such elements whose function is to conduct water and mineral nutrients.

Vermiform larva. Relatively long and slender larva; worm-shaped larva; worm-like larva typical of some Diptera.

Vertical resistance. A type of resistance effective against certain specific biotypes of the insect, but not against others; also called as specific resistance.

Vestigial. Having the nature of a degenerate or atrophied organ, more fully functional in an earlier stage of development of the individual or species.

Virulence. State of being pathogenic.

Viruliferous. A vector containing a virus and capable of transmitting it.

Viviparity. A type of reproduction in some species of insects in which the embryonic development is completed within the body of the female parent and hence instead of laying eggs young larvae or nymphs are laid.

Volatile compound. A compound, which evaporates or vaporizes (changes from a liquid to a gaseous stage) at ordinary temperatures on exposure to air.

Water-soluble powders and concentrates (WSP and WSC). Insecticidal powders or liquids that will go into solution when mixed with water for dilution before spraying without being dispersed as in the case of wettable powders or emusifiable concentrates

Weevils. Insects belonging to Order- Coleoptera with prominent snout in the adult insects.

Wettable powders (WP). Insecticide formulations consisting of finely powdered solid particles of active ingredient and inert carriers, which form a suspension when mixed with water for spraying

White ears. White-colored, dried up, chaffy earheads invariably without any normal grains, produced as a result of stem borer attack, which cuts off the inner portion of the culm. The white ears come off quite easily when pulled; also called as 'white head'.

Wrigglers. The aquatic larval stage of mosquitoes.

Xylem. A plant tissue consisting of tracheids, vessels, parenchyma cells and fibers; wood.

Yellow fever. An acute infectious disease caused by a virus transmitted by *Aedes aegyptii* and marked by fever, jaundice and albumin and globulin in the urine.

Yellows. A type of plant disease characterized by pronounced chlorosis; a plant disease characterized by yellowing and stunting of the host plant.

Zoophagous. Feeding on material of animal origin.

Zygote. A fertilized egg.